AIR QUALITY – MONITORING AND MODELING

Edited by **Sunil Kumar** and **Rakesh Kumar**

Air Quality – Monitoring and Modeling
Edited by Sunil Kumar and Rakesh Kumar

Published by InTech

Janeza Trdine 9, 51000 Rijeka, Croatia

Publishing Process Manager Maja Bozicevic

Technical Editor Teodora Smiljanic

Cover Designer InTech Design Team

Additional hard copies can be obtained from orders@intechopen.com

Air Quality – Monitoring and Modeling, Edited by Sunil Kumar and Rakesh Kumar
p. cm.
ISBN 978-953-51-0161-1

For EU product safety concerns:
IN TECH d.o.o., Prolaz Marije Krucifikse Kozulić 3, 51000 Rijeka, Croatia,
info@intechopen.com or visit our website at intechopen.com.

Meet the editors

Er. Kumar has been working in the field of Environmental Engineering and Management in general and Solid and Hazardous Waste Management in particular at National Environmental Engineering Research Institute (NEERI), Council of Scientific and Industrial Research, Government of India. He has published many research papers in journal of international repute with high impact factor covered referred journals and many papers are under review/ submission stage in various international journals. He is the Editorial Board member of International Journal of Environmental Monitoring and Assessment besides working as Guest Editors for many international journals in the domain of Environmental Engineering in general and solid waste management in particular.

Dr. Rakesh Kumar is a Chief Scientist and Head at Mumbai Zonal Centre of Council of Scientific and Industrial Research-National Environmental Engineering Research Institute (CSIR-NEERI). He has been closely associated with many prestigious studies carried out by different funding agencies in the domain of air quality monitoring and modeling which could lead toward policy formulation in this sector in India. He has numerous publications in the field of air pollution engineering. He is a member of various committees relating to air quality.

Contents

Preface

There is a great challenge for monitoring of Air quality specially for formulating abatement plans which will decrease the burden of environmental pollution. The management of polluted and fungal contaminated air is required to be dealt with different science and technological applications to reduce its impact on the environment.

There are many publications dealing with air quality monitoring and its health impact but still very few documents are available which could reveal the information relationship between fungal contamination of indoor air and health problems of some residuals. Microbial pollution is a key element of indoor air pollution. It is caused by hundreds of species of bacteria and fungi, in particular filamentous fungi (mould), growing indoors when sufficient moisture is available. In indoor environments, the air and surface fungal contaminations are influenced not only by temperature, relative humidity and air flux/drafting but also by the weather conditions, hygiene standards and human activity.

In the present book, chapters covering a methodology for estimation of air pollution and its health effects in large Asian cities, a qualitative study of air pollutants from road traffic and air pollution, air quality in low-energy buildings, particulates in indoor environment, evaluation of dry atmospheric deposition at sites in the vicinity of fuel oil fired power, PM 10 and its chemical composition, etc. are discussed in detail. Each chapter of the book carries valuable and updated information from basics to apex which help readers to understand more precisely.

Sunil Kumar
Scientist Council of Scientific and Industrial Research-National Environmental Engineering Research Institute (CSIR-NEERI), India

Rakesh Kumar
Chief Scientist and Head CSIR-NEERI Mumbai Zonal Laboratory
India

1

Relationship Between Fungal Contamination of Indoor Air and Health Problems of Some Residents in Jos

Grace Mebi Ayanbimpe[1],
Wapwera Samuel Danjuma[2] and Mark Ojogba Okolo[3]
[1]Department of Medical Microbiology University of Jos Nigeria,
[2]Department of Geography and Planning University of Jos,
[3]Department of Microbiology, Jos University Teaching Hospital Jos, Nigeria

1. Introduction

The problem of indoor air quality has in recent times attracted more concern than was the case several decades ago. Morey, (1999) presented some probable reasons why this is so. Construction materials have markedly changed from stones, wood and other 'natural' products to synthetic, pressed wood and amorphous cellulose products, some of which may be less resistant to microorganisms than the earlier ones. The indoor environment is invariably the closest to humans in terms of daily interaction, as a large proportion of man's time is spent indoors, especially in the modern cities and urban societies. There are indications however, that in many parts of the world, our homes, schools and offices are heavily contaminated with airborne molds and other biological contaminants. Biological contaminants of the indoor air include fungi, bacteria, viruses, pollen etc.; and many species of fungi are able to grow wherever there is moisture and an organic substrate. Building materials such as ceiling tiles, wood, paint, carpet rugs, etc., present very good environment for growth of fungi (Robin et al, 2007; Morey, 1999; Heseltine and Rosen, 2009). Water accumulation and moist dirty surfaces also encourage growth of fungi within a building, and contribute to human health problems. Fungi are very successful organisms capable of survival in diverse environments. This is possible because of their physiological versatility and genetic plasticity. Their spores are produced in large quantities and easily spread over a wide area. These spores can remain dormant for a long period of time during unfavorable conditions. Fungal spores and products of their metabolism are able to trigger allergic reactions which include hyper-sensitivity pneumonitis, allergic rhinitis and some types of asthma. Infections such as influenza, respiratory fungal infections, mycotoxicoses, and eye irritation by fungal volatile substances, among others, are health risks and challenges to susceptible individuals. Children, the elderly and immuno-compromised persons are at greater risks (Simoni et al, 2003).

1.1 Medically important fungi

Molds are part of the kingdom Fungi. Fungi are a diverse group of organisms within a wide range of species that include mushrooms, bracket fungi, molds and yeasts. Fungi, are eukaryotic, heterotrophic organisms which may be filamentous, i.e. multicellular (molds) or single celled (yeasts). They are found everywhere around the human environment and more than 1, 500,000 species of molds exist in the world (Hawksworth, 1991; Cannon & Hawksworth, 1995; Hawksworth, 1999). Fungi are not photosynthetic and hence obtain nutrients by absorption. Molds are composed of linear chains of cells (hyphae) that branch and intertwine to form the fungus body called mycelium. All fungal cell walls contain (1-3)-beta-D-glucan, chitin and mannan (Hawksworth, 1998). They exist as saprobes, mutualists or parasites in their relationships with other organisms. They reproduce by forming sexual or asexual spores which are in most instances, their dispersal propagules. Fungi grow very easily in damp and wet conditions; especially on materials and products containing moisture and an organic substrate. When these conditions are met, molds will grow and reproduce by forming spores that are released into the air. Molds are very adaptable and can grow even on damp inorganic materials such as glass, metal, concrete or painted surfaces if a microscopic layer of organic nutrients is available. Such nutrients can be found on household dust and soil particles (Robins and Morell, 2007). Fungi are capable of extracting their food from the organic materials they grow on and the ability to reproduce by way of minute spores makes them ubiquitous (DeHoog et al., 2007). Fungi are a part of nature's recycling system and play an important role in breaking down materials such as plants, leaves, wood and other natural matter (Robins and Morell, 2007). Because various genera grow and reproduce at different substrate water concentrations and temperatures, molds occur in a wide range of habitats. Microbial growth may result in greater numbers of spores, cell fragments, allergens, mycotoxins, endotoxins, β-glucans and volatile organic compounds in indoor air (Nielson, 2003). Persistent dampness and microbial growth on interior surfaces and in building structures should be avoided or minimized, as they may lead to adverse health effects. When mold growth is observed remediation should be given top priority (Zaslow, 1993; Douwes, 2009).

1.2 Fungi and the Indoor environment

Fungi grow on any material that provides a carbon source and adequate moisture. Fungi can grow on building and other materials, including: ceiling tiles; wood products; paint; wallpaper; carpeting; some furnishings; books/papers; the paper on gypsum wallboard (drywall); clothes; and other fabrics (Robins and Morrell, 2007). Mildew most often appears on natural fibers, such as cotton, linen, silk, and wool. It can actually rot the fabric. Mold can also grow on moist, dirty surfaces such as concrete, fiberglass insulation, and ceramic tiles (North Carolina Cooperative Extension Services 2009; Robins and Morell, 2007). Fungal growth in building materials is more dependent on the moisture content of the substrate than on atmospheric relative humidity. The minimum moisture content of building materials allowing fungal growth is near 76 %. Wood, wood composites (plywood, chipboard), and materials with a high starch content are capable of supporting fungal growth, at the lowest substrate moisture content. Plasterboard reinforced with cardboard and paper fibers, or inorganic materials coated with paint or treated with additives that offer an easily-degradable carbon source, are excellent substrates for fungal growth when substrate moisture content reaches 85-90 % (Pasanen et al, 1992; Dubey et al, 2011).

1.2.1 Common fungal contaminants and factors responsible for indoor contamination

Several fungi have been isolated from the indoor environment in various parts of the world and there seems to be relatedness in the types found from different places. Common air-borne fungi in indoor environments include: *Alternaria sp, Aspergillus glaucus group, Aspergillus versicolor, Aspergillus fumigatus, A sydowii, Penicilium aurantiogriseum, Penicillium chrysogenum, Stachybotrys chartarum, Chaetomium globosum, Cladosporium cladosporoides, Ulocladium, Aspergillus nidulans,* and *Alternaria alternata.* Others are *Mortierella sp., Trichoderma sp.Penicilium lividum, P. arenicola, P. verrucosum P. citrinum, P. bilai, P. cyaneum, P. granulatum, P. sublateritium, Cladosporium sphaerospermum, C. herbarum, Fusarium sp. Trichoderma sp. Memnoniella sp. Epicoccum nigrum, Scopulariopsis fusca and Chrysosporium queenslandicum.* There seems to be some variation in the abundance of fungi at different seasons of the year (Shelton et al., 2002; Mirabelli et al, 2006; Sautour, 2009) while others occurred irrespective of the season (Horner et al, 2004). *Cladosporium species* are abundant in the summer, but *Penicillium* concentrations are higher in the wetter months (Verhoeff, 1993; Flannigan, 1997; Dubey et al, 2011).

Mold types and concentrations indoors are primarily a function of outdoor fungi (Shelton et al., 2002). They presented an exhaustive study of the mycoflora of outdoor and indoor atmospheres in all USA regions. More than 12,000 samplings were carried out, both in outdoor and indoor atmospheres in more than 1,700 buildings. The moisture content of the substrate (related to indoor humidity level) has also been considered. Higher concentrations of outdoor molds and other fungi occur where trees, shrubs and landscape irrigation occur close to exterior building walls (McNeel and Krautzer, 1996). Humidity is the most important factor determining fungal growth in indoor environments (Got et al, 2003). Atmospheric relative humidity influences directly the release of conidia from conidiophores, and concomitantly, the concentration of spores in the atmosphere. The pattern may differ for different types of fungi. Pasanen et al., 1991, showed that spore release by *Cladosporium* was favored by low humidity unlike in *Penicilium* where the reverse was the case. These differences influence the seasonal patterns of outdoor fungi.

In general, the types and concentrations of molds that affect indoor air quality are similar to those found in outdoor air and molds may occur in homes without dampness problems (Gots et al, 2003; Horner et al, 2004; O'connor et al, 2004; Codina et al, 2008). However, background mold numbers may shift whenever water accumulates in buildings. Damage caused by floods, plumbing leaks, poor understanding of moisture dynamics and careless building design and construction lead to structures that are more susceptible to water intrusion. Also, lack of good maintenance practices in some buildings can lead to moisture buildups that, when left alone, can result in microbial contamination and higher levels of bioaerosols (BCBS, 2007; Stetzenbach et al, 2004). Dampness has usually been associated with moldiness indoor and its attendant ill health conditions (Dales et al, 1997; O'connor et al, 2004; BC Non-profit Housing Association,2007; Vocaturo et al, 2008). Damp buildings often have a moldy smell or obvious mold growth; some molds are human pathogens (Kuhn et al, 2003). Fungal colonization of buildings is especially common in low income communities in developing countries, where buildings are situated indiscriminately without consideration for environmental sanitation, urban planning and building regulations (Ahiamba et al, 2008; Ayanbimpe et al, 2010). Among factors responsible for indoor air contamination in developing countries are poor locations of refuse dumps, standing water bodies, water damaged materials and absence of proper urban planning. According to Heseltine &Rosen, (2009) indicators of dampness and microbial growth include the presence of condensation on surfaces or in structures, visible mould, perceived mould odour and a history of water damage, leakage or penetration. Thorough inspection and, if necessary, appropriate measurements can be used to confirm indoor moisture and microbial growth. Similarly, contaminated air handling systems can become breeding grounds for molds and mildews and also distribute spores all over the home. Excessive humidity and high water content of building materials, lack of thermal insulations and incorrect behaviours of residents may also contribute (Ahiamba et al, 2008).

1.3 Indoor molds and health

A healthy indoor environment is a necessity for every human being. A number of studies have examined the health risks related to the presence of fungi indoors and shown positive associations between fungal levels and health outcomes. Evidences abound on the problems of dampness which consequently contribute to mold colonization of the indoor environment (European Environment and Health Information System, 2007). To control diseases caused by fungi the primary avenues of their introduction into the indoor environment should be attended to and knowledge of these avenues is required. Identifying conditions which contribute to fungal presence in the indoor environment and correction of such is a necessary step to preventing ill-health and resultant losses. Various suggestions have been made to this effect (Buck, 2002).

Some reports have highlighted the importance of housing conditions as determinants of mental and physical health and the importance of housing as a medium for promoting universal health efforts (BC Non-profit Housing Association, 2007; Shenasa et al, 2007; BCBS, 2007; Canadian Center for Occupational Health and Safety, 2008). Surveys of indoor environments have shown that symptoms of ill-health associated with air-borne fungi and their products are higher in households with molds and mildews and that are characterized by moist or damp surfaces, damp walls, wet surfaces, damaged materials, standing water and refuse dumps (Shelton et al, 2002; O'connor et al, 2004; Ahiamba et all, 2008; Ayanbimpe et al, 2010;). Pathogenic infections may be elicited by some molds which are primarily found in soil or on vegetation but are carried into homes by human foot wears or animals. When spores of some air-borne fungi are inhaled by immunocompromised persons, they may develop disease processes, for example Aspergillosis, and histoplasmosis. Respiratory illnesses are the major reported problems among persons living in homes colonized by molds (Dales et al., 1991). Persons at greater risk of effects of indoor air fungal colonization are the young, elderly, immunocompromised individuals, persons undergoing intensive surgical procedures like transplant patients and those on immunosuppressive therapy (Simoni et al, 2003; Ritz et al, 2009;). Fungal components implicated in pathogenesis include spores; β-1,3-D-glucan; Hydrophobins; Volatile organic compounds (VOCs) and Dihdroxynaphthalene (DHN) Melanin.

1.4 Symptoms of ill health associated with poor indoor air quality

There are suggestions that even very low levels of some common air pollutants can gradually cause health problems. Such condition have been reported to include; allergic reactions in which mild and occasional, to severe and chronic symptoms caused by most species of fungi may develop in susceptible individuals who may present with hay fever, runny nose, breathing problems, and sinusitis (Bryant, 2002; Moloughney, 2004, Wu et al, 2007; Kuhn and Ghannoum,, 2004; Mirabelli et al, 2006). Some people could develop skin diseases and skin rashes as an allergic reaction to toxic mold. (Gots et al, 2003; http://www.cleanwaterpartners.org/mold/black-mold.html). Most people can recognize and react to the moldy smell negatively with symptoms like headaches, vomiting, nausea, blocked noses, and asthma. Volatile products of indoor molds have also been suggested to cause toxic illnesses or mycotoxicoses (Abbott, 2002; Straus and Wilson, 2006). Mycotoxins can be inhaled by people through the air, ingested through food, and can also come in contact with skin, resulting in many illnesses. Different molds produce different mycotoxins, which also depends on what kind of material the toxic mold is growing on. Among the illnesses caused by mycotoxins are gastrointestinal, respiratory, and reproductive disorders (O'connor et al, 2004; Mirabelli et al, 2006). Also reported are eye irritations, poor taste sensations, sore throat and tiredness (U.S. Environmental Protection Agency, 2003; Worksafe Bulletin, 1999). Even severe neurologic conditions such as depression have been linked to indoor mold colonization (Shenassa et al, 2007). Under certain metabolic conditions, many fungi produce mycotoxins, natural organic compounds that initiate a toxic response in vertebrates. Molds that are important potential producers of toxins indoors are certain species of *Fusarium, Penicillium,* and *Aspergillus.* In water-damaged buildings *Stachybotrys chartarum* and *Aspergillus versicolor* may also produce toxic metabolites. A large body of information is available on the human and animal health effects from ingestion of certain mycotoxins (Sorenson, 1989; Smith and Henderson, 1991), Two classes of mycotoxins have been isolated from house dust samples: aflatoxins from some strains of *Aspergillus flavus* and trichothecenes from some species and strains of *Fusarium, Cephalosporium, Stachybotrys* and *Trichoderma* (Kuhn and Ghanoum, 2004). The fungus cell wall component, (1-3)-beta-D-glucan, a medically significant glucose polymer that has immunosuppressive, mitogenic (i.e. causing mitosis or cell transformation) and inflammatory properties. This mold cell wall component also appears to act synergistically with bacterial endotoxins to produce airway inflammation following inhalation exposure in guinea pigs.

1.5 Environmental pollution and regulations

The problem of environmental pollution is enormous and various attempts have been made to establish facilities for its control and regulation in various parts of the world. In Nigeria, the Federal Military Government set up the Federal Environmental Protection Agency (FEPA) Act 1988, established under Decree 58 of 30 December 1988 with statutory responsibility for overall protection of the environment and setting and enforcing ambient and emission standards for air, water and noise pollution (Federal Government of Nigeria (FGN), 1988). The Environmental Impact Assessment (EIA) Act 1992 reaffirmed the powers of the FEPA and defined the minimum requirements for an EIA (Environmental Policy, 1999). The democratic government created a Ministry of Environment in 1999, thus bringing agencies such as the FEPA and the Environmental Health Unit of the Ministry of Health under one administrative system. Further effort to ensure the maintenance of a healthy environment was made by instituting the National Environmental Standards and Regulations Enforcement Agency (NESREA) charged with the responsibility for protection, and development of the Nigerian environment (Federal Republic of Nigeria, 2007). Among numerous initiatives advanced to attain the United Nations Millennium Development Goal is the relationship between the physical environment and human health, and this is a top priority of the WHO. Here the need to ensure that people, especially children, live in environments with clean air was emphasized. The WHO recognized healthy indoor air as a basic right because people spend a large part of their time each day indoors; in homes, offices, schools, health care facilities, or other private or public buildings, therefore quality of the air they breathe in those buildings is an important determinant of their health and well-being (World Health Organization, 2009). Regulations by relevant authorities to control the environmental air pollution have focused on industrial emissions and chemical contaminants but less on microbiological contaminants, especially of the indoor air.

1.6 Relationship between indoor air fungi and health of residents

Indoor air pollution, such as from dampness and mould, chemicals and other biological agents, is a major cause of morbidity and mortality worldwide (Environmental Protection Agency, 2010). Biological pollutants, volatile organic compounds and indoor molds have been associated with symptoms such as rhinitis, wheezing, conjunctiva irritation, skin rashes, tightness in the chest, difficulty in breathing, runny nose, fever, muscle aches, headaches, diarrhea, nausea and nose bleeding. Aggravation of respiratory symptoms has been linked to increased concentration of airborne particles (McNeel and Kreutzer, 1996; Wong et al, 1999; Petrescul, et al, 2011). Interventions to reduce the colonization of indoor environment by contaminants which may serve as triggers to respiratory illnesses, especially asthma in various places yielded positive results. Follow-up from one of such interventions (Morgan et al. 2004), showed that the subjects experienced fewer symptom days for 2 years, and there were significant reductions in dust mite and cockroach allergens, also for 2 years. Another study showed the rate of doctor-diagnosed asthma was higher (25%) in children residing in deteriorated public housing compared with only 8% in those in other houses. A similar observation was made by Howell et al. (2005), on the self-reported health status of some individuals. Petrescu1 et al., (2011) related increase in chronic respiratory symptoms to increased concentrations of particles in the air. Shenasa et al, (2009) reported a strong association between depression and indoor mold. In their study, 9% of the respondents reported 3 or 4 depressive symptoms and older women and unemployed persons were more prone depressive symptoms. This was irrespective of dampness of the indoor environment.

1.7 Health effects of mycotoxins and volatile organic compounds

Molds produce acute health effects through toxin-induced inflammation, allergy, or infection. Some reports have associated overgrowths of trichothecene-producing fungi with human health effects such as cold and flu-like symptoms, sore throats, headache and general malaise (Croft et al., 1986; Johanning et al., 1993; Pasanen et al., 1996). Molds also produce a large number of volatile organic compounds (VOCs) which are responsible for the musty odors produced by growing molds. These VOCs have not concretely been associated with disease, although ethanol, the most common VOC, synergizes many fungal toxins. When fungal spores are inhaled, they may reach the lung alveoli and induce an inflammatory reaction, creating toxic pneumonitis. Severe toxic pneumonitis can cause fever, flu-like symptoms and fatigue (organic toxic dust syndrome). Hypersensitivity pneumonitis, a particular form of granulomatous lung disease, is a syndrome caused by inhalation of large concentrations of dust containing organic material including fungal spores.

Mycotoxicosis may also be acquired through exposure to toxins produced by some fungi as a result of their secondary metabolism, elaborated through various metabolic pathways (Wang et al, 2002). Mycotoxins have been detected from crude building materials and include Acetyl-T-2 toxin; Aflatoxin A; DON, HT-2 toxin; OchratoxinA; citrinine; Nivalenol, indoor environments (Nielson 2003). These mycotoxins may target various organs of the human body such as the liver, lungs, kidney, nervous system, immune system and the endocrine system. The severity of the mycotoxicoses depends on the target organ, nutritional and health status of the individual, type of toxin, and the synergistic effect of the mycotoxins with other chemicals or mycotoxins in the target organ.

1.8 The general challenge of indoor air quality

Indoor air pollution has been identified as one of the biggest health threats facing Americans in recent times (US Environmental Protection Agency, 2003). This may be even more in the developing countries but are underreported because of the near absence of research into this aspect. Knowledge of indoor air quality, its health significance and the factors that cause poor quality is key to enabling action by relevant stakeholders, including building owners, developers, users and occupants (Heseltine &Rosen, 2009). Maintaining a clean indoor air involves every one of the above. It is clear that elucidating the actual state of indoor air and its relationship to health of residents is an important step that would enable proper action, by regulation and policy to be taken by appropriate authorities. The building users on their part would also play a role in this very important issue. There are very few reported investigations in the study area on the health effects of indoor air fungi on the residents and because of this, little or no data is available.

2. Materials and methods

This household survey and examination of health data of residents of houses with indoor air fungal contamination was carried out to establish the association between microbial contaminations of indoor air and the health of residents. Some aspect of this was highlighted in a previous report by us (Ayanbimpe et al, 2010), thus the present study is a follow up on subjects previously identified in order to establish a relationship, if any, between the indoor air fungal colonization and the health of residents. The criteria considered for determination of a relationship between ill-health symptoms in residents and indoor air were as follows:

1. The presence of fungus growth indoor and isolation of fungi from indoor air of the residence.
2. Dampness of the indoor environment and presence of other parameters considered as risk factors (see Table 3).
3. The experience of symptoms among residents while in the specific indoor environment.
4. The reduction of symptoms following mold remediation.

2.1 Description of the study area

The Jos metropolis is the capital city of Plateau State of Nigeria located at 1900m above sea level in the North Central part of the country. The Jos Plateau is bounded approximately by latitudes 8° 0′55″N and 10° 0″N, and longitude 8 ⁰′S 22″ E and 9 ⁰′ 30″E. Tin mining which began around 1904 was a major occupation of people of this area. This activity has led to creation of ponds; gullies and water logged ditches (some of which are used as dumping sites by residents) around residential areas that may contribute to dampness. The rock materials are used extensively for various forms of building construction in and around Jos (Solomon, et al 2002). The study area is characterized by several unplanned settlements, with haphazardly located buildings.

2.2 Evaluation of indoor air quality in the study area

A previous study had highlighted the poor indoor air quality of the study area (Ayanbimpe et al, 2010). The following fungi were isolated: *Aspergillus versicolor, Eurotium sp, Stachybotrys chartarum, Chaetomium globosum, Stachybotrys alternans, Aspergillus versicolor, Alternaria alternata, Aspergillus fumigatus, Xylohypha bantianum, Rhizopus sp, Stemphylium sp, Penicillium sp, Sepedonium sp* and *Phoma sp.* (see Table 1). More than 75% of the population in that report had complaints of respiratory symptoms; frequent headache, eye irritation, skin rash etc. (see Table 2). 17 households out of the 150 had molds growing in one or more parts of the house. The relationship between poor housing conditions and indoor air contamination was also established.

2.3 Investigation of factors in the study area

Using structured questionnaires and one on one interviews the conditions of the buildings considered as risk factors were determined. This was done by obtaining responses for the following questions:

a. How old is the building?
b. What type of accommodation is it?
c. How many rooms are there in the house?
d. How many persons live in a room?
e. Did you experience flooding at any time or are there leaking areas in the house?
f. How many households share the building?
g. Do you have moist walls and moldy spots/areas within the house?

The following problems were identified as risk factors for the different households in the study area:

- Flooding
- Bad house roofs.
- Fungal growth on asbestos ceiling sheets
- Non - painted house walls
- Over crowding (families of 5 to 8 persons living in one room)
- Fungal growth on some food stuff stored in -doors.

Residents with health problems related to airborne fungi had been identified and specific symptoms recorded. The present study assumed that the presence of moldy surfaces indoors contributed to the concentration of airborne spores. By a random selection, 8 of the 17 households with indoor mold colonization and high concentration of airborne fungi were visited for follow-up. Each of the 8 households revisited underwent mold clean up and remediation or correction of conditions / factors supposedly responsible for indoor mold colonization.

Fungus	Average number	percent
Chaetomium globosum	64	16.8
Stachybotrys alterans	53	13.9
Aspergillus versicolor	53	13.9
Alternaria alternata	53	13.9
Aspergillus fumigatus	37	13.9
Xylohypha bantianum	29	7.6
Penicilium sp	16	4.2
Rhizopus sp	28	7.8
Stemphylum sp	24	6.3
Sepedonium sp	13	3.4
Phoma sp	10	2.6
Total	380	100

Table 1. Fungi isolated from indoor air in the study area (Adopted from Ayanbimpe et al., 2010)

Symptoms	0-10	11-20	21-30	31-40	41-50	51-60	61-70	71-80	Total
Cough,	7	8	1	0	10	4	0	1	31
Nocturnal cough	2	2	0	0	1	1	0	1	7
Dyspnea	0	1	0	0	8	0	0	0	9
Sore throat	1	12	2	0	8	2	0	0	25
Hoarseness	1	4	2	0	3	0	0	1	26
Rhinitis	7	6	3	0	7	2	0	1	26
Nasal bleeding	1	4	0	0	0	0	0	0	5
Sinusitis	0	0	0	0	1	1	0	0	2
Skin rash	2	4	1	0	3	0	0	1	11
Headaches	3	9	3	0	9	2	0	1	27
Nausea	2	8	0	0	8	0	0	0	18
Diarrhea	4	7	1	0	4	0	0	0	16
Tightness in the chest	2	4	0	0	3	0	0	0	9
Fever	7	8	2	0	4	2	0	0	23
Muscle aches	2	8	1	0	12	1	0	1	25
Eye irritation	4	9	2	0	4	4	0	1	24

Table 2. Symptoms reported among residents of various age groups before mold remediation

2.4 Fungal clean up and remediation

Because of the seriousness of the problem of indoor air contamination, the WHO guideline recommended that remediation of the conditions that lead to adverse exposure should be given priority to prevent an additional contribution to poor health in populations who are already living with an increased burden of disease (WHO Regional Office for Europe, 2006). The various conditions that supposedly contributed to mold colonization were physically assessed and the necessary remediation procedure determined. Methods proposed in standard manuals and guidelines (Kennedy; Buck, 2002; New York City Department of Health and Mental Hygiene; 2008; North Carolina University Cooperative Extension Services, 2007) were followed, with slight modifications, under close expert monitoring. The steps adopted for each house depended on the extent of mold contamination and the part of the building affected. Two of the households completely relocated to new accommodations. The process of remediation essentially involved:

1. Flooded apartment - Only one of the households experienced flooding which affected two of the rooms. Here the rugs were removed, washed with detergent solution and rinsed with water and Sodium hypochlorite disinfectant solution (diluted according to Manufacturer's instructions). These were dried in the sun. The rooms were also washed, disinfected and allowed to dry before re-occupation.
2. Bad Roofs - Leaky roofs were repaired by replacing punctured sheets and boards. This was done by a carpenter.
3. Fungal growth on asbestos and wooden ceiling sheets - These were removed using absorbent cotton pads soaked in disinfectant, held in place by a wooden handle over a disposable plastic collector. The spot was then scrubbed with detergent solution and dabbed dry, after which paint was applied. Adequate protective measures were observed with every procedure. A physical inspection of each household was done to ascertain the absence of moldy areas following remediation.

2.5 Questionnaires to determine link between indoor air fungi and health symptoms

From these households, 47 residents, made up of 20 males and 27 females, ages between 0 and 80 years old, with one or more complaints of respiratory and other symptoms before remediation, were closely monitored after remediation for a period of 3 to 6 months to determine their state of health with respect to the previously observed symptoms. This was done with the aid of structured questionnaires designed to obtain personal information on age, sex, location before and after remediation and other salient features considered in determining the state of health of the residents post remediation. The following questions were asked:

a. What symptom(s) or complaint(s) have you experienced and when did it begin?
b. Do symptoms occur only when you are in the home, or at other places?
c. Do symptoms disappear when you are out of the house?
d. How frequent was the symptom
e. Does any other person in your home experience similar symptoms?
f. What steps did you take to solve the problem?
g. After the repairs and mold clean-up of your home did you notice a difference?

i. Frequency of symptoms
ii. Type of complaint
iii. Moldy odor in the home?

The symptoms included cough, nocturnal cough, dyspnea, sore throat, hoarseness, rhinitis, nasal bleeding, sinusitis, skin rashes, headaches, nausea, diarrhea, tightness in the chest, fever and muscle aches. The questionnaire responses were analyzed to establish the state of health of subjects post remediation. The Pearson correlation test was applied to determine significant association between mold presence or absence in homes and ill health symptoms experienced by residents.

3. Results

Table 3 shows outcome of questionnaire responses with respect to the eight houses revisited. Mold growth, Arthropod infestation and presence of suggestive symptoms were reported in all (100%) of the houses before remediation. All the households inspected, post remediation, had complete absence of moldy spots. There was significant correlation between reduction in frequency of symptoms of some health problems among residents and mold remediation ($r = 0.577$; $P>0.05$). Significant reduction of the following symptoms,

Features of responses	Yes (%)	No (%)
Location unplanned	72	28
Males	100	0
Females	100	0
Age of building (old)[1]	62.5	37.5
Leaking	50	50
Flooding	12.5	87.5
Dampness indoors	75	25
Mold growth indoor	100	0
Musty odor indoors	55	45
Arthropods infestation	100	0
Suggestive symptoms	100	0

Table 3. Features of Questionnaire responses with respect to the eight residences

cough (83.3%), dyspnea (100%), nasal bleeding (96.9%), skin rash (85.1%), tightness in the chest (83.3%), rhinitis (85.1), muscle aches (92.3%) and eye irritation (92.3%), were observed (Table 4). Reduction in frequency of occurrence of symptoms with respect to age and gender is shown in table 3. 72.3% of individuals experienced significant reduction or total

[1] Building above 25 years old was considered old

disappearance of symptoms of ill-health. More females (77.8%) experienced reduction in frequency of symptoms than males. Among the age groups examined, there was no significant difference in the rate of reduction of symptoms. There was no obvious change in the condition of some residents even six months beyond remediation and clear absence of moldy spots in their homes. Some symptoms ranked very low in reduction and were not significantly affected by mold remediation. These included sinusitis (Percentrank =0), nocturnal cough (Percentrank = 0.066), and fever (Percentrank = 0.133). Headache and nausea reduced appreciably but there was no strong association between mold absence and their reduction. Residents also reported relief from some of the symptoms like sneezing, cough and headaches when they left the home but reoccurred on their return. Exposure to moldy environment was influenced by location of building, dampness of the indoor environment and age of building (Student ttest =0.5; P<0.05).

Symptoms of ill-health	**Frequency before (days/year)**	**Frequency after (days/year)**	**Difference (days/year)**	**Actual reduction (%)**	**Percentrank**
Cough,	90	15	75	83.3	0.533
Nocturnal cough	60	35	25	41.7	0.066
Dyspnea	33	0	33	100	0.933
Sore throat	90	20	70	77.8	0.333
Hoarseness	66	12	54	81.8	0.466
Rhinitis	87	13	74	85.1	0.666
Nasal bleeding	65	2	63	96.9	0.866
Sinusitis	94	94	0	0	0
Skin rash	67	10	57	85.1	0.666
Headaches	94	40	54	57.4	0.2
Nausea	45	0	45	100	0.933
Diarrhea	52	15	37	71.2	0.266
Tightness in the chest	30	5	25	83.3	0.533
Fever	87	45	42	48.3	0.133
Muscle aches	90	20	70	77.8	0.333
Eye irritation	65	5	60	92.3	0.8

Table 4. Frequency of symptoms of ill-health experienced by residents before and after mold remediation

Age group (Years)	Males		Female		Total	
	N[2]	R[3] (%)	N[1]	R[2] (%)	N[1]	R[2] (%)
0 - 10	4	2(50.0)	3	3(100.0)	7	5(71.4)
11 - 20	6	3(50.0)	14	11(78.6)	20	14(70.0)
21 - 30	2	2(100.0)	1	1(100.0)	3	3(100.0)
31 - 40	0	0	0	0	0	0
41 - 50	4	4(100.0)	8	5(62.5)	12	9(75.0)
51 - 60	4	2(50.0)	0	0	4	2(50.0)
61 - 70	0	0	0	0	0	0
71 - 80	0	0	1	1(100.0)	1	1(100.0)
Total	**20**	**13(65.0)**	**27**	**21(77.8)**	**47**	**34(72.3)**

Table 5. Reduction of symptoms among residents with respect to age and sex

4. Discussion

Indoor air fungi may be responsible for some of the health problems experienced by residents of contaminated homes. There was persistent cough, headache, skin rashes and sneezing in some of the residents before mold cleanup. This did not reoccur after relocation of cleanup of molds. The major factors suggested to contribute to indoor air fungi were observed in a significant percentage of the households. Mold growth was seen in all the houses, at one point of the other with varying abundance and locations. There was significant reduction in frequency of symptoms of some health problems among residents post remediation. This implies that there may be a link between indoor air contamination by fungi and occurrence of such health problems. Jacobs et al., (2007) reported that housing interventions and remediation effected reduction of symptoms of asthma among subjects. Dales et al, (1999) and Shenasa et al, (2007) had, independently, related some ill-health conditions to indoor mold contamination. Some health conditions however barely changed, suggesting that sources other than fungi may be responsible. Symptoms like headaches and fever are also indicators of several other disease entities such as malaria, typhoid and other endemic diseases in the study area. Others such as sinusitis may have assumed a chronic status and hence the little or no reduction recorded. Frequencies of occurrence of symptoms among residents recorded before and after remediation showed remarkable difference between males and females. Shenasa et al, (2007) made a similar observation persons with symptoms of depression. This may reflect the level of exposure to sources of fungi outside the home, and the difference in reaction to triggers of such symptoms in relation to proximity to source of fungi. A similar observation was made by O'connor, (2004) and Mirabell et al, (2006) relating outdoor exposure with asthma in children. Males are more

[2] Number of persons in each group

[3] Number of persons with significant reduction of symptoms

involved in outdoor activities either for relaxation or work. There was no significant difference in the rate of reduction of symptom among the age groups. Of note, however is the only elderly female subject, who suffered prolonged respiratory symptoms (cough and runny nose) prior to remediation, which resolved barely three weeks after remediation. It was clear that the source of the symptoms was indoor mold since the old woman spent most of her days indoors.

Residents are gradually taking interest in the relationship between their residential environment and recurring health problems. Other researchers have suggested that effective communication with building occupants is an important component of all remedial efforts. Awareness of the causes and sources of agent of ill health would enable individuals who believe they have mold-related health problems to seek medical attention.

The health symptoms reduced in frequency after leaky roofs and other risk factors that promote mold growth were attended to. This agrees with the numerous reports which emphasize that the health risks of biological contaminants of indoor air could thus be addressed by considering dampness as the risk indicator. Several widely acknowledged global trends contribute to the conditions associated with increased exposure to dampness and mould (Canadian Center for Occupational Health and Safety, 2009). Conditions that favoured presence of indoor molds was well established in the residences studied and may have contributed to the high occurrence of symptoms. The high relative humidity of the study area is another factor that may encourage growth of fungi. Mshelgaru and Olonitola (2010) in considering the post occupancy contamination of timber buildings in Nigeria, found that fungi were predominant in the rain forest belts which included Plateau State. He found that 64% of the workers on such sites suffered symptoms of the sick building syndrome.

To our understanding, most people in the study area, both in administration and planning positions, as well as residents who are the direct victims of a moldy indoor environment, are still very much ignorant of the enormity of the problem. There is a serious need for more research, enlightenment and implementation of research findings by all stakeholders if a better indoor air quality is to be achieved.

5. Conclusions

Dampness in indoor environment encourages mold colonization and indoor air contamination. Molds and moisture indoor were associated with symptoms of ill health among residents. Some adverse health effects of indoor fungi on residents were established in the study area. The young children and the elderly were particular prone to mold-related symptoms and remediation provided relief. Reduction of symptoms was easily noticeable among these individuals when molds were removed from the home. The need for collaborative efforts to tackling the problem of indoor contamination is further highlighted. All stakeholders should participate in engaging resources at our disposal to prevention and control of contamination of the indoor environment with pathogenic fungi.

6. Acknowledgement

The authors wish to acknowledge the following: All member of the households who agreed to be enlisted in this work, the data analyst and family members.

7. References

Abbot, S. P. (2002). Mycotoxins and indoor moulds. *Indoor Environment CONNECTIONS* Vol. 3, No. 4, pp. 14-24.

Ahiamba J.E.; Dimuna, K.O. & Okogun, G.R.A. (2008). Built Environment Decay and Urban Health in Nigeria. *Journal of Human Ecology*, Vol. 23 No. 3 pp. 259-265.

Ayanbimpe, G. M.; Wapwera, S.D.& Kuchin D. (2010). Indoor air mycoflora of residential dwellings in Jos metropolis. *African Health Sciences* Vol. 10, No. 2, (June 2010), pp. 172-176.

British Columbia Non-Profit Housing Association (2007). Determining Good Health as Part of Housing Solution inBritish Columbia. In: *Housing Affects Health Affects Housing.* May 2007. Accessed January 2009.

Buck, K. M. (2002). Fighting Fungi Environmental Contaminants in Healthcare.

Canadian Center for Occupational Health & Safety (2006). Canadian Centre for Occupational Health & Safety (2006).

Canadian Center for Occupational Health and Safety (2009). Biological Hazards. Indoor Aie Quality Molds and Fungi. Available from: www.ccohs.ca

Cannon, P. F. and Hawksworth, D. L. (1995)The Diversity of Fungi Associated with Vascular Plants: the known, the unknown and the need to bridge knowledge gap. Advances in Plant Science vol. 11, pp. 277-302.

Codina, R.; Fox, R. W.; Lockey, R. F.; Demarco, P. & Bagg, A. (2008). Typical levels of airborne fungal spores in houses without obvious moisture problems during a rainy season in Florida, USA. *Journal of Investigative Allergology and Clinical Immunology* Vol. 18, No. 3, pp. 156-162.

Croft W. A., Jarvis B. B. and Yarawaya C. S. (1986). Airborne outbreak of Trichothecene toxicosis. Atmosphere and Environment vol. 20. Pp. 549-552.

Dales, R. E.; Zwanenburg H.; Burnett R. & Franklin C. A. (1991). Respiratory health effects of home dampness and mold among Canadian xhildren. American Journal of Epidemiology Vol. 134, pp. 196-203.

Dales, R. E.; Miller, D. & Mc Mullen, E. D. (1997). Indoor Air Quality and Health: Validity and Determinants of Reported Home Dampness and Moulds. *International Journal of Epidemiology* Vol. 26, No. 1, July 1996, pp. 120-125.

DeHoog, G. S.; Guarro, J.; Gene, J. & Figueras M. J. (2004). Atlas of Clinical Fungi. Atlas Version 2004.1, CD realization by Weniger, T. Computer Science ii University of Wurzburg Germany.

Douwes, J. (2009). Building dampness and its effect on indoor exposure to biological and non- biological pollutant, In: Dampness and Mould, vol. 7, pp.7-29; WHO Europe, ISSN 978-92-890-4168-3, Copenhagen.

Dubey S.; Lanjewar S.; Sahu, M.; Pandey, K. and Kutti, U. (2011). The Monitoring of Filamentous Fungi in the Indoor , Air Quality and Health. Journal of Phytology vol. 3, No. 4, pp. 13-14.

Olowoporoku D. (2007). Air Quality Management in Lagos. Air Quality Management Resource Centre, UWE, Bristol 9 May 2007.

BCBS, (2007). Key Questions: Environment and Housing Quality. Version 1.1, Accessed July 2009, Available from www.designforhealth.net

Environmental Protection Agency (2010). Indoor Air Pollution: An introduction for Health Professionals. Available from: http://epa.gov/iaq/pubs/hpguide.html

European Environment and Health Information System (2007). Children Living in Homes with Problems of Damp. Fact Sheet No. 3.5- May 2007- Code RPG3_Hous_Ex2.

Enhis (2008). Children living in homes with problems of dampness. In Enhis Version 1.8, 21 October 2008.

Federal Republic of Nigeria (2007). National Environmental Standard and Regulation Agency (Establishment) Act, 2007. Federal Republic of Nigeria Official Gazette vol. 94, No. 92., 31 July 2007.

FEPA, (1999). National Master Plan for Public Awareness on environment and Natural Resource conservation in Nigeria, FEPA Garki, Abuja.

FGN (1988). Federal Environmental Protection Agency Decree 58, 1988,Federal Ministry of Information and Culture, Lagos, Nigeria.

Gots, R. E.; Layton, N. J. (2003). Indoor Health: Background Levels of Fungi. Journal of Occupational Health and Hygiene vol. 64, No. 4, 427-438.

Hawksworth, D. L. (1991). The fungal dimension of biodiversity: magnitude, significance, and conservation. *Mycological Reasearch* Vol. 95, pp. 641-655.

Hawksworth, D. L. (1998). Kingdom Fungi: Fungal Phylogeny and Systematics. In: Collier L., Balows, A., Sussman, M. (eds.). Topley and Wilson's Microbiology and Microbial Infections , 9th edition. Arnold, London. 1988; pp. 43-54.

Heseltine E. and Rosen, J. (2009). WHO Guidelines for Indoor Air Quality: Dampness and Mold. WHO Europe Damness and Molds. Available from: www.euro.who.int ISBN: 978 92 870 4168 3.

Housing Health and Safety Rating Systems London Communities and Local Government (2007). Accesses July 2009. A vailable from: http://www.communities.gov.uk/index.asp?id=1152820

Horner, W. E.; Worthan, A. G. & Morey P. R. (2004). Air- and dustborne mycoflora in houses free of water damage and fungal growth. *Applied Environmental Microbiology*. Vol. 70, No. 11, November 2004, pp. 6394-6400.

Howell E.; Harris, L. J. & Popkin S. J. (2005). The health status of Hope VI public housing residents. *Journal of Health Care Poor & Underserved* Vol. 16, pp. 273–285.

http://www.epa.gov/mold/moldresources.html

Husman T. (1996). Health effects of indoor-air microorganisms. *Scand J Work Environ Health* Vol. 223, pp. 5-13.

Jacobs, E. D.; Kelly, T. & Sobolewsky, J. (2007). Linking Public Health, Housing, and Indoor Environmental Policy: Successes and Challenges at Local and Federal Agencies in the United States. *Environmental Health Perspectives* Vol. 115, No. 6, June 2007, pp.976-982.

Johaninning E., Biagini R., Hull D., Morey P. R., Jarvis, B., and Landsbergis, P. (1996). Health and Immunology Studies following exposure to toxigenic fungi (Starchybotrys chartarum) in a water-damaged office environment. Int. Arch. Occup. Environ Health. Vol. 68, pp.207-218.

Kennedy, R. (n.d). Strategies for Preventing or Removing Mold Growth After Contamination, Handout 4. In: US Department of Agriculture,University of Akansas, Cooperative and Extension Services.

Kuhn, D. M. & Ghannoum, M. A. (2003). Indoor mold, toxigenic fungi, and Stachybotrys chartarum: infectious disease perspective. *Clinic Microbiology Reviews* Vol. 16, No.1, 144-172.

McNeel, S. V.; and Kreutzer R.A. (1996). Fungi and Indoor Air Quality. Health and Environment Digest vol. 10, No. 2, pp. 9-12.

Mirabell, M. C.; Wing, S.; Marshall, S. W. & Wilkosky, T. C. (2006). Asthma Symptoms Among Adolescents Who Attend Public Schools That Are Located Near Confined Swine Feeding Operations. *Paediatrics* Vol. 118, No. 1, July 2006, pp. e66-e75.

Moloughney, B. (2004). Housing and Population Health: The State of Current research Knowledge. Ottawa: Canadian Institute for Health Information.

Morey P. R. (1999). Indoor Air Quality, Fungi, and Removal of Fungal Colonization. Presented at: Mealy Sick Building Litigation Conference November 1999. West Palm Beach, Florida.

Morgan W. J.; Crain, E.F.; Gruchalla, R.S.; O'Connor, G.T. Kattan, M.; Evans, R. III et al. (2004). Results of a home-based environmental intervention among urban children with asthma. *New England Journal of Medicine* Vol. 351, pp.1068–1080.

Mshelgaru I. H. and Olonitola O.S. (2010). Health and Safety Conditions of Buikding Maintenance Sites in Nigeria: Evaluating the Post-Occupancy Contamination of Timber Buildings buy Microorganisms. African Journal of Environmental Science and Technology vol. 4, No. 1, pp. 013-020.

New York City Department of Health and Mental Hygiene (2009). Guidelines on Assessment and Remediation of Fungi in Indoor Environments November 2008. Available from: www.nyc.gov/health

North Carolina Extension Service (2009). Mildews Prevention in the Home. Housing and Home Furnishing Specialists. Revised by Kirby S. D. Available from: http://www.ces.ncsu.edu

O'Connor, G. T.; Walter, M.; Mitchell, H.; Kattan, M.; Morgan, W. J.; Gruchalla R. S.; Pongracic, J. A.; Smartt, E.; Stout, J. W.; Evans, R.; Crain, E. F. & Burge, H. A. (2004). Airborne fungi in homes of children with asthma in low-income urban communities: The inner-City asthma study. *Journal of Allergy and Clinical Immunology* Vol. 114, No. 3, September 2004, pp. 599 - 606.

Pasanen A.L., Lapalainen S., and Pasanen P. (1996). Volatile organic Metabolites Associated with some toxic fungi and their mycotoxins. Analyst vol. 12, pp. 1949-1953.

Pasanen A. L., Korip, A., Kasanen, J.P. and Pasanen P. (1996). Critical aspects on significant of microbial volatile metabolite as indoor air pollutants. Environment International vol. 24, No. 7, pp. 703-712.

Petrescu, C.; Suciu, O.; Lonovici R.; Herbarth, O.;Franck U. and Schlink U. (2011). Respiratory Health Effectof Air Pollution and Climate Parameters in the Population of Drobeta Turnu-Severin, Romania. In: Air Pollution-New Developments, Moldoveanu A. M. (Ed.), ISBN: 978-953-307-527-3, InTech, Available from: http://www.intechopen.com/article/show/title/respiratory-health-effects-of-air-pollution-and-climate-parameters-in-the-population-of-population-of-drobeta-turn

New York City Department of Health and Mental Hygiene (2008). Preventing and Cleaning Mold Growth, Fact Sheet for Building Owners and Managers. Accessed July 2011. Available from: www.nyc.gov/health

Nielson, K. F. (2003). Mycotoxin Production by Indoor Molds. Fungal Genetics and Biology vol. 39, pp. 103-117.

Robins C. & Morell J. (2007). Mold Housing and Wood. In: Western Wood Products Association 2007. Available from: Â© 2007 Western Wood Products Association.

Sautour, M.; Sixt, N.; Dalle, F.; L'Ollivier, C.; Fourquenet, V.; Calinon, C.; Paul, K.; Valvin, S.; Maurel, A.; Aho, S.; Couillault, G.; Cachia, C.; Vagner, O.; Cuisenier, B.; Caillot, D. & Bonnin, A. (2009). Profiles and seasonal distribution of airborne fungi in indoor and outdoor environments at a French hospital.Science of the Total Environment, vol. 407, No. 12, pp.3766-3771.

Shelton, B. G.; Kirkland K. H.; Flanders W. D. & Morris, G. K. (2002). Profiles of airborne fungi in buildings and.outdoor environments in the United States. Applied Environmental Microbiology, Vol. 68, No. 4, April 2002, pp. 1743-1753.

Shenassa, E. D.; Liebhaber, A.; Daskalakis, C.; Braubach, M. & Brown M. (2007). Dampness and Mold in the Home and Depression: An Examination of Mold-Related Illness and Perceived Control of One's Home as Possible Depression Pathways. *American Journal of Public Health* Vol.97, No.10, 1893-1899.

Solomon A. O.; Ike E. E.; Ashaano E. C. and Jwanbot D. N. (2002). Natural Background Radiation on the Jos Plateau and use of the Rock for House Construction. AfricanJournal of Natural Sciences vol. 5.

Stetzenbach, L. D.; Amman, H.; Johanning, E.; King, G & Shaughnessy, R. J. (2004). Microorganisms, Mold, and Indoor Air Quality. In: American Society for Microbiology (2004), Indoor Air Quality, December, 2004, pp. 1-20.

Straus, D. C. & Wilson, S. C. (2006). Reespiratory Trichothecene Mycotoxins can be Demonstrated in the air of Stachybotrys chartarum contaminated buildings buildings. *Journal of Allergy and Clinical Immunology* Vol. 118, pp. 760.

U.S. Environmental Protection Agency (2003). An Introduction to Indoor Air Quality (IAQ). Biological Pollutants. Accessed, April 2011.

Verhoeff, A. P. & Burge, H. A., (1997). Health risk assessment of fungi in home environments. *Annals of Allergy, Asthma and Immunology.* Vol. 78, No. 6, pp.544-556.

Vocaturo, E.; Kunseler, E.; Slovakova G.; Ruut, J.; Cavoura O. & Otorepec P. (2008). Children living in homes with problems of dampness. In: *ENHIS, World Health Organisation,* Version 3.5, Menu. Bilthoven, accessed 22 October 2008. Available from: RIVM, Home\ Environment and health issues\ Housing

Workplace Health, Safety and Compensation Commission of New Brunswick, (2000): Microbials and indoor air quality Moulds and Bacteria. Published: December, 2000

Worksafe Bulletin (1999). Building Related Moulds. Bulletin No. 194.

Wu, F.; Biksey, T. & Karol, M. H. (2007). Can Mold Contamination of Homes Be Regulated? Lessons Learned from Radon and Lead Policies. *Environmental Science Technology,*Vol.41, No. 14, pp. 4861-4867.

Zaslow S. A.; and Genter M. B. (1993). Mold, Dust Mites, Fungi, Sporesand Pollen: Bioaerosols in the human Environment. North Carolina Extension Service FCS-360-5. Available from: www.ces.ncsu.edu

2

A Qualitative Study of Air Pollutants from Road Traffic

Gabriela Raducan* and Ioan Stefanescu
The National Research and Development Institute for Cryogenics and Isotopic Technologies, Rm. Valcea, Romania

1. Introduction

Air quality in urban areas continues to represent a major concern, taking into account the impact of pollution on environment and human health. Air quality study involves researchers of several scientific disciplines and requires the development of theoretical, practical and methodological concepts, always new, for understanding, controlling and combating urban air pollution. Together with monitoring, modeling air pollution is a very important tool in the management of air quality. It enables efficient verification in order to reduce emissions and make prediction of pollution levels associated with changes in urban infrastructure.

In recent decades many experimental and modeling studies have been conducted, in order to obtain information on the dispersion of pollutants and transformations suffered by them in the street canyon. Depending on the purpose, different techniques were used for monitoring and modeling. Some of these studies are purely experimental, which means they rely only on small-scale measurements. Others are purely theoretical, concentrating on investigating the various schemes of atmospheric fluid flows and dispersion of pollutants, using mathematical models. The most common are those that combine mathematical modeling with field measurements, as the models for street canyon.

This work presents a series of studies for the correlation of emissions from road traffic, meteorology and concentrations of pollutants at a local scale. It is a qualitative analysis meaning that the study compares the concentration of fictitious values. The aim of this study has been to enrich and understand how to make an analysis of the spatial and temporal variability of air pollution in urban areas.

2. Urban air pollutants

The most important urban air pollutants in terms of emissions and the impact on the environment and human health which could be taken into account are: carbon monoxide (CO), nitrogen oxides (NO and NO_2), volatile organic compounds (VOCs), ozone (O3) and aerosol particles with a diameter less than 2.5μm ($PM_{2,5}$).

---

* Corresponding Author

Carbon monoxide is found in abundance in urban air pollution in quantities that no other toxic gases are. Carbon monoxide is formed, usually by one of the following three processes: incomplete combustion of fuel containing carbon, reactions of carbon dioxide and materials containing carbon at high temperatures and dissociation of carbon dioxide at high temperatures.

The main nitrogen oxides are nitrogen monoxide (NO) and nitrogen dioxide (NO_2), and together are called noxious NO_x. About the effects on health, NO are considered harmless, at least at concentrations that usually exist in the street. Instead, NO2 can cause serious problems. Only a small fraction of the gases emitted by engines is represented by NO2, the principal amount being NO. This NO2 in ambient air is mainly due to further oxidation NO. Chemistry of nitrogen oxides is quite complex, but due to the small time the pollutants reside in the street canyon, the reactions of practical interest are: oxidation of nitrogen monoxide to nitrogen dioxide, nitrogen dioxide decomposition in nitrogen monoxide and oxygen and the forming of ozone resulting from the reaction between the oxygen molecule and the oxygen root.

In the category of volatile organic compounds (VOCs) there are a large number of air pollutants, emitted from industrial sources and urban traffic, as well as from other sources. In terms of chemistry, volatile organic compounds include aliphatic and aromatic hydrocarbons, halogenated hydrocarbons, some alcohols, esters and aldehydes. The importance of these compounds is obvious, considering the fact that the agency EPA (Environmental Protection Agency) has designated them as belonging to the six types of critical air pollutants.

Both natural sources and anthropogenic sources contribute to VOC emissions. Natural sources include oil, forest fires and some chemical reactions to produce these compounds. The main anthropogenic sources include fossil fuel combustion at high temperature, the obtaining of oil refined and unrefined, incineration, burning of crops and other debris after collection.

Ozone is formed and destroyed chemically in the atmosphere, the stratosphere or the low troposphere. The complicated processes of training and distribution of ozone in the stratosphere are of great importance for the amount of ultraviolet radiation reaching the earth's surface and for the energy balance of earth-atmosphere system. Ozone is formed under the influence of direct solar radiation from volatile organic compounds and nitrogen oxides. Both the training and the destruction of ozone are photochemical reactions. Ozone concentration in the troposphere has increased lately, especially in the northern hemisphere, due to increasing anthropogenic emissions. Ozone is a strong oxidizer and high concentrations at this level can cause strong negative effects on the environment and human health.

Aerosol particles cover a wide dimensional field, from the cluster of a few molecules to particles greater than 100 μm in diameter. Dimensional distribution of aerosol particles presents a certain regularity that is the result of production of nuclei from different sources and loss through different processes. Each type of particles, taken individually, has its own characteristics, related to its size, geometry, concentration, chemical composition and physical properties.

The particles with less than 2.5μm diameter are a real concern because they remain suspended in the atmosphere, where, depending on size, can be settled relatively difficult.

Aerosol particles, both those produced naturally, as well as those produced by anthropogenic sources are involved in the extinction of solar radiation, can spread the light and reduce visibility and, therefore, can influence, both direct and indirect (cloud formation), the global or regional climate. Inhalation by humans and animals may have adverse effects on health. The small particles, with the diameter less than 1μm, behave like gases; they are involved in the Brownian movement, follow the lines of fluid current near the obstacles, are able to coagulate and are deposited on the surface of the earth very hard. Particles larger than 10μm are strongly affected by gravity, being involved in processes of dry and wet settling.

3. Methods of monitoring atmospheric pollutants

The preliminary assessment of atmospheric pollution in urban area is a very important step in finding the locations where to deploy fixed monitoring stations. Such locations are selected according to information which includes: distribution of pollutants sources in the urban area, the expected maximum concentration points (hot points), the meteorological and topographical conditions, the model applications and other suitable features. The monitoring sites have to be representative of a sufficiently large area in the vicinity, so that the sampling station can be considered representative of a larger area or representative of sites characterized by similar environmental conditions.

The examination of modern methods used for air monitoring permits to answer the question: how air monitoring has to be performed. The variety of analytical methods offers a wide selection of procedures, which can be carried out by means of static, mechanized or automatic devices. The choice will depend upon the use of the monitoring data and the aim which has to be reached.

Static devices, such as the ones employed to measure the amount of deposited particles (dust fall) are used for mapping, for definition of special problem areas and for general survey.

Mechanized bubbler devices are used to collect nitrogen dioxide, mercury, and other gases and vapors. These samplers, although typically designed for collecting 24 h integrated samples, can be modified to collect 1 or 2 h samples in sequence, and thus allow definition of diurnal variations.

In automatic sampler—analyzers, collection and analysis are combined in a single device. These automatic instruments produce continuous analysis, with the output in a machine-readable format or in a suitable form to a central data-acquisition facility.

The approach of the various instruments described to monitor a pollutant either by laboratory analysis or by analyzers is the same. These instruments can be called 'point' sensors as they measure the concentration of the given pollutant at a single point. Another approach for monitoring is the 'remote sensing'. This term indicates the use of instruments which can provide the average concentration of a pollutant in a certain area either by looking at the emissions as they exit at the stack output or by sampling an optical volume at a point within the plume and conducting a spatially integrated measurement across the diameter of the plume. Remote sensing can be performed also by means of a 'long-path sensor'; this term indicates any device which permits one to measure extended or diffuse sources, such as oil refineries and chemical complexes between two points.

For some pollutants however, there are no methods of measurement based on sensors and monitors, and in that case the system requires a periodic transfer of this data.

Choosing compounds to be measured is related to the type of monitoring stations. These, according to the standards, may be:

1. Urban Background Station (for nitrogen oxides, ozone, aerosol particles and sulfur dioxide (optional));
2. Industrial / Residential Station (for nitrogen oxides, sulfur dioxide, aerosol particles and possibly other compound-specific ammonia or volatile organic compounds);
3. Traffic Station (for nitrogen oxides, carbon monoxide, aerosol particles, volatile organic compounds (benzene) and sulfur dioxide (optional));
4. Regional Background Station (for nitrogen oxides, sulfur dioxide, ozone, aerosol particles, volatile organic compounds).

In recent years the focus has been on the use of modern techniques, easy to use, inexpensive and having a good accuracy. Thus, modern information systems for monitoring air quality were made, which enable integrated approach and direct and rapid access to data. They can be used for the purpose of evaluating and planning future actions. Such a surveillance system could be consisted of:

- Sampling systems, sensors and monitors installed in special monitoring stations;
- Data package transfer systems and programs related to quality assurance and control data;
- Database for the air quality, weather and the emission of pollutants;
- Numerical or statistical models (including models of dispersion and meteorological forecasting models);
- Modules for graphic presentation of the results of tests;
- Distribution systems and communication of results through network in order to inform users and the public or for preparing strategies to reduce pollution by the authorities.

Automatic analyzers can provide continuous measurements of concentrations of pollutants without further laboratory analysis. Monitoring procedures are fully automated and use analysis techniques such as chemiluminescence or infrared absorption. Concentration determined by the analyzer is recorded in a data storage device that can be internal or external. The monitoring stations usually operate with multiple analyzers (eg CO, NO, NO2, NOx, O3), because the price is much smaller than the individual analyzers, taken together. A typical monitoring station, along with the analyzer, implies the existence of a room weatherproof and highly secure, air-conditioning system, sample collection lines, a powerful vacuum pump to absorb quantities of sample in ambient air, calibration gas, a data storage system and another data transfer.

There are several types of measurement devices with various features. It is very important to choose them properly.

To measure CO and CO_2 concentrations devices operating on the principle of non-dispersive IR analysis could be used. The IR radiation is absorbed by the CO and CO_2 at specific wavelengths. Gas sample found in the room where measurements are made, is exposed to IR radiation coming from a powerful source. Absorbed energy is measured. The camera is filled with sample gas and clean air, in turn, and the difference of energy absorbed is calculated.

Analyzer for nitrogen oxides (NO, NO_2,) works on the principle of chemiluminescence. Gas sample is first filtered through a filter that can have a Teflon membrane. Nitric oxide present in the sample reacts with ozone generated within the analyzer, producing nitrogen dioxide in the excited state. It emits electromagnetic radiation that is detected and measured by photomultiplier tube, being proportional to the concentration of NO in the sample. A catalytic converter that reduces NO_2 to NO is used to measure NO_2, allowing measurement of total amount of nitrogen oxides (NOx). The measuring cycle of the NO and NO_2 concentrations is very important for the NOx analyzer.

Low levels of O3 could be measured using a spectrophotometer. Ozone absorbs UV radiation of 254 nm wavelength, according to Beer-Lambert law (Schneider et al, 1997). The analyzer determines the concentration of ozone in the sample by measuring the attenuation of UV radiation due to ozone absorption wavelength above mentioned. The gas sample to be measured is led into an analyzer using a diaphragm pump. It passes first through a filter for aerosol particles or by a catalytic converter. The converter selectively changes the ozone sample in O2, thus generating a reference gas to contain no ozone. When the reference gas passes through the cell is set "0" of light intensity. This value is kept as a reference, by the microprocessor. When the sample gas replaces the reference gas the luminous intensity is measured and then the system calculates the difference which is proportional to the mass of ozone contained in the cell. To obtain a relationship of proportionality between light intensity and amount of light absorbed, the ozone analyzer was equipped with a temperature and pressure compensator which includes a temperature converter and a pressure converter.

To measure the particulate matters (PM), the element oscillating microbalance (Tapered Element Oscillating Microbalance- TEOM) could be used or a device based on beta radiation. Both methods are based on basic physical principles.

TEOM is a device used to measure real-time mass concentration of particles with sizes smaller than 2,5 μm (or 10 μm), indoors and outdoors. It records: mass concentration, mass ratio and total mass accumulated in the filter cartridge.

Beta radiation attenuation monitor (BAM-Beta Attenuation Monitor): the beta radiation flow, emitted by a radioactive source located inside the monitor, is directed to the filter containing deposited particles and is attenuated in proportion to the mass of collected particles. This system also allows continuous hourly (or 30 minutes) measurements of aerosol particles of different sizes.

4. The dispersion process

In the initial dispersion of pollutants from the emission sources, they are dispersed into the environment by movements that depend on weather and surveying conditions. In the dispersion process, significant weather phenomena are related to wind speed and direction, turbulence and atmospheric stability.

The wind speed plays an important role in transport and dilution of pollutants. If the emission speed is relatively constant, a doubling of wind speed will result in halving the amount of pollutants, so the concentration is inversely proportional to wind speed.

The speed of the horizontal wind is affected by friction, which is proportional to surface roughness. The last one depends on local natural features: mountains, valleys, rivers, lakes, forests, cultivated fields and buildings. Wind speed over soft surfaces (cultivated areas or lakes, etc.) is on average higher than the wind speed over rough surfaces (mountains, buildings, etc.). The effect of surface roughness on wind speed is a function of height above the earth's surface: the wind speed is lower near the surface.

Dispersion of pollutants is also affected by wind direction variability. If wind speed is relatively constant, the same area will be exposed to high levels of pollutant concentrations. On the other hand, if the wind direction will change constantly, it will disperse pollutants over a larger area, and concentrations in any of the exposed areas will be much lower. Wind direction and wind speed can be represented by a type of graph named wind rose. The spokes' length indicates wind direction frequency. This kind of graph can be used to predict dispersion from a point source or surface.

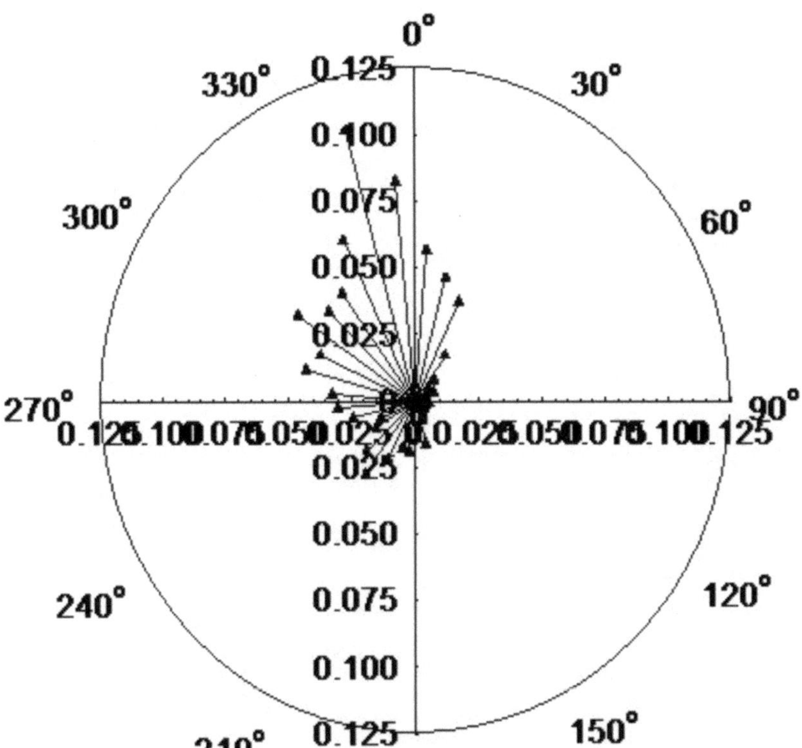

Fig. 1. Wind Rose

The air flow is not uniform near the surface of the earth, but presents a series of movements called vortices. The vortices are produced by two specific processes: thermal turbulence (turbulence caused by the atmospheric heat) and mechanical turbulence (resulting from the movement of air near various obstacles). Thermal turbulence is dominant in bright and sunny days with weak winds. Although mechanical turbulence is produced in various weather conditions, it is dominant in windy, in neutral atmospheric conditions. Turbulence has the effect of increasing the dispersion process, even when there is mechanical turbulence, on in certain areas, high concentrations of pollutants could be recorded, (Godish, 1997).

In large urban centres there are a multitude of fixed and mobile pollution sources. This urban community forms a heated and polluted island which is one of the biggest problems of the contemporary world. In an air quality analysis the ventilation rate of the entire region must be taken into account, because the dispersion can be influenced by both atmospheric mesoscale and macroscale movements. The atmospheric conditions as the anticyclones or thermal inversions slow the dispersion, resulting in high concentrations of pollutants. On the other hand, high weather variability such as moving cyclones, associated with strong winds, will enhance the dispersion of pollutants in urban areas. The pollution can move on, over very large areas, affecting air quality within hundreds of kilometers distances (Briggs, 1969).

Dispersion of car emissions is a complex process that acts at different scales of time and space and is strongly influenced by weather variability and air dynamics around vehicles and buildings. The amount of the exhaust gas varies over time, and it moves in time and space. Immediately after being discharged, the gases are dispersed due to turbulence that occurs around the vehicle, a phenomenon influenced by the non-stationary flow of air masses. A few seconds later, the dispersion is further accelerated by turbulence from other vehicles, moving within the street area.

In parallel with the mixture due to movement of vehicles, emissions are dispersed by air masses that have a dynamic movement because of the weather and topographical conditions (the buildings, trees, open areas, etc.) (Dobre et al, 2005).

Movement of vehicles creates uneven air movements (Plate, 1998) which mix the pollutants emitted from the exhaust pipe. After moving, the vehicle vortices arise. Basically, there isn't a database of local-scale air movements produced by motor vehicles in urban areas (Tate, 2004).

The influence of driving mode on vehicle emissions is of particular importance in such a study as higher emission rates are associated with certain traffic characteristics. (Tate, 2004)

The urban environment is characterized by different widths streets and buildings with different geometries. Dispersion of emissions from vehicles in narrow streets from urban areas has become a very intensely studied research theme (Chan et al, 2002, Xie et al, 2003) because it is known that very high levels of pollutants can be found within these streets, which means, most often, very high levels of pollution to which citizens are exposed.

The term "street canyon" refers to a relatively narrow street with buildings built by both sides, continuously (Nicholson, 1975). Dimensions of the street canyons are expressed usually by the characteristic ratio H / W, where H is the average height of a building located on both sides of the street, and W is the width of the street. A street canyon can be considered regular if the characteristic ratio is approximately equal to 1 and there are no major gaps in buildings on both sides of the canyon, which are like walls. If the wind blows obliquely to the direction of street-canyon, air flow reflection (wind in the roof) in the canyon wall which is exposed to the wind flow, induces a spiral flow along the canyon (Johnson and Hunter, 1999).

Atmospheric fluid dynamics inside the street canyon have been extensively studied by all types of experiments. Historically, most studies were based on idealization of the canyon with buildings placed evenly on both sides of the canyon, considering a wind perpendicular to the street (Berkowicz 1997, Chan et al, 2002). In reality, the geometry and weather conditions are statistically very unlikely.

When analyzing measurements of air pollution in street canyon it is important to have basic knowledge about the dynamics of air and wind effects on dispersion. Also, the moving vehicles generate turbulence which is preferable to be taken into account, if possible. A serious limitation of many studies of urban pollution is the lack of good data set to describe the synoptic weather conditions. Wherever possible, these critical parameters (wind speed, direction, and turbulent flows) must be measured in the area.

5. A study of spatial and temporal variability of air pollution in urban area

The purpose of this work is to understand how the examination of spatial and temporal variations of the pollutants' concentration in a street canyon could be done. It has to be linked with the traffic data, with emissions and weather conditions. To do this, the frequency of road traffic, the generation of emissions and the pollutants dispersion phenomenon must be taken into account.

A long term database is preferred, to also make an analysis of the pollutants trend. The database must be created using the raw measurements and eliminate negative values or strange data which could appear because of the device malfunctioning. If the measurement period is long, the devices must be calibrated from time to time (e.g. one time a month) and a few data could be lost. Time series plots must be done, pollutants together with traffic flow and meteorological data, to make complex analysis of the monitored data. The graphs must be analyzed, to find the correlations between different pollutants and the traffic flow. So, carbon monoxide, nitrogen oxide and particulate matter are emitted by vehicle engines during usage - primarily pollutants (Raducan and Stefan, 2009).

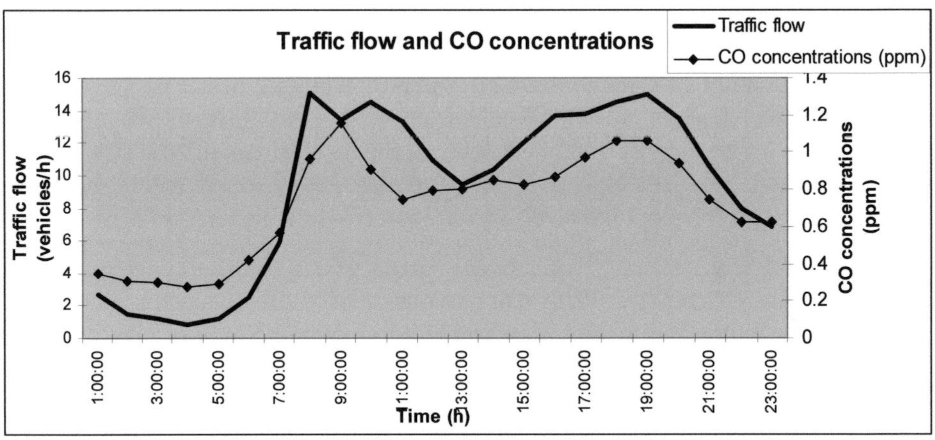

Fig. 2. Traffic flow - CO concentrations (primarily pollutant) – almost the same trend

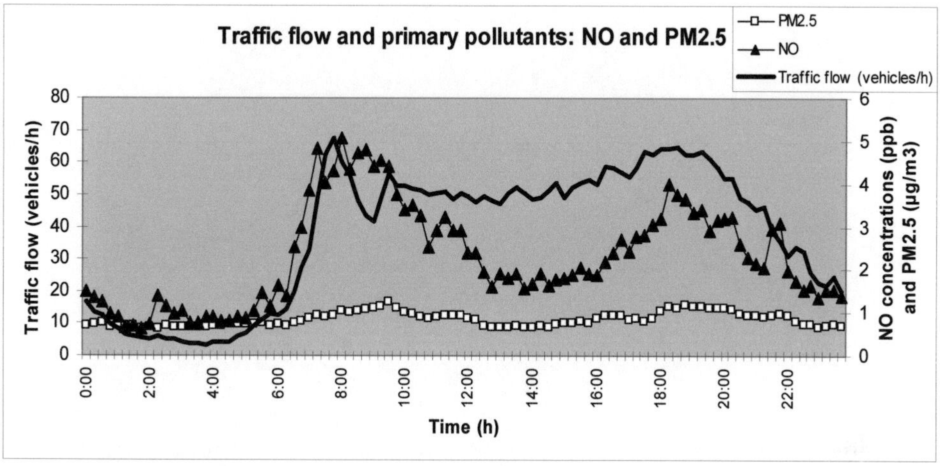

Fig. 3. Traffic flow - NO and PM $_{2.5}$ concentrations (primarily pollutants) – almost the same trend

The graphs show a relatively large amounts of traffic flow and concentration of pollutants around two moments of the day: in the morning, when people go to work and in the afternoon, when they come back, and a good correlation between the above mentioned pollutants, which demonstrates that their main source is the traffic.

Also, relatively high concentrations are recorded near the street, compared with the higher points or the more distant points from the street.

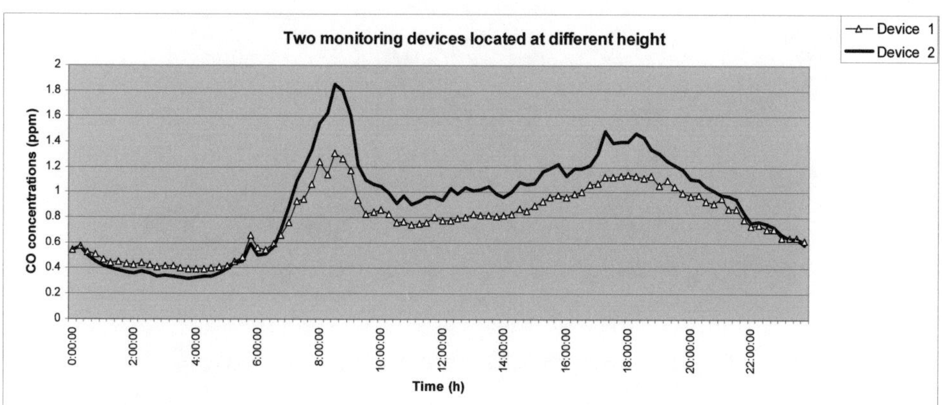

Fig. 4. The device 2 was closer by the street and it recorded higher level of pollution

When there are many devices which measure the same pollutant, installed at different heights, they must record the data as it can be seen in the graph below: the device which is set lower, records higher concentration values.

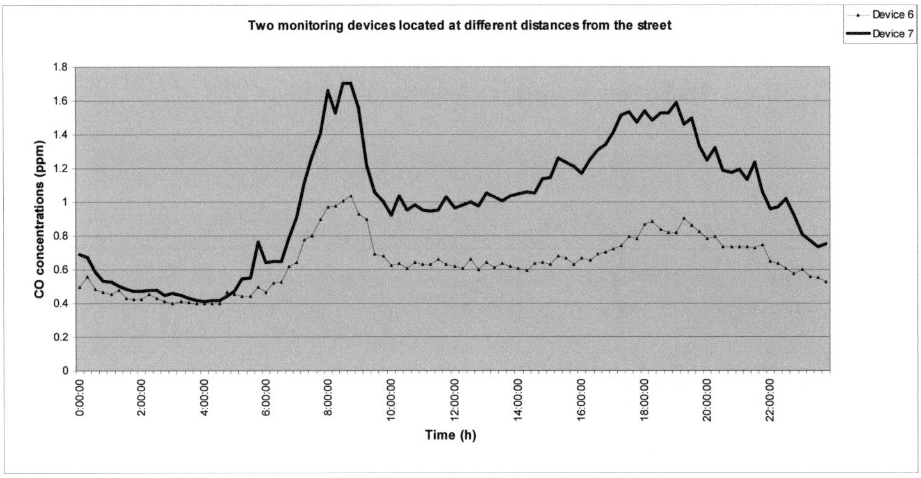

Fig. 5. The devices 6 and 7 ware situated in the same point near the street. Device 6 was located above

Meteorology has a key role; the results clearly show the importance of wind direction and speed: increasing wind speed cause a good dispersion of pollutants, so the registration of low concentration of pollutants. On the contrary, stable atmospheric conditions cause an accumulation of pollutants in the area of the street, considerably increasing concentrations.

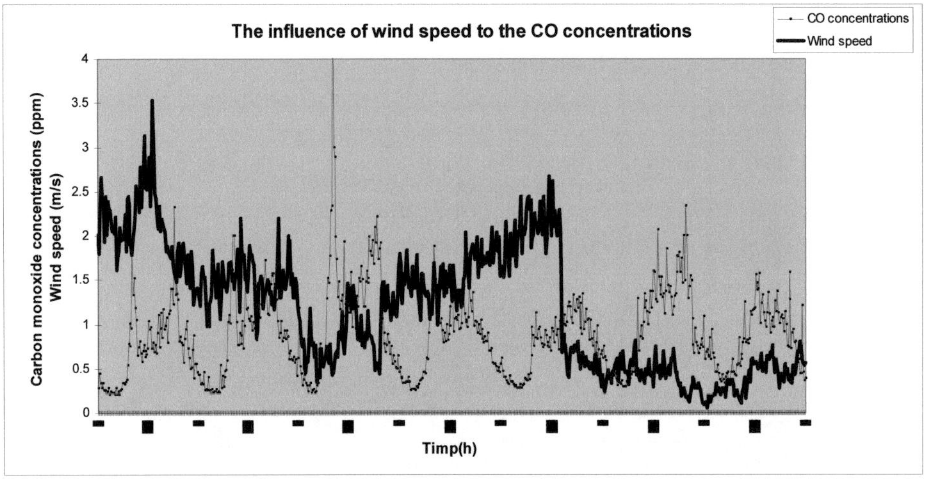

Fig. 6. When the wind speed is lower, the CO concentrations are higher

When the wind direction is perpendicular to the street, a vortex is created in the street, carrying pollutants towards the wind. If the wind direction is parallel to the street, the atmospheric fluid transports the pollutants along the street increasing the rate of concentrations in this direction. When the direction of the wind is skew toward the street the atmospheric fluid is moving helicoidally, carrying the pollutants both to the other side of the road, and along the street.

6. Air quality modeling

Mathematical modeling is one of the most powerful modern tools used in most studies of air pollution. Since the monitoring urban pollutants is very expensive, the modeling of the pollutants dispersion is an alternative that offers good results in the study of air quality.

A simple definition of the concept of "model" may be as follows: the model is a mathematical representation of facts, factors and interpretation of data quantity or situations.

The models are used mainly for: generalization and interpretation of measurement, the pollution prediction and the simulation of the pollution scenarios (answers to questions: what will happen if ...?)

Within the operational plans, the models are now widely used, the main focus being (Zanetti, 1993):

- The establishment of legislative measures to control emission of pollutants into the atmosphere, by determining the maximum emission of pollutants that can take place without exceeding air quality standards;
- The assessment of the proposed emission control techniques (air cleaning technologies);
- The selection of positions for the future pollution sources, in order to minimize environmental impact;
- The control of pollution accidents by defining the measures and strategies to be taken immediately after the accidental pollution episodes (warning systems and real-time strategy to reduce emissions);
- The assumption of the responsibilities, for higher levels of pollution, by establishing the source-receptor relationship.

Types of models

Models currently used can be classified as follows:

- Physical models;
- Empirical models;
- Mathematical models;

Physical models are small-scale representations of "real world", e.g. representations and reproductions of the phenomena of nature in the laboratory. An example of such common representations is the wind tunnel or water where they can reproduce the dispersion experiments in the presence of miniature buildings, relief, vegetation - forests, etc. Another example is the fog chambers where chemical reactions can be studied under the same atmospheric pressure, temperature, humidity and solar radiation. There are several limitations in this small-scale reproduction of the phenomena from the nature: the conditions of turbulence in wind tunnels or water can not fully simulate the real conditions, many chemical reactions occur in fog room at their walls, etc.

Physical models have shown very interesting results, revealing some mechanisms and providing validation data, then used in developing mathematical models.

Empirical models generalize the relationships between different parameters, experimentally determined.

Mathematical models, the most complex models currently developed, can be divided into:

- Statistical models based on semi-empirical statistical relationships between various data and available measurements.
- Deterministic models based on mathematical description of atmospheric processes, where the effects (e.g. air pollution) are generated by causes (e.g. pollutant emissions).

The models of type receiver could be good examples of statistical models. Receptor type models use measured data for different concentrations of pollutants, for a long time. They can identify the impact of the contribution from different sources using statistical analysis of the measured concentrations at a certain point (receiver). Receptor type models are used mainly for the interpretation of air quality measurements.

An example of a deterministic model could be a diffusion model in which the output data generated by the model (pollutant concentrations) is calculated based on mathematical computations performed on the input data (emissions, atmospheric parameters, etc.).

Deterministic models are most suitable for practical applications, because if they are correctly calibrated and used, they generate very accurate and consistent results regarding the source-receiver relationship. Establishing accurate source-receptor relationships is the target of any study which is made to improve the air quality or to keep the current levels of concentration for future development of industrial or urban areas. In other words, only deterministic models can generate clear assessment of the responsibility of each source of pollution from each receiver, this thing being absolutely necessary in the design and implementation of emission control strategies.

In terms of mathematical models, the modeling techniques used in the deterministic models must be applied to all aspects and phenomena related to air pollution. The main phenomena which are mathematically modeled:

- atmospheric transport,
- turbulent diffusion in the atmosphere,
- atmospheric photochemical and chemical reactions,
- deposition to the soil surface.

Turbulent diffusion theory is focused on the behavior of pollutant particles which follow trajectories that allegedly are caused by the wind field. Due to the randomness of atmospheric dynamics it is impossible to determine the exact distribution of pollutants concentration in the atmosphere due to the pollution sources. Although the basic equations describing turbulent diffusion can be written without difficulty, there is no single mathematical model that can be used to calculate concentrations field in all possible states of the atmosphere.

There are currently two theories developed that address turbulent diffusion: the Eulerian approach and the Lagrangian approach.

The Eulerian method is based on solving the equation of mass conservation in a fixed region of space, while the Lagrangian method is based on tracking the behavior of pollutant particles, carried along the flow lines. Each approach has a number of shortcomings that make it virtually impossible to determine the exact diffusion problems. For example, the

Eulerian approach, although capable to include complex physical phenomena in the diffusion equation (dry deposition, wet deposition, topographic effects and building, chemical reactions), has serious mathematical problems when the equations must be solved, using different parameterizations for this purpose.

Lagrangian formalism, by the statistical description of particle displacement along current lines, is mathematically easier to approach, but the results are limited by the impossibility of defining a complete statistic for a system of particles. Furthermore, Lagrangian formalism can't include a number of important physical phenomena such as nonlinear chemical reactions.

7. OSPM – A street canyon model

Operational Street Pollution Model (OSPM) has been tested many times by comparison with data obtained in different measurement campaigns (Hertel and Berkowicz, 1989a, b, c, Berkowicz, 2000a, b). Test results have helped to improve the model parameterization.

Quality of results obtained with the model depends on the quality of input data. Thus, studies by Schaedler et al. (1996, 1999) and Roeckle and Judges (1995) examined the influence of quality input data on modeled concentrations. As with any model of urban pollution, the most important parameters of the model used as input data are meteorological and background data, traffic data and emission factors that are needed to estimate emissions and topographical conditions (Pavageau et al., 1997).

Pollution generated by road traffic in a street canyon is characterized by a strong temporal but also spatial variability, both horizontally and vertically, which not only depends on diurnal variation of traffic, but also on weather conditions. Very little things are known about the vertical distribution of traffic pollution. Some of them are obtained from experiments in wind tunnels (Pavageau et al., 1999) or from direct measurements (Vachon et al., 1999). The analysis of measurements of turbulence in the street is shown in Vachon et al. (2001) and Louka et al. (2001). The dependence of urban pollutant concentrations on traffic and thermal effects, and comparison of experimental results with those modeled by OSPM was realized in Berkowicz (2000a,b).

OSPM is a semiempiric, parameterized model, where the concentrations of exhaust gas are calculated using a combination of a plume model for direct contribution and a box model for the recirculated part of the pollutants from the street. OSPM uses a very simplified parameterization of atmospheric fluid dynamics and dispersion of pollutants in street canyon. This parameterization was deducted from the analysis of experimental data and testing of the model. The results of these tests were used to improve the model's performance, especially in connection with various configurations of the street and different weather conditions. On the other hand, OSPM is based on many empirical assumptions that may not be applicable in all urban environments. The comparison data measured in the street with the OSPM model proves that a model with a simple parameterization like this can be applied to predict pollution from traffic. An important feature is the ability of the model to determine the concentrations of pollutants on a very weak wind, when it records data of severe pollution.

The results obtained with WinOSPM can be saved in a file that can be text, Excel or Access. The easiest to use is the .xls file, because it allows an easily analysis of the correlation between modeled and measured data. The modeled data is compared with those measured by calculating a set of statistical indicators used to validate the model. By this method, the quality of the output data can be determined, and so the model performance.

This data set includes the averages, the medians and the standard deviations for measured and simulated values, the mean error (bias), the normalized residue (FB), the normalized square error (NMSE), the correlation coefficient (R2) and the fraction of predicting the measured values up to a factor 2 (FA2) (Raducan, 2008b).

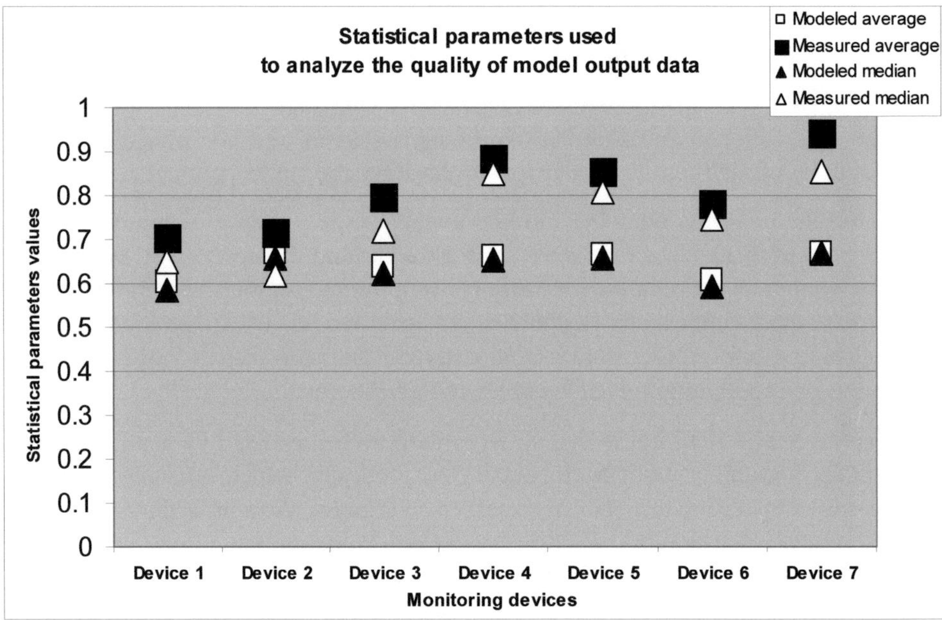

Fig. 7. The statistical parameters: the averages, the medians calculated for measured and modeled data

If the values of the averages and medians for both sets of data (modeled and measured) are close, it means that there are not very large extreme values which could be wrong.

The standard deviation shows how much variation or "dispersion" there is from the average. A low standard deviation indicates that the data points tend to be very close to the mean, whereas high standard deviation indicates that the data points are spread out over a large range of values.

If Bias <0 the model underestimates. Otherwise, it overestimates. Also, if FB>0, the model underestimates. Otherwise, it overestimates. So, figure 8 shows that the model underestimates.

When the normalized square error (NMSE) is low, it indicates that the model correctly describes the processes in time and space.

Analyzing the correlation coefficients between the simulated and measured data (Raducan, 2008a) there is good correlation if $R^2 > 0,5$ and a weak correlation if $R^2 < 0,5$.

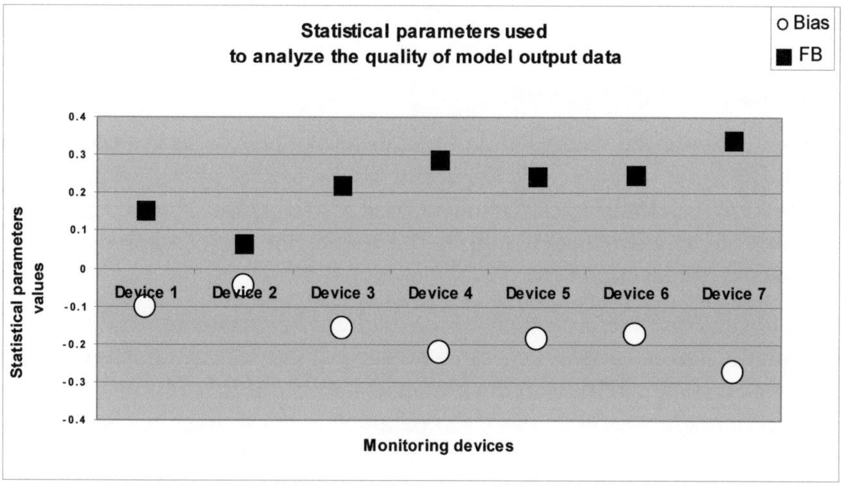

Fig. 8. The statistical parameters: the mean error (bias), the normalized residue (FB)

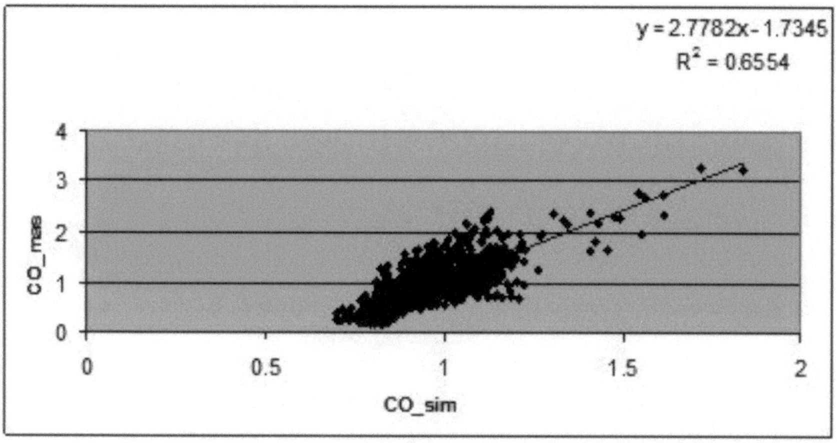

Fig. 9. A good correlation between simulated and measured data (R^2> 0,5)

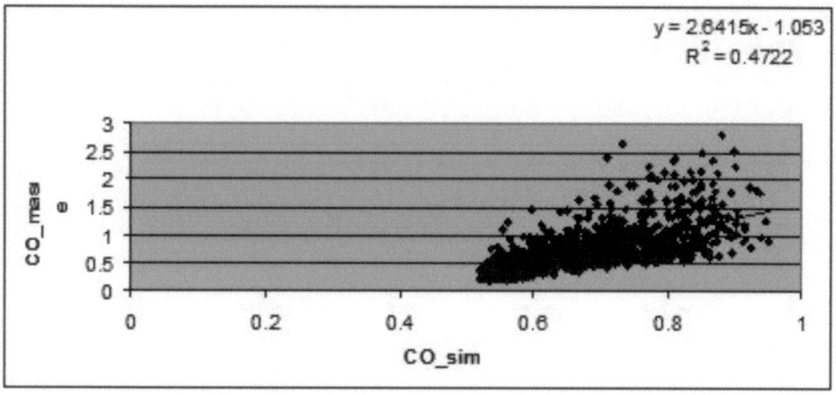

Fig. 10. A weak correlation between simulated and measured data (R^2< 0,5)

The values of the fractions of predicting (FA2) must be between 0,5 and 2.

8. The model's parameterization

The correlation between the simulated and measured data could be improved if the errors are corrected. The input data are very important for the quality of output data. In the input data there are a few parameterization mathematical relationships which could be improved. The model must be run with the new inputs. The results must be compared as it can be seen in the graph below (Raducan, 2008a)

If the values of the correlation coefficients are higher, the parameterization has improved the model outputs and it is validated for that set of data. This parameterization must be tested for many others sets of data in the same location and if it works properly, than it can be used within that street canyon. Also, the parameterization must be tested from time to time, using real data.

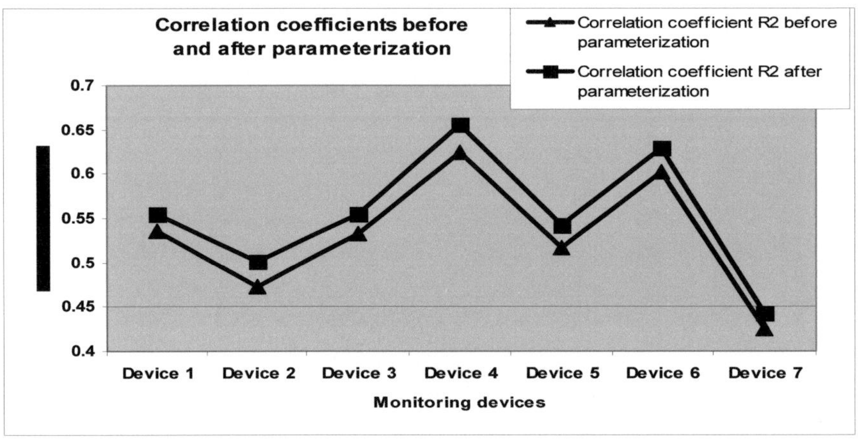

Fig. 11. A comparison between two sets of model outputs, before and after a parameterization; the model was run in 7 points in the street canyon

In many studies with OSPM, the results are compared with those obtained with other models to improve physical concepts of the model parameterization. Yet, the numerical models do not describe well enough some important phenomena and they require various parameterization relationships. These relationships are very important in the case when the wind is weak and the turbulence produced by traffic has a great influence.

It is noted that OSPM has been successfully applied to analysis of data obtained from routine monitoring in order to predict trends in traffic emissions (Palmgren et al., 1999).

Because OSPM, as software, does not require significant resources of the computer system, several years' data were processed within a few minutes.

9. Conclusions

The aim of the study has been enriching and understanding how to make an analysis of the urban air pollution.

A complex analysis of urban air quality must contain an observational part that examines the concentrations of pollutants recorded in the street and a dispersion modeling part.

First, the monitored pollutants must be established. Also, the monitoring devices must be chosen and the location in the urban area. Then, from the multitude of pollutants dispersion models, the most appropriate must be chosen (could be more then one model in order to compare the models output data).

Data analysis is done taking into account both the measured concentrations in the street, and emissions, traffic, weather conditions and topography, all being related to one another. The best way to compare data is to use some types of graphs to highlight the features that interest us.

The results should be stated clearly, visibly and after careful and thorough analysis. It is preferable to highlight both the results and shortcomings after a practical study. The latter could become research topics for those who made that study but also for other researchers.

10. References

[1] Berkowicz, R., 1997, Modelling Street Canyon Pollution: Model Requirements and Expectations, International Journal of Environment and Pollution 8 (3–6), 609–619.

[2] Berkowicz, R., 2000a, OSPM - A Parameterized Street Pollution Model, Environment Monitoring Assessment, 65, 323–331.

[3] Berkowicz, R., 2000b, A simple model for urban background pollution. Environmental Monitoring and Assessment 65, 259–267.

[4] Briggs G.A., 1969, Plume rise, ABC Critical Review Series, Washington DC

[5] Chan, T.L., Dong, G., Leung, C.W., Cheung, C.S., Hung, W.T., 2002, Validation of a Two-Dimensional Pollutant Dispersion Model in an Isolated Street Canyon. Atmospheric Environment 36, 861–872.

[6] Dobre A., Arnold S.J., Smalley R.J., Boddy J.W.D. Barlow J.F., Tomlin A.S., Belcher S.E., 2005, Flow Field Measurements in the Proximity of an Urban Intersection in London, UK, Atmospheric Environment 39, 4647-4657.

[7] Godish, T., 1997, Motor Vehicle Emissions Control. Air Quality. Boca Raton: CRC Press.

[8] Hertel, O. and Berkowicz, R., 1989a, Modelling Pollution from Traffic in a Street Canyon. Evaluation of Data and Model Development, DMU Luft A-129, 77p.

[9] Hertel, O. and Berkowicz, R., 1989b, Modelling NO2 Concentrations in a Street Canyon, DMU Luft A-131, 31p.

[10] Hertel, O. and Berkowicz, R., 1989c, Operational Street Pollution Model (OSPM). Evaluation of the Model on Data from St. Olavs Street in Oslo, DMU Luft A-135.

[11] Johnson, G.T., Hunter, L.J., 1999, Some Insights into Typical Urban Canyon Air Flows. Atmospheric Environment 33, 3991–3999.

[12] Louka, P., Vachon, G., Sini, J.F., Mestayer, P.G., Rosant, J.M., 2001, Thermal Effects on the Air Flow in a Street Canyon—Nantes'99 Experimental Results and Model Simulations, Third International Conference on Urban Air Quality, Loutraki, Greece.

[13] Nicholson, S.E., 1975, A Pollution Model for Street-Level Air, Atmospheric Environment 9, 19–31.

[14] Palmgren, F., Berkowicz, R., Ziv, A., Hertel, O., 1999, Actual Car Fleet Emissions Estimated from Urban Air Quality Measurements and Street Pollution Models. Science of the Total Environment 235, 101–109.

[15] Pavageau, M., Rafailidis, S. and Schatzmann, M., 1997, A Comprehensive Experimental Databank for the Verification of Urban Car Emission Dispersion Models, Int. J. Environ. Pollut. 8, 738–746.

[16] Pavageau, M., Schatzmann, M., 1999, Wind Tunnel Measurements of Concentration Fluctuations in an Urban Street Canyon. Atmospheric Environment 33, 3961–3971.

[17] Plate, E.J. and Kastner-Klein, P.,1998, Wind Tunnel Modelling of Traffic Induced Pollution in Cities, Proceedings of the 2nd German-Japanese meeting "Klimaanalyse für die Stadtplanung", 25-27.9.97, to appear in the series Report of Research Center for Urban Safety and Security in March 1998.

[18] Raducan, G., 2008a, Study of the air pollution regional variability, PhD Thesis.

[19] Raducan, G., 2008b, Pollutants Dispersion Modeling with OSPM in a Street Canyon from Bucharest, Romania Reports in Physics, Romania, no.4, 1099-1114.

[20] Raducan, G., Stefan, S., 2009, Characterization of Traffic-Generated Pollutants in Bucharest, Atmosfera, Mexic, Vol. 22, No.1, 99-110.

[21] Roeckle, R., C.J. Judges, 1995, Ermittlung des Strömungs- und Konzentrationsfeldesim Nahfeld typischer Gebäudekonfigurationen.- Modellrechnungen-. Abschlußbericht PEF 92/007/02, Forschungszentrum Karlsruhe, English Version, http://bwplus.fzk.de

[22] Schaedler, G., W. Baechlin, A. Lohmeyer, T. Van Wees, 1996, Vergleich und Bewertung derzeit verfügbarer Strömungs- und Ausbreitungsmodelle. In: Berichte Umweltforschung Baden-Württemberg (FZKA-PEF 138), (English Version), http://bwplus.fzk.de

[23] Schaedler, G., W. Baechlin, A. Lohmeyer, 1999, Immissionsprognosen mit meikroscaligen Modellen. Vergleich von berechneten und gemessenen Größen. Förderkennzeichen PEF 2 96 004. FZKA-BWPLUS 14, (English Version), http://bwplus.fzk.de

[24] Schneider, T., Koning, H. W. and Brassed, L. J., 1997, Air Pollution Reference Measurement Methods and Systems, Studies in Environmental Science 2. Elsevier, New York, 1998.

[25] Tate J.E., 2004, A Study of Vehicular Emissions and Ambient Air Quality at the Local-Scale, PhD Thesis

[26] Vachon, G., Rosant, J.-M., Mestayer, P.G. and Sini, J.F.,1999, Measurements of Dynamic and Thermal Field in a Street Canyon, URBCAP Nantes'99', Proceedings of the 6th Int. Conf. on Harmonisation within Atmospheric Dispersion Modelling for Regulatory Purposes, 11–14 October, Rouen, France, Paper 124.

[27] Vachon, G., Louka, P., Rosant, J.M., Mestayer, P., Sini, J.F., 2001, Measurements of Traffic-Induced Turbulence within a Street Canyon during the Nantes '99 Experiment. Third International Conference on Urban Air Quality, Loutraki, Greece.

[28] Xie, S., Zhang, Y., Qi, L. and Tang, X., 2003, Spatial Distribution of Traffic-Related Pollutant Concentrations in Street Canyons. Atmos. Env., 37, 3213-3224.

[29] Zanetti, P., 1993, Air Pollution Modelling: Theories, Computational Methods, and Available Software. Computational Mechanics Publications, Southampton, UK.

3

Estimation of Ambient Air Quality in Delhi

Kaushik K. Shandilya[1,*], Mukesh Khare[2] and A. B. Gupta[3]
[1]*Civil Engineering Department, The University of Toledo, Toledo, OH,*
[2]*Civil Engineering Department, Indian Institute of Technology, Delhi,*
[3]*Civil Engineering Department, Malaviya National Institute of Technology, Jaipur,*
[1]*USA*
[2,3]*India*

1. Introduction

For decades, it has been well known that air pollution causes adverse health effects on humans (Thurston et al., 1989, Shandilya and Khare, 2012). Since the late 1980s, epidemiological studies were published relating the mortality and morbidity to the ambient levels of fine particles without showing the expected results (Borja-Aburto et al., 1997). Most emphasis was then given to the particles with an aerodynamic diameter smaller than 10μm (PM_{10}) particularly to the class of the fine particulates. A special study carried out by United Nations Sources (UNECE, 1979) shows that the frequency of finding a particle of diameter range 2-4μm is 27.5%, which is higher in an overall range of 1-12μm, so the danger of the fine particulates of diameter ranges 2-4μm is higher than the danger by any particle which may lie in the range of 4-12μm. Therefore, there is a rising awareness about further fine fraction i.e., particles with an aerodynamic diameter smaller than 2.5μm ($PM_{2.5}$). Aerosols, which originate directly from the sources, are termed as "Primary aerosols". There may be the conversion of gaseous matter into the particulate due to many Physico-Chemical reactions. Such types of particulates are known as "Secondary Aerosols" (Perkins, 1974, Shandilya et al., 2009).

Delhi, the Capital and third largest city of India, ranks third in population among other Indian cities, estimated among 8.5 million (Crowther et al., 1990). Rapid urbanization and the unprecedented industrial and economic development during the last three decades have increased the vehicular population of Delhi by several folds. Consequently, harmful emission such as Pb, PAH, SO_2, NOx, particulate matter and carbon monoxide have also registered a sharp increase in these intervening years. As per the WHO report, Delhi is the fourth most populated city in the world. A 1997 study by the Center for Science and Environment (CSE) revealed that at least one person died prematurely every hour in Delhi in 1995 strictly because of Suspended Particulate Matter (SPM). A 1998 study by All India Institute of Medical Sciences (AIIMS), New Delhi, showed how emergency visits and deaths due to respiratory and heart problems are the highest when particulate levels peak during the winter (Down to Earth, 2000). It is estimated that in Delhi, the incidence of chronic

*Corresponding Author

bronchitis is 6-14 times more than the average elsewhere in India because of the thermal power plants and the vehicular pollution (Rao & Rao, 1989).

The objective of this study was to monitor NOx, SO_2, PM_{10}, $PM_{2.5}$, and Non Respirable Suspended Particulate (NRSP) levels in South Delhi at different land use patterns; to compare it with the levels found in the crowded residential site of East Delhi; and to compare the observed SPM,NOx, and SO_2 levels with the prescribed CPCB limits (except $PM_{2.5}$ compared with USEPA primary standards), which serves as the guidelines for the environmental planning of the region. Another objective was to find out the spatial and diurnal variation of NRSP, the Organic Matter level in NRSP, and the 12 hour variation of NOx and SO_2. The final objective was the determination of Air Pollution Index at different places.

2. Materials and methods

A background to the Study Area is outlined in detail in Shandilya et al. (2007). In order to assess ambient pollution level, particulate samples were taken with the help of respirable dust sampler and fine particulate sampler. For the high volume Measurement of RSP, air was drawn through a 20.3X25.4cm Whatman GF/A glass fiber filter at a flow rate which was kept around 1-1.1m^3/min. Detailed Procedures of Equipment (APM 460DX) are specified in operating manuals, available from equipment manufacturers (Envirotech Instruments Private Limited). The gases were determined with APM 411. The entire weighing procedure was done under controlled conditions (room temperature) in a balance room at Envirotech Instruments Pvt. Ltd., New Delhi using a Dhona-100DS microbalance. For $PM_{2.5}$ monitoring, ambient air enters APM550 system through an omni directional inlet designed to provide a clean aerodynamic cut-point for particles >10μm.

2.1 Sampling sites and details

The air quality monitoring in Delhi for present investigation was carried out from October 19, 2000 to November 29, 2000, during which PM_{10}, $PM_{2.5}$, and non-respirable suspended particulate concentrations were measured at four monitoring sites. In order to obtain a clear picture of a source-based pollution, it was required that the monitoring must be carried out at different places of different characteristics in New Delhi. It was important to consider residential, industrial, and commercial areas for carrying out such source differentiation. Based on the above considerations, four sites were chosen for setting up the sampling sites, namely Envirotech Instruments Pvt. Ltd, Okhla, II Floor Roof (Industrial Area); Indian Institute of Technology, Delhi, V Block IV Floor Roof (Residential Area); Vasant Vihar Bus Depot II Floor Roof (Commercial Area); Shakarpur a Residence, III Floor Roof (Residential Area). More details and figures of sampling locations can be obtained from Shandilya et al., (2007, 2012).

Out of the four sites chosen, two were in residential areas, one each in industrial and commercial areas. The criterion that was taken into account was the presence of local sources. The Shakarpur site was of particular importance, since the results obtained were used to compare the SPM level in South Delhi. Simultaneously, when we analyzed the respirable suspended particulates in respect of heavy metals, it was interesting to know the concentration near prominent sources like the Indraprastha power station. The site was located just on the bank of the Yamuna, when on the other bank was the Indraprastha power station. But the site location was mainly a residential area. Except for Shakarpur, where the Indraprastha power station is situated nearby, the sources of the local emissions were considered.

SPM was collected on a filter paper weighed before and after exposure. It was dried in an oven at 80°C for 45 minutes before each weighing. The size classification of PM was achieved through a cyclone installed in a Hi-Vol sampler, which separates the respirable PM_{10} and non-respirable fraction in the case of the Hi-Vol sampler. The particles sized <10μm were collected on the filter, and >10μm were collected in a separate sampling cup, also known as a dust collector. Flow rates were monitored before and after every sampling period with manometer, which was factory calibrated with an accuracy of 1.3% full scale. The flow rate checks were made before and after each sampling event to ensure that filter holders were not leaking and to determine that fitter clogging had not occurred.

These gases were monitored with the help of APM411 assembly attached with APM460DX (Envirotech Instruments Pvt. Ltd.) using the impingers. The modified Jacob-Hochheiser method (or arsenate method) have determined nitrogen dioxide (IS-5182, 1975). For the 24 hour field sampling, it is preferable to use this method. The West–Gaeke spectrophotometric method is the standard method for SO_2 monitoring of 0.0005-5 ppm SO_2 in the ambient air.

2.2 Air quality index determination

Based on the monitoring data of various pollutants at different sites on different days, an index called "Air Quality Index" was calculated using the following equation, thereby assessing the present air quality status of Delhi. As per the EPA guidelines, pollutants such as O_3, CO, SO_2, NO_2, PM_{10}, and $PM_{2.5}$ are used for calculating the AQI for an area.

$$lp = \frac{l_{Hi} - l_{Lo}}{BP_{Hi} - BP_{Lo}}(Cp - BP_{Lo}) + l_{Lo} \quad (1)$$

lp=the index for pollutant P
Cp=the rounded concentration of pollutant P
BH_{Hi}=the breakpoint greater than or equal to Cp
LHi=the AQI value corresponding to BP_{Hi}
l_{Lo}=the AQI value corresponding to BP_{Lo}
BP_{Lo}=the breakpoint less than or equal to Cp

In the present study, the pollutants used for calculating the AQI were NO_2, SO_2, and RSP. AQI was calculated for individual pollutants simultaneously at a place, and the maximum value corresponding to one particular pollutant was reported as the AQI of that area. The same procedure was adopted at all sites on each day, with AQI finally reported along with the responsible pollutants.

3. Results and discussion

For all sites, we have calculated the mean, standard deviation, and maximum, and we also calculated the geometric means. The geometric mean is the basis of preparing the air quality standards. The mean, standard deviation, maximum value, and geometric mean values are listed as below:

	RSP	FSP	NRSP	TSP	$PM_{2.5}/PM_{10}$	RSP/TSP
OKHLA						
GEO MEAN	268.6	126.71	405.35	687.71	0.47	0.39
MEAN	271.11	129.7	425.2	696.3	0.49	0.40
STD DEV.	39.1	28.6	139.1	117.4	0.11	0.11
MAXIMUM	320.3	157.3	599.2	827.6	0.56	0.53
IIT						
GEO MEAN	202.1	139.02	289.0	492.1	0.69	0.41
MEAN	204.3	144.1	290.7	495.0	0.71	0.41
STD DEV.	35.4	38.6	32.9	59.3	0.20	0.04
MAXIMUM	274.7	170.6	326.7	601.3	0.91	0.46
VV						
GEO MEAN	256.9	152.58	251.3	514.6	0.59	0.50
MEAN	260.7	159.5	260.3	521.1	0.65	0.50
STD DEV.	49.7	45.0	77.9	97.3	0.26	0.09
MAXIMUM	322.6	198.1	394.5	717.2	0.94	0.64
SKP						
GEO MEAN	226.8	174.05	192.3	427.6	0.77	0.53
MEAN	233.0	175.0	199.4	432.4	0.78	0.54
STD DEV.	64.5	21.2	67.1	73.7	0.18	0.11
MAXIMUM	321.2	198.4	299.7	497.4	0.95	0.66
Delhi						
GEO MEAN	238.1	144.8	283.3	532.0	0.61	0.45
MEAN	243.1	150.0	302.5	545.6	0.65	0.46
STD DEV.	51.1	37.0	117.6	130.3	0.21	0.10
MAXIMUM	322.6	198.4	599.2	827.6	0.95	0.66
ITO Delhi						
GEO MEAN	245.65		371.35	628.99		0.39
MEAN	257.62		384.62	642.24		0.40
STD DEV.	86.28		101.04	130.81		0.10
MAXIMUM	586.00		659.00	984.00		0.63

Table 1. Statistical Analysis of Results.

3.1 General discussion

In general, for $PM_{2.5}$ the secondary formation is more relevant than the primary emissions, but Wilson and Suh (1997) stated that some natural particles also occur in size range <2.5µm. A typical study (Pekkanen et al., 1997) reports that a considerable fraction of the road dust consisted of particles<2.5µm in diameter. The primary particles are emitted by a source as particles and are dispersed in the atmosphere without any major chemical transformation (Shandilya and Kumar, 2010a, 2010b, 2011).

PM_{10} mainly consisted of three distinct components: Primary vehicular emissions, secondary pollutants, and windblown dust. Specifically, major contributors to excessive PM_{10} levels were combinations of area-wide emitters such as fugitive dust from roads, construction, and agriculture; smoke from residential wood combustion or prescribed burning; directly emitted exhaust from motor vehicle engines; and secondary sulfates, nitrates, and organics formed from gaseous ammonia, sulfur dioxide, nitrogen oxides, and reactive organic gas emissions (Sharma et al., 2004, Shandilya et al., 2012). The emissions inventories for these "nontraditional" types of sources were found to be much less accurate than those obtained for "traditional" ducted emissions from the industrial sources. The coarse fraction was mainly composed of the re-suspended dust. Farming operations (e.g., agriculture tillage, harvesting, and travel on unpaved roads) account for large portions of the primary PM_{10} in many emissions inventories. PM_{10} non-attainment areas are suspected of having RWC as a major contributor. The prescribed burning of agricultural fields and forest slash is suspected of being a major contributor to PM_{10} in other areas. PM_{10} is considered more a local and urban scale problem, whereas $PM_{2.5}$ has a much longer residence time and is considered a regional or trans-boundary problem. The spatial and temporal distribution of fine particle levels may vary substantially.

Natural background levels, excluding all anthropogenic sources in the US and elsewhere, were 4-11-μgm^{-3} and 1-5-μgm^{-3} for PM_{10} and $PM_{2.5}$, respectively. PM_{10} mean concentrations were typically from 10-80-μgm^{-3} around the world. 10-20-μgm^{-3} is found at remote sites with no local sources (i.e., even in clean air). In urban areas, 60–220-$\mu g/m^3$ is typical, and in a heavily polluted area, levels may approach 2000-$\mu g/m^3$. More than 50-μgm^{-3} was found at heavily polluted sites with high traffic density and/or local sources in the surroundings, where heavily polluted urban areas may reach 100-μgm^{-3} on average. In the heavily polluted cities of southern Europe, Latin America, and Asia, average PM_{10} concentrations have reached 100-μgm^{-3}, and more. $PM_{2.5}$ mean concentrations were between 7-80-μgm^{-3}. PM_{10} trends were decreasing in the last decade, for emission and ambient levels, in the developed and developing world. The ratio of $PM_{2.5}$ to PM_{10} at the urban sites usually averaged 0.55-0.6.

Many studies indicated that TSP and PM_{10} concentration in the ambient air would be affected by various meteorological factors such as wind speed, wind direction, solar radiation, relative humidity, rainfall, and source conditions. Apart from the emission factors of the vehicles, SPM concentrations on the street level would be mainly affected by the mechanical turbulence (created by moving vehicles and wind) and by thermal turbulence produced by the hot vehicle exhaust gas (Shandilya and Kumar, 2009) .

24-hour average concentrations of the FSP, RSP, and TSP were calculated (See Methodology for a sample calculation) from samples collected at four stations at Southern Delhi and East Delhi and from samples taken from Satna. Our study is also not limited by the fact that, during winter, mixing depth in the early morning is lower than in the summer, thereby concentrating primary crustal aerosol emissions (e.g., from road dust). Since our study did not take place in the summer or winter, this condition is not applicable and the purpose of our study is not forfeited.

3.2 Comparison of SPM level in Delhi

The sources prevalent at VV are Diesel Buses, CNG Buses, Traffic, and Mechanical Disturbances, while at IIT the sources are Traffic, Vegetative Burning, construction work, and residential sources like cooking. At Okhla, the sources are Vegetative Burning, Diesel Generators, Mechanical Disturbances, Traffic, and Industrial Activities. Concentrations of fine particles are close to main roads. Road transport and diesel exhaust are major sources of fine particles in urban areas (Shandilya and Kumar, 2009, Kumar et al., 2011). PM_{10} and $PM_{2.5}$ concentrations are elevated close to main roads compared to background sites. More monitoring details can be found in Shandilya et al (2006, 2011).

3.2.1 Variation of $PM_{2.5}$

With a minimum of one daily sample at a site, 24-hour $PM_{2.5}$ concentration at the urban-industrial Delhi site varies from 85.1-157.3μgm^{-3}. 24-hour $PM_{2.5}$ concentration at the urban-commercial Delhi site varies from 78.6-198.1μgm^{-3}, and 24-hour $PM_{2.5}$ concentration at the urban-residential Delhi sites varies from 84.6-198.4μgm^{-3}. The value at Delhi is well above 24-hour $PM_{2.5}$ USNAAQS (65μgm^{-3}). More details of this monitoring can be found in Shandilya et al., (2011). Variation of $PM_{2.5}$ value is highest on SKP, which suggests that the contribution from different sources varies to a great extent. The Standard Deviation values suggest that $PM_{2.5}$ source emission rate at VV varies greater than the emission rate at any place in South Delhi. The large variation is most likely due to the differences in air mass, emission characteristics, and meteorology.

3.2.2 Peak day composition

The maximum $PM_{2.5}$ concentration in Delhi, 198.4-μg/m^3, was at SKP on November 26, 2000, which may be due to a wash out caused by the rain. The lowest $PM_{2.5}$ value of 78.6-μg/m^3 was observed at VV bus depot (November, 24, 2000). When there was a strike and there was less traffic on the road, there was a significant decrease in $PM_{2.5}$ fraction. It shows an important relationship between $PM_{2.5}$ fractions with the traffic sources. Maximum $PM_{2.5}$ concentration in South Delhi, 198.1-μg/m^3, was at VV on November 22, 2000. Maximum $PM_{2.5}$ concentration at Okhla, 157.3-μg/m3, was on October 21, 2000. Maximum $PM_{2.5}$ concentration at IIT, 170.6-μg/m^3, was on November 6, 2000.

3.2.3 Variation of PM_{10}

With a minimum of one daily sample at a site, 24-hour PM_{10} concentration at the urban-industrial Delhi site varies from 205.0-320.3-μgm^{-3}. 24-hour PM_{10} concentration at the urban-commercial Delhi site varies from 210.0-322.6-μgm^{-3}, and 24-hour PM_{10} concentration at the urban-residential Delhi sites varies from 174.5-321.2-μgm^{-3}. The PM_{10} concentration value was always greater than the 24-hour NAAQS (150-μg/m^3 in industrial and commercial areas and 100-μg/m^3 in residential areas).

3.2.4 $PM_{2.5}/PM_{10}$ ratio

The most common ratio of $PM_{2.5}/PM_{10}$ is 0.60, but the prevalence of the local sources, with either a dominating fine or coarse fraction, may shift this ratio. $PM_{2.5}/PM_{10}$ concentration at the urban-industrial Delhi site varies from 0.27-0.56 with an average±standard deviation of 0.47±0.11. $PM_{2.5}$/RSP concentration at the urban-commercial Delhi site varies from 0.25-0.94 with an average±standard deviation of 0.59±0.26. $PM_{2.5}$/RSP concentration at the urban-residential South Delhi sites varies from 0.43-0.91 with an average±standard deviation of 0.69±0.20. $PM_{2.5}$/RSP concentration at the urban-residential East Delhi sites varies from 0.62-0.95 with an average±standard deviation of 0.77±0.18. This shows that at Okhla, $PM_{2.5}$

contribution to PM_{10} is 47%, which increases to 59% at VV, further increases to 69% at IIT and achieves its maximum (77%) at SKP. $PM_{2.5}/PM_{10}$ ratio standard deviation is highest (0.26) at the VV site, suggesting the heterogeneous contribution of secondary particle formation sources. It has been observed that a part of PM_{10} is $PM_{2.5}$ but not as much as in the case of IIT in South Delhi. The standard deviation ($PM_{2.5}$/RSP) low value at Okhla (0.11) shows that the contribution of $PM_{2.5}$/RSP is more consistent compared to other sites, and $PM_{2.5}$ sources contributing to RSP remains the same. The ratios of $PM_{2.5}/PM_{10}$ are generally between 20 and 30%, with abnormal values from industrial particulate (both high-50% in Cynon- and low-17% in Port Talbot). The highest contribution of $PM_{2.5}/PM_{10}$ strongly shows that industrial sources are predominant in the SKP area. Comparison with higher ratios generally found in large urban areas (Birmingham-QUARG report) indicates that the finer traffic particulate stick together or deposit on surfaces less rapidly than the coarse fraction of PM_{10}, so they tend to build up in larger urban areas or still conditions. It has been observed that a part of PM_{10} is $PM_{2.5}$, but not as much as in the case of SKP. It shows that the condensation products of the chemical reactions primarily contribute PM_{10} fraction that originated in this area. Data ($PM_{2.5}/PM_{10}$) shows that PM_{10} fraction that originated in industrial areas contains the condensation products of the chemical reactions and some primary particulate.

3.2.5 Peak day composition

The maximum PM_{10} concentration in Delhi, South Delhi and VV 322.6-µg/m^3 was on November 17, 2000 during our study; the lowest PM_{10} of 174.5-µg/m^3 was observed at Shakarpur. This can be due to some change of wind pattern since there is a reduction in $PM_{2.5}$ concentration (171.8-µg/m^3), which eradicates the possibility of contribution from secondary particles and is further confirmed by an increase in NRSP fraction. Maximum PM_{10} concentration at Okhla, 320.3-µg/m^3, was on October 29, 2000 next to Deepawali. This high concentration can be attributed to crackers; this is well confirmed by the reduction of $PM_{2.5}$ concentration (85.1-µg/m^3), which is primarily caused by vehicles and is further confirmed by decline in the contribution of $PM_{2.5}$ to PM_{10} suggested by the value of $PM_{2.5}/PM_{10}$ (0.27, second lowest in all). Maximum PM_{10} concentration at IIT, 274.7-µg/m^3, occurred on November 3, 2000, the day when a celebration was held, so there was a great rush of vehicles in IIT. The increase in PM_{10} was due to the resuspension of dust and vehicle exhaust. It was further confirmed by the increase in NRSP fraction on that day and an increase in $PM_{2.5}$. The $PM_{2.5}$ value on the second day remained on the same level (168.7-µgm^{-3}), showing that the residence time $PM_{2.5}$ was much higher, which was not the case with PM_{10} and NRSP. Maximum PM_{10} concentration at SKP, 321.2-µg/m^3, took place on November 26, 2000, which may have been due to the wash out caused by the rain.

3.2.6 Variation of TSP

With a minimum of one daily sample at a site, 24-hour TSP concentration at the urban-industrial Delhi site varies from 553.4-827.6-µgm^{-3}. 24-hr TSP concentration at the urban-commercial Delhi site varies from 453.2-717.2-µgm^{-3} and 24-hr TSP concentration at the urban-residential Delhi sites varies from 345.2-601.3-µgm^{-3}. The TSP concentration value was always greater than the 24-hr NAAQS of TSP i.e., 200-µg/m^3 in the residential area. The TSP concentration value was always greater than the NAAQS of TSP i.e., 500-µg/m^3 in the industrial and commercial area while it crossed the limit (NAAQS, 500-µg/m^3) only once on November 17, 2000 at the VV commercial area.

3.2.7 RSP/TSP ratio

RSP/TSP at the urban-industrial Delhi site varies from 0.27-0.56 with an average±standard deviation of 0.39±0.11. RSP/TSP concentration at the urban-commercial Delhi site varies from 0.42-0.64μgm^{-3} with an average±standard deviation of 0.50±0.09. RSP/TSP concentration at the urban-residential South Delhi sites varies from 0.37-0.46 with an average±standard deviation of 0.41±0.04. RSP/TSP concentration at the urban-residential East Delhi sites varies from 0.40-0.66 with an average±standard deviation of 0.53±0.11. This shows that at Okhla, RSP contribution to TSP is 39%, increasing to 50% at VV, decreasing to 41% at IIT, and again achieving its highest of 53% at SKP. The standard deviation (RSP/TSP) low value at IIT (0.04) shows that the contribution of RSP to TSP is consistent, while it varies equally at SKP and at Okhla. Therefore, the RSP emitting sources in the SKP area were contributing equally throughout the study, while these sources' emission pattern was changed at the other places.

3.2.8 Peak day composition

The maximum TSP concentration in Delhi, South Delhi and at Okhla 827.6-$\mu g/m^3$ occurred on October 23, 2000. The maximum TSP concentration at IIT, 601.3-$\mu g/m^3$, occurred on November 3, 2000. It is further confirmed by an increase in NRSP fraction on that day. The maximum TSP concentration at VV, 717.2-$\mu g/m^3$, occurred on November 17, 2000, which may be due to some change of wind pattern since there is a reduction in $PM_{2.5}$ concentration (171.8-$\mu g/m^3$) while there is an increase in NRSP fraction. The maximum TSP concentration at SKP, 489.4-$\mu g/m^3$, was on November 26, 2000, which may be due to the wash out caused by the rain. In short, TSP concentrations at IIT and at SKP sites are lower than the other values observed in Delhi, but $PM_{2.5}$ concentrations are higher than those in the Okhla industrial location.

Thus, industrial emissions have considerable impact on the ambient non-respirable particulate matter concentrations, while residential emissions have considerable impact on $PM_{2.5}$ concentration. Similarly, PM_{10} concentrations at the Okhla site are higher than the values found at any other site. This suggests that the industrial emissions contribute more to $PM_{2.5\text{-}10}$ fraction.

Our observations at Okhla in South Delhi showed that $PM_{2.5}$, RSP, and TSP concentrations were 126.71±28.6, 268.6±39.1, and 687.71±117.4-μgm^{-3}, respectively; IIT showed that $PM_{2.5}$, RSP, and TSP concentrations were 139.02±38.6, 202.1±35.4, and 492.1±59.3-μgm^{-3}, respectively; and VV showed that $PM_{2.5}$, RSP, and TSP concentrations were 152.58±45.0, 251.3±49.7, and 514.6±97.3-μgm^{-3}, respectively. As stated before, SKP in East Delhi showed that $PM_{2.5}$, RSP and TSP concentrations were 174.05.0±21.1, 226.8±64.5, and 427.6±73.6-μgm^{-3}, respectively.

Our results could be compared with the results of Kanpur, a study carried out by Sharma et al. (2003). They found at the background site, IIT Kanpur that PM_{10} levels were 62±11.22-$\mu g/m^3$ and at Agricultural University, the residential area, PM_{10} levels were 71.46-335.83-$\mu g/m^3$. At the commercial area Naveen Market, PM_{10} levels were 114.22-634.68-$\mu g/m^3$, and at the industrial area, Lajpat Nagar, PM_{10} levels were 221.77-650.73-$\mu g/m^3$. On comparison, it can be concluded that PM_{10} levels in all areas (industrial, commercial, residential) found the same range at both places (Delhi and Kanpur).

Our results could also be compared with the results of Jaipur, a study carried out by Das et al. (1998). They found at the industrial site that TSP levels were 384.61-µg/m^3 and, in another study by Gajraj et al. (submitted) in the Vishwakarma industrial area, mean PM_{10} concentration during May 1999 was 242-µg/m^3; during June 1999, the mean concentration was PM_{10} (208.5-µg/m^3). Our TSP levels at Okhla are certainly higher than TSP levels in Jaipur, as obtained by Das et al. (1998).

3.2.9 Explanation of spatial variation

The terminal settling velocity is an important factor for the spatial variation of SPM (Willeke & Baron, 1993). For very fine particles, such as a particle of 0.1-µm d_{ae} setting velocity, is $8.65x10^{-5}$-cms^{-1} for a particle of 1-µm $3.48x10^{-3}$ and for a particle of 10-µm in size $3.06x10^{-1}$-cms^{-1}. Primary particles (Aitken particles <0.1-µm) and large particles (e.g.>10-µm) are expected to have a bigger spatial variability than the particles in the accumulation mode (0.1–1-µm). Coagulation (for very fine particles <0.01-µm) and gravitational settling (for particles>1-µm) are the underlying mechanisms, which cause spatial heterogeneity. The distribution of $PM_{2.5}$ might be uniform in a situation where the secondary formation is important; in cities, however, with large emissions from heavy duties (diesel exhaust), $PM_{2.5}$ can also exhibit significant spatial variation. Secondary aerosols that contribute to the accumulation mode (0.1-2-µm) show quite homogeneous spatial distribution. In general, small-scale spatial variations of $PM_{2.5}$ were described to be smaller than the spatial variations of PM_{10} (Tuch et al., 1997). The spatial variation was small for $PM_{2.5}$ but larger for PM_{10}. Our results are in agreement with Monn (2001). With respect to the small-scale spatial variation in the urban areas, the largest variations occurred in the ultrafine (0.1-µm) and coarse mode ($PM_{10-2.5}$) (Burton et al., 1996). Ultrafine particle number counts have large spatial variations, and they are not well correlated to mass data, so spatial variation is much more reflected in PM_{10} than in $PM_{2.5}$.

The spatial variation was small for $PM_{2.5}$ but larger for PM_{10}. Spatial correlations for $PM_{2.5}$ are found to be lower than 0.9 and around 0.6 for PM_{10}. The contribution of coarse particle fraction ($PM_{10-2.5}$) on PM_{10} can be important. Conclusively, we can say that SPM is complex with respect to its sources in Delhi. With the exception of the geological material, $PM_{2.5}$ source contribution estimates are not similar to those for PM_{10}. A deep insight in the source contribution could be obtained if we consider emission factors of SPM from different sources.

3.3 Oxides of sulfur and nitrogen

In order to understand the sources, behavior, and mechanism of particle formation in the atmosphere, it is important to measure the precursors like SOx and NOx; this primarily contributes to the formation of the secondary particulate, which again constitutes a big fraction of total particulate matter in the atmosphere. This type of precursor characterization for the continental, marine background, and urban influenced aerosols has been reported from various sites throughout the world (Garg et al., 2001). 24-hour average concentrations of SO_2 and NO_2 (See Methodology) from the samples collected at four stations at Southern Delhi and East Delhi are given in the table below; they were calculated by taking the average of the day and night value.

For all sites, we have calculated the mean, standard deviation, and minimum and maximum values. We have also calculated the geometric means to show a homogeneous distribution of the data. Geometric mean concentrations are better representations of the data population as the elements are log-normally distributed. The geometric mean, standard deviation, maximum and minimum values, and geometric mean values are listed as below:

Okhla	Day Value	Night Value	Sox 24-hr	Day Value	Night Value	NOx 24-hr	Sox/NOx 24-hr	Sox/NOx Day	Sox/NOx Night
Geo.Mean	5.86	25.47	15.76	32.48	107.85	70.68	0.22	0.18	0.24
Mean	6.13	25.55	15.84	32.58	110.57	71.58	0.23	0.19	0.24
Std Dev	1.89	2.32	1.76	2.73	24.20	11.68	0.06	0.06	0.08
Maximum	8.42	30.12	18.45	36.89	136.62	84.87	0.35	0.28	0.42
Minimum	3.50	23.58	13.54	28.20	64.55	50.72	0.18	0.11	0.19
IIT									
Geo.Mean	4.30	14.45	9.40	24.56	57.29	40.95	0.23	0.18	0.25
Mean	4.37	14.49	9.43	24.72	57.83	41.27	0.23	0.18	0.25
Std Dev	0.83	1.21	0.81	3.06	8.67	5.67	0.02	0.03	0.03
Maximum	5.78	16.32	10.48	29.85	70.37	50.11	0.26	0.22	0.29
Minimum	3.21	12.96	8.45	21.00	47.13	35.12	0.20	0.14	0.23
VV									
Geo.Mean	5.48	26.67	16.12	37.86	97.48	68.36	0.24	0.14	0.27
Mean	5.51	26.97	16.24	38.21	99.98	69.09	0.24	0.15	0.28
Std Dev	0.64	4.38	2.15	5.53	24.71	11.07	0.04	0.02	0.06
Maximum	6.32	33.18	19.12	44.60	139.19	85.49	0.30	0.16	0.38
Minimum	4.90	20.65	12.81	31.05	68.45	54.55	0.20	0.11	0.22
SKP									
Geo.Mean	5.32	29.66	17.49	42.88	133.91	88.45	0.20	0.12	0.22
Mean	5.40	30.00	17.70	42.99	134.91	88.95	0.20	0.12	0.22
Std Dev	1.07	5.26	3.13	3.47	19.36	11.10	0.02	0.01	0.02
Maximum	6.56	36.84	21.70	46.92	162.02	104.47	0.21	0.14	0.24
Minimum	4.46	24.52	14.52	39.45	116.09	78.43	0.17	0.11	0.19
Delhi									
Geo.Mean	5.22	22.83	14.11	32.98	92.48	63.19	0.22	0.16	0.25
Mean	5.38	23.81	14.59	33.81	98.28	66.04	0.23	0.16	0.25
Std Dev	1.36	6.65	3.69	7.51	33.25	19.08	0.04	0.04	0.06
Maximum	8.42	36.84	21.70	46.92	162.02	104.47	0.35	0.28	0.42
Minimum	3.21	12.96	8.45	21.00	47.13	35.12	0.17	0.11	0.19
ITO									
Geo.Mean			17.30			78.42	0.22		
Mean			17.81			80.57	0.22		
Std Dev			4.40			18.83	0.04		
Maximum			28.00			126.00	0.30		
Minimum			11.00			42.00	0.10		

Table 2. Statistical Analysis of the Result Obtained

3.3.1 Sources of sulfur oxides and nitrogen oxides

The natural sources of nitrogen oxides are the geothermal activity and bacterial action. Natural sources include the soil microbes, vegetation, biomass, slum, burning, and lightning. Natural sources include the biological emissions (e.g. vegetation and soil microbes) and abiotic processes (e.g., geogenic, lightning, and biomass burning). Globally, lightning strikes occur about 100-times/second (Borucki and Chameides, 1984). Lighting strikes are characterized by a highly ionized and high temperature channel, in which molecular nitrogen is dissociated and reacts with oxygen to form NO (Guenther et al, 2000).

The primary sources of NOx are the motor vehicles, electric utilities, and other industrial, commercial, and residential sources that burn fuels. Other sources include the chemical and nitric acid manufacture, the detonation of nitrate-containing explosives, and electric sources. Man made sources of nitrogen oxide include the automobile, railways (coal fired and diesel locos), airways (landing and takeoff operations), industrial processes like chemicals (nitric acid plants, processes where nitrogenous compounds are used), metal processing, fertilizer (nitrogenous fertilizers), and other miscellaneous activity like domestic combustion, forest fires and managed fires, industrial boilers, furnaces, and commercial combustion. Although most NO is produced by the natural bacterial action, it must be emphasized that the rate of emission from man-related sources has, by estimate, increased by one-fourth between 1966 and 1968. Certainly, man's greatest contribution to the family of nitrogen oxides arises from the combustion of fuels. The contribution to the total NOx can be categorized as Utilities 27%; Motor Vehicles 49%; Other Sources 5%; and Industrial/ Commercial/Residential 19%. Nearly 2/3 of NOx emissions originate at the Earth's surface from fossil fuel and biomass combustion. Vehicle exhausts are known to increase the inputs of NO_2 to the atmosphere (see, e.g., Hargreaves et al., 2000). The burning of natural gas (which produces negligible SOx, CO, and HC) does produce a very significant quantity of oxides of nitrogen.

Man caused nitrogen emissions as equivalent NO_2 are estimated as $53x10^6$ TPA over the world, 95% of which are emitted in the Northern Hemisphere. Half of these are from the combustion of coal. Natural sources of NOx are estimated as between 15 and 17-times pollutant emissions. (Sawyer et al, 1998). The major source of NOx is the combustion when the fixation of atmospheric N occurs at the high flame temperature. The oxides are emitted mainly as NO, which is normally rapidly oxidized to NO_2 by atmospheric O_2 and O_3. Motor vehicle exhaust contributes a sizable fraction of the total emissions of NOx. As well as the stationary combustion sources, the manufacture of HNO_3 and nitrate fertilizer is the source of NO & NO_2. On a global basis, natural emissions of oxides of nitrogen (such as releases from soil and the ocean, lightning strikes, oxidation of ammonia, and stratospheric oxidation) all make substantial contributions to the total emissions.

Garg et al (2001) has reported the sectoral shares of NOx emissions for India in 1995. According to them, a major part of NOx emissions is contributed from 28% by the non-energy sources, 28% by electric power generation, 2% by road transport, 19% by Biomass burning, 4% by other transports, and 19% by industries. Delhi tops the table of top 10 NOx emitting districts in the terms of total emissions by emitting total NOx 66-Gg in 1990 and 82-Gg in 1995; in the terms of emissions per unit area, NOx/area (tons km^{-2}) was 44 in 1990 and 55 in 1995. The largest NOx emission chunk in Delhi is from the oil sources.

The major atmospheric man made sources of sulphur dioxide (SO_2) are the burning of fossil fuel in stationary combustion and industrial process, viz. petroleum, chemical, metallurgical and mineral industries, domestic emission and mineral industries, domestic emission, and fires, while the natural source are the volcanoes (wherever these are present, but in India there are none), geothermal activity, the bacterial decomposition of organic matter, forest fires, and managed fires. Motor vehicles are a relatively minor source of SO_2, since refined motor fuel normally has low sulfur content.

Ambient SO_2 results largely from the stationary sources (that burn the coal and oil, refineries, pulp, and paper mills) and from the nonferrous metal smelters. The sources of the pollution include the emissions from the on-road vehicles, non-road vehicles (like planes, ships, trains and industries), and small businesses and households where polluting products are used.

Nitrogen oxides (NOx) are produced when fossil fuels are burned in the motor vehicles, power plants, furnaces, and turbines. Other gasoline powered mobile sources (motorcycles, recreational vehicles, lawn, garden, and utility equipment) have high emissions on per quantity of the fuel consumed basis, but their contribution to the total emissions is small. Heavy-duty diesel vehicles are the dominant mobile source of the Oxides of nitrogen. Oxides of nitrogen emissions on a fuel-consumed basis are much greater from the diesel mobile sources than from the gas.

Another common source of SO_2 in the atmosphere is the metallurgical operations. Many ores, like Zn, Cu, and Pb are primarily sulfides. During the smelting of these ores, SO_2 is evolved in the stack concentrations of 5-10% (SO_2). Among miscellaneous operations releasing SO_2 into the atmosphere are the sulphuric acid plant, paper-manufacturing plants, open burning of the refuse, and municipal incinerators which contribute some amount of SO_2 to the atmosphere (Peavy et al, 1985).

In the entire sulphur entering atmosphere, $1/3^{rd}$ is by the anthropogenic activities, and $2/3^{rd}$ come from natural sources such as Hydrogen Sulfide (H2S) or SOx. The problem with the manmade sources is the distribution, rather than the amount. Delhi, in this instance, is being recognized as a city with a potentially high traffic load on its roads. Therefore, the combustion of fossil fuels like diesel and petrol can be considered as a major source of the SO_2 in Delhi. Besides this thermal power plant, which is located in the Southeast about 1-1.5 km far from the SKP station, it is also contributing SO_2 in the atmosphere because the coal is used as a main fuel for the heat and the power generation. However, SO_2 concentration in the air also depends upon the sulfur content of fuels that vary from less than 1% of good quality anthracite to over 4% for bituminous coal.

Most crude petroleum products contain less than 1% sulfur; a few contain up to 5%. In the coming years, Delhi is supposed to have less concentration of SO_2 in its atmosphere because the fuel with less S content has already been introduced into the market. Another common source of the sulfur dioxide in Delhi is metallurgical operation. Among the miscellaneous operations releasing releasing SO_2 into the atmosphere are sulfuric plants and paper manufacturing plants. The open burning of refuse and municipal incinerator also contribute some amount of SO_2.

Garg et al (2001) also reported the sectoral shares of SO_2 emissions for India in 1995. According to them, a major part of SO_2 emissions is contributed to 46% by non-energy sources, 36% by electric power generation, 8% by road transport, 6% by biomass burning, 3% by other transports, and 1% by industries. The details of industrial SO_2 emissions shows that 40%, 27%, 14%, 10%, 7%, and 2% are shared by non-energy industries, steel industry, fertilizer, cement, other industries, and refineries, respectively.

In Delhi, SO_2/area (tkm^{-2}) in 1990 was 47.9, in 1995 it was 46.7, and the average annual concentrations (µg/m^3) were 16 in 1990 and 24 in 1995. The largest contribution to SO_2 levels is by the transport sector. Our study results should be considered by keeping the points of the study done by Kamyotra et al (2000) and Jain (2001). These studies have pointed out two important points: by increasing the sampling period, the concentration shown becomes less (Kamyotra et al., 2000) and the diurnal pattern over 24 hours changes with time, local conditions, and activities (Jain, 2001).

The nighttime high values can be attributed to the phenomenon of the inversion. According to Kamyotra et al (2000), the lower values found during the day can be ascribed for the evaporation of the liquid, but this evaporation is absent during the night time which, in turn, increases the night time values, so high values obtained can be attributed to the combined effect of these facts.

3.3.2 Variation in SO_2

With a minimum of one 12-hour daily sample at a site, 12-hour a day SO_2 concentration at the urban-industrial Delhi site varies from 3.50-8.42-µgm^{-3}, 12-hour a night SO_2 concentration varies from 23.58-30.12-µgm^{-3}, and the average 24-hour SO_2 concentration varies from 13.54-18.45-µgm^{-3}. With a minimum of one 12-hour daily sample at a site, 12-hour a day SO_2 concentration at the urban-commercial Delhi site varies from 4.90-6.32-µgm^{-3}, 12-hour a night SO_2 concentration varies from 20.65-33.18-µgm^{-3}, and average 24-hour SO_2 concentration varies from 12.81-19.12-µgm^{-3}. With a minimum of one 12-hour daily sample at two sites, 12-hour a day SO_2 concentration at the urban-residential Delhi site varies from 3.21-6.56-µgm^{-3}, 12-hour a night SO_2 concentration varies from 12.96-36.84-µgm^{-3}, and average 24- hour SO_2 concentration varies from 8.45-21.70-µgm^{-3}.

On Okhla, the highest day SO_2 concentration variation (1.89-µgm^{-3}) suggests that the source emission rate variation at Okhla is higher comparatively. On the VV site, day SO_2 concentration variation was lowest in all (0.64-µgm^{-3}), suggesting that the source diversity and source emission rate variation are lower comparatively. On the IIT site, day SO_2 concentration variation is 0.83-µgm^{-3}, and on the SKP site day SO_2 concentration variation is 1.07-µgm^{-3}. In all of Delhi, day SO_2 concentration varies from 3.21-8.42-µgm^{-3}.

On SKP, Night SO_2 concentration variation was the highest in all (5.26-µgm^{-3}), suggesting that the source emission rate variation at SKP is higher comparatively. On the IIT site, night-SO_2 concentration variation was the lowest in all (1.21-µgm^{-3}), suggesting that the source diversity and source emission rate variation are lower comparatively. On the VV site, night SO_2 concentration variation is 4.38-µgm^{-3}, and on the Okhla site night SO_2 concentration variation is 2.32-µgm^{-3}. In all of Delhi, night SO_2 concentration varies from 12.96-36.84-µgm^{-3}. On SKP, 24-hour SO_2 concentration variation was highest in all (3.13-µgm^{-3}), suggesting that the source emission rate variation at SKP is higher comparatively. On the IIT site, 24-hour SO_2 concentration variation was the lowest in all (0.81-µgm^{-3}), suggesting that the source diversity and source emission rate variation are lower comparatively. On the VV site,

24-hour SO_2 concentration variation is 2.15-µgm^{-3}, and on the Okhla site 24-hour SO_2 concentration variation is 1.76-µgm^{-3}. In all of Delhi, 24-hour SO_2 concentration varies from 8.45-21.70-µgm^{-3}.

3.3.3 Peak day composition

The maximum day SOx concentration in Delhi, 8.42-µg/m^3, was at Okhla on October 25, 2000. The lowest day SOx value 3.21-µgm^{-3} was observed at IIT on November 2, 2000. The maximum day SOx concentration in SKP, 6.56-µg/m^3, was on November 27, 2000. The maximum day SOx concentration at VV, 6.32-µg/m^3, was on November 22, 2000. The maximum night SOx concentration in Delhi, 36.84-µg/m^3, was at SKP on November 27, 2000. The lowest night SOx value 12.96-µgm^{-3} was observed at IIT on November 5, 2000. The maximum night SOx concentration in Okhla, 30.12-µg/m^3, was on October 23, 2000. The maximum night SOx concentration at VV, 33.18-µg/m^3, was on November 23, 2000.

The maximum 24-hour SOx concentration in Delhi, 21.70-µg/m^3, was at SKP on November 27, 2000. The lowest 24-hour SOx value, 8.45µg/m^3, was observed at IIT on November 2, 2000. The maximum 24-hr SOx concentration in Okhla, 18.45-µg/m^3, was on October 23, 2000. The maximum 24-hour SOx concentration at VV, 19.12-µg/m^3, was on November 23, 2000. Summarizing our observations at Okhla in South Delhi showed that day SOx, night SOx, and average 24-hour concentrations were 5.86±1.89, 25.47±2.32, and 15.76±1.76-µgm^{-3}, respectively. IIT showed that day SOx, night SOx, and average 24-hour concentrations were 4.30±0.83, 14.45±1.21, and 9.40±0.81-µgm^{-3}, respectively. VV showed that day SOx, night SOx, and average 24-hour concentrations were 5.48±0.64, 26.67±4.38, and 16.12±2.15-µgm^{-3}, respectively. SKP in East Delhi showed that day SOx, night SOx, and average 24-hour concentrations were 5.32±1.07, 29.66±5.26, and 17.49±3.13-µgm^{-3}, respectively, and overall in Delhi were 5.22±1.36, 22.83±6.65, and 14.11±3.69-µgm^{-3}.

3.3.4 Variation in NO_2

With a minimum of one 12-hour daily sample at a site, 12-hour per day NO_2 concentration at the urban-industrial Delhi site varies from 28.20-36.89-µgm^{-3}, 12-hour per night NO_2 concentration varies from 64.55-136.62-µgm^{-3} and average 24-hour NO_2 concentration varies from 50.72-84.87-µgm^{-3}. With a minimum of one 12-hour daily sample at a site, 12-hour per day NO_2 concentration at the urban-commercial Delhi site varies from 31.05-44.60-µgm^{-3}, 12-hour per night NO_2 concentration varies from 68.45-139.19-µgm^{-3}, and average 24-hour NO_2 concentration varies from 54.55-85.49-µgm^{-3}. With a minimum of one 12-hour daily sample at two sites, 12-hour per day NO_2 concentration at the urban-residential Delhi site varies from 21.00-46.92-µgm^{-3}, 12-hour per night NO_2 concentration varies from 47.13-162.02-µgm^{-3} and average 24-hour NO_2 concentration varies from 35.12-104.47-µgm^{-3}.

On VV, day NO_2 concentration variation was highest in all (5.53-µgm^{-3}), which suggests that the source emission rate variation at VV is higher comparatively. On the Okhla site, day NO_2 concentration variation was the lowest in all (2.73-µgm^{-3}), which suggests that the source diversity and source emission rate variation is lower comparatively. On the IIT site, day-NO_2 concentration variation is 3.06-µgm^{-3}, and on the SKP site, the day NO_2 concentration variation is 3.47-µgm^{-3}. In all of Delhi, the day NO_2 concentration varies from 21.00-46.92-µgm^{-3}.

On the VV site, night-NO_2 concentration variation was the highest in all (24.71-µg/m^3), which suggests that the source emission rate variation at SKP is higher comparatively. On the IIT site, night NO_2 concentration variation was the lowest in all (8.67-µgm^{-3}), which suggests that the source diversity and source emission rate variation is lower comparatively. On the Okhla site night NO_2 concentration variation is 24.20-µgm^{-3}, and on the SKP site night NO_2 concentration variation is 19.36-µgm^{-3}. In all of Delhi, night NO_2 concentration varies from 47.13-162.02-µgm^{-3}.

On Okhla, 24-hour NO_2 concentration variation was highest in all (11.68-µgm^{-3}), which suggests that the source emission rate variation at Okhla is higher comparatively. On the IIT site, 24-hour NO_2 concentration variation was the lowest in all (5.67-µgm^{-3}), which suggests that the source diversity and source emission rate variation are lower comparatively. On the VV site, 24-hour NO_2 concentration variation is 11.07-µgm^{-3}, and on the SKP site, 24-hour NO_2 concentration variation is 11.10-µgm^{-3}. In all of Delhi, 24-hour NO_2 concentration varies from 35.12-104.47-µgm^{-3}.

3.3.5 Peak day composition

The maximum day NOx concentration in Delhi, 46.92-µg/m^3, was at SKP on November 27, 2000. The lowest day NOx value, 21.00-µg/m3, was observed at IIT on November 4, 2000. The maximum day NOx concentration in Okhla, 36.89-µg/m^3, was on October 25, 2000. The maximum day NOx concentration at VV, 44.60-µg/m^3, was on November 15, 2000. The maximum day NOx concentration at IIT, 29.85-µg/m^3, was on November 6, 2000.

The maximum night NOx concentration in Delhi, 162.02-µg/m^3, was at SKP on November 27, 2000. The lowest night NOx value, 47.13-µgm^{-3}, was observed at IIT on November 2, 2000. The maximum night NOx concentration in Okhla, 136.62-µg/m^3, was on October 23, 2000. The maximum night NOx concentration at VV, 139.19-µg/m^3, was on November 23, 2000. The maximum night NOx concentration at IIT, 70.37-µg/m^3, was on November 6, 2000.

The maximum 24-hour NOx concentration in Delhi, 104.47-µg/m^3, was at SKP on November 27, 2000. The lowest 24-hour NOx value, 35.12-µgm^{-3}, was observed at IIT on November 2, 2000. The maximum 24-hour NOx concentration in Okhla, 84.87-µg/m^3, was on October 23, 2000. The maximum 24-hour NOx concentration at VV, 85.49-µg/m^3, was on November 23, 2000. The maximum NOx concentration at IIT, 50.11-µg/m^3, was on November 6, 2000.

Our observations at Okhla in South Delhi showed that day NOx, night NOx, and average 24- hour concentrations were 32.48±2.73, 107.85±24.20, and 70.68±11.68-µgm^{-3}, respectively. Observations at IIT showed that day NOx, night NOx, and average 24-hour concentrations were 24.56±3.06, 57.29±8.67, and 40.95±5.67-µgm^{-3}, respectively. VV showed that day NOx, night NOx, and average 24-hour concentrations were 37.86±5.53, 97.48±24.71, and 68.36±11.07-µgm^{-3}, respectively. SKP in East Delhi showed that day NOx, night NOx, and average 24-hour concentrations were 42.88±3.47, 133.91±19.36, and 88.45±11.10-µgm^{-3}, respectively, and the overall concentrations in Delhi were 32.98±7.51, 92.48±33.25, and 63.19±19.08-µgm^{-3}.

As evident from the table 2, the concentration of NOx was the highest at SKP, followed by Okhla, VV, and IIT. The trend is attributed to the prevailing activities at the sites. 24-hour average NOx concentration is mainly contributed by the nighttime values. These nighttime higher NOx concentrations at the different sites also reflect the local source conditions.

3.3.6 SOx to NOx day ratio

The day ratio at an urban-industrial Delhi site varies from 0.11-0.28 and at the urban-commercial Delhi site varies from 0.11-0.16; at two urban-residential Delhi sites, it varies from 0.11-0.22. At SKP, variation of the ratio (0.01) is lowest, while it is highest at the Okhla (0.06). The variation of ratio at VV was 0.02, and at IIT it was 0.03, and in Delhi the ratio ranged from 0.11-0.28. The variation in Delhi (4%) shows that the sources liberating NOx are also releasing SOx. Summarizing our observations at Okhla in South Delhi showed that the day ratio was 0.18±0.06. IIT in South Delhi showed that the ratio was 0.18±0.03. VV in South Delhi showed that the ratio was 0.14±0.02. SKP in East Delhi showed that the ratio was 0.12±0.01, and overall the Delhi ratio was 0.16±0.04.

3.3.7 SOx to NOx night ratio

The night ratio at the urban-industrial Delhi site varies from 0.19-0.42, and at the urban-commercial Delhi site varies from 0.22-0.38, and at two urban-residential Delhi sites varies from 0.19-0.29. At SKP, the variation of the ratio (0.02) is the lowest, while it remains the highest at Okhla (0.08). The variation of ratio at VV was 0.06, and at IIT it was 0.03, and in Delhi the ratio ranges from 0.19-0.42. The variation in Delhi (6%) shows that sources liberating NOx are also releasing SOx. Summarizing our observations at Okhla in South Delhi showed that the night ratio was 0.24±0.08. IIT in South Delhi showed that the ratio was 0.25±0.03. VV in South Delhi showed that the ratio was 0.27±0.06. SKP in East Delhi showed that the ratio was 0.22±0.02, and overall in Delhi the ratio was 0.25±0.06.

3.3.8 Sox to NOx 24-hr ratio

The 24-hour ratio at the urban-industrial Delhi site varies from 0.18-0.35, and at the urban-commercial Delhi site varies from 0.20-0.30, and at two urban-residential Delhi sites varies from 0.17-0.26. The SKP and IIT variation of the ratio (0.02) is the lowest, while the highest is at Okhla (0.06). The variation of ratio at VV was 0.04, and overall in Delhi the ratio ranges from 0.17-0.35. The variation in Delhi (4%) shows that the sources liberating NOx are also releasing SOx.

Summarizing our observations at Okhla in South Delhi showed that the 24-hour ratio was 0.22±0.06. IIT in South Delhi showed that the ratio was 0.23±0.02. VV in South Delhi showed that the ratio was 0.24±0.04. SKP in East Delhi showed that the ratio was 0.20±0.02, and overall in Delhi the ratio was 0.22±0.04.

Many studies indicated that NOx and SOx concentration in ambient air would be affected by the various meteorological factors such as the wind speed, wind direction, solar radiation, relative humidity, rainfall, and source conditions. Our results could also be compared with the values obtained by the CPCB personnel at ITO during the study. These were taken from their site (http://www.envfor.nic.in/). Our values are in good agreement with the CPCB values. During comparison, it was kept in mind that these values are of those samples, which were taken for eight-hour duration as per CPCB rules. But these values are normalized for 24-hours to represent the daily mean value (as per CPCB personnel).

The SOx levels were always lesser than the 24-hour NAAQS of SOx, i.e., 80-$\mu g/m^3$ in the residential area and 120-$\mu g/m^3$ in the industrial area. The NOx levels were lesser than 24-

hour NAAQS of NOx i.e. 80-µg/m^3 in the residential area and 120-µg/m^3 in the industrial area, except on November 23, 2000 at VV and on November 26, 27, and 28, 2000 at Shakarpur (80-µg/m^3).

The ITO site can be considered as a commercial site in East Delhi, so we will compare our results of VV with it. We will also compare our results of SKP with it. The levels obtained at ITO in East Delhi showed that NOx, SOx concentrations and SOx/NOx ratio were 78.42±18.83, 17.30±4.40-µgm^{-3}, and 0.22±0.04, while at VV levels were 68.36±11.07, 16.12±2.15-µgm^{-3}, and 0.24±0.04. Our data also did not show any weekend or holiday effect. The variation is most likely due to differences in the air mass, emission characteristics, and the meteorology.

3.3.9 Comparison with other studies

Das et al (1997) reported in NO_2, SO_2, and NO_2/SO_2 levels in Jaipur for the sampling time of two-hours in the evening peak traffic hours at 49-stations. At residential areas, NO_2 and SO_2 values varied between 4.61-332.89-µg/m^3 and 115.84-296.74-µg/m^3; at the commercial areas, between 1.07-974.44-µg/m^3 and 136.96-912.64-µg/m^3; and at the industrial areas, between 26.16-78.81-µg/m^3 and 141.44-262.02-µg/m^3. Gajraj et al (2000) reported for May-June 1999 SO_2 and NOx levels during the morning and the evening for the rush hours in Jaipur as 56.7-184.4-µg/m^3 and 70.8-139-µg/m^3. Sharma et al. (2003) reported SO_2 levels in Kanpur for 24-hours at five stations. They found at IIT Kanpur SO_2, the levels were 5.44±1.58-µg/m^3, at GT road were 28.6±9.9-µg/m^3, at Agricultural University were 11.0±2.7-µg/m^3, at Naveen Market were 14.6±3.8-µg/m^3, and at Lajpat Nagar were 17.73±4.38-µg/m^3.

CPCB (http://www.envfor.nic.in/divisions/cpoll/delpoll.html) have reported an increase in the annual SO_2 level over the years (1989-1997), such as 8.7, 10.2, 13.3, 18.4, 18.5, 19.5, 19.0, 19.0, and 16.2-µg/m^3, and in annual NO_2 level such as 18.5, 22.5, 27.2, 30.4, 33.2, 33.0, 34.1, 33.7, and 33.0-µg/m^3. Though the annual mean value of SO_2 (15-26-µg/m^3) and NOx (28-46-µg/m^3) remain within the prescribed limits of 60-80-µg/m^3, there is a rising trend. Compared to 1989, SO_2 atmospheric concentrations in 1996 have registered a 109% rise, and NOx an 82% rise. Glikson et al (1995) reported in Brisbane, Australia for the period of the study that the maximum 1-hour average for NO_2 was 6.7-ppm. Cheng and Lam (2000) reported in Hong Kong the seasonal daily mean SO_2 concentrations for the entire period (1983-1992) were 28.2, 38.3, 20.0, and 23.5-µgm^{-3} for spring, summer, autumn, and winter, respectively; their standard deviations were 30.8, 42.7, 16.9, and 23.6-µgm^{-3}. The corresponding figures of the overall mean NOx concentrations were 129.7, 110.4, 121.0, and 148.6-µgm^{-3} with standard deviations of 54.2, 49.7, 40.1, and 74.5-µgm^{-3}. Mondal et al (2000) reported the ground level concentration of NOx at 19-important traffic intersection points in Calcutta ranging from 55-222-µgm^{-3}. Lebret et al (2000) reported small area variations in ambient NO_2 concentrations in four European areas, i.e., in Amsterdam (Netherlands), Huddersfield (UK), Poznan (Poland), and Prague (Czech Republic) found NO_2 concentrations as 24-72, 10-79, 12-48, and 9.6-18.4-µgm^{-3}. In NOx emission Delhi holds second position, and in SO_2 Delhi is not in the top 10-districts of India (Garg et al., 2001). India is the second largest contributor of SO_2 emissions in Asia, after China. For India, the estimates of national SO_2 emissions (Gg/yr.) in the period 1985-97 shows a rising trend as the emissions were 3402.0, 3609.1, 3829.9, 4085.2, 4256.1, 4437.2, 4684.3, 4863.1, 5039.6, 5289.3, 5609.5, 6106.3, and 6276.6-Gg/yr in the years 1985, 1986, 1987, 1988, 1989, 1990, 1991, 1992, 1993, 1994, 1995, 1996, and 1997, respectively (Streets et al., 2000). Our study is in good agreement with all of these studies.

3.4 Air quality index

In order to assess the degree of air pollution in Delhi, an index called air quality index (AQI) has been used as recommended by USEPA (EPA-454/R-99-010). More details are given in the literature review section, and the calculation method adopted is given in the methodology section. Calculated AQI has been presented in the table below.

DATE	RSP	SOx	NOx	AQI RSP	AQI SOx	AQI NOx
Okhla						
20/10	259.06	14.41	77.48	153.03	20.59	70.10
21/10	297.51	15.66	71.03	172.04	23.03	67.20
23/10	273.07	18.45	84.87	159.96	27.13	73.42
24/10	271.74	17.74	50.72	159.27	26.10	58.07
29/10	320.31	16.07	62.90	183.32	23.63	63.54
9-Nov	204.99	13.54	72.67	125.75	19.91	67.93
IIT						
2-Nov	179.09	8.45	35.12	112.93	12.43	51.05
3-Nov	274.65	9.09	42.62	160.75	13.37	54.53
4-Nov	189.11	9.57	36.17	117.88	14.07	51.53
5-Nov	196.80	8.76	38.94	121.69	12.88	52.77
6-Nov	200.86	10.22	50.11	123.72	15.03	57.79
7-Nov	185.06	10.48	44.69	115.90	15.41	55.36
VV						
15/11	245.62	12.81	63.90	145.84	18.84	63.99
16/11	258.31	17.68	62.80	152.63	26.00	63.50
17/11	322.63	15.26	77.01	184.46	22.44	69.89
22/11	210.57	16.34	70.80	128.52	24.03	67.09
23/11	209.97	19.12	85.49	128.23	28.12	73.70
24/11	317.32	16.24	54.55	181.84	23.88	59.79
SKP						
25/11	174.54	16.08	78.43	110.65	23.65	70.52
26/11	321.24	18.48	88.07	183.77	27.18	74.86
27/11	238.72	21.70	104.47	142.43	31.91	82.23
28/11	197.64	14.52	84.82	122.09	21.35	73.40

Table 3. Daily AQI along with Respective Parameter Concentrations

The category of air quality based on AQI for RSPM at Okhla was always unhealthy except on one day i.e. November 9, when it was unhealthy for sensitive people. The category of air quality based on AQI for RSPM at IIT was always unhealthy for sensitive people except on one day i.e. November 3, when it was unhealthy for the general public. The category of air quality based on AQI for RSPM at VV was unhealthy on November 16, 17, and 24, 2000, while it was unhealthy for sensitive people on November 15, 22, and 23, 2000. The category of air quality based on AQI for RSPM at SKP was always unhealthy for sensitive people except on one day i.e. November 26, 2000, when it was unhealthy for the general public as well. The category of air quality based on AQI for SOx and NOx at all sites was always good for the public.

It can be concluded that PM is the critical parameter for the deterioration of ambient air quality in Delhi.

3.4.1 Comparison with other studies

In the industrial area of Jaipur, Gajraj et al (2000) found 24 observations of RSPM, where the air quality was unhealthy for 14 days and unhealthy for sensitive groups for the remaining days. For 32 observations of RSPM in Jaipur, Jain (2001) found that the air quality was good on six days, moderate on 20 days, and unhealthy for sensitive people on six days. She also found for 24 observations of SO2 that the air quality was good on all days.

4. Conclusions

The $PM_{2.5}$ value at Delhi is well above 24-hour $PM_{2.5}$ USNAAQS (65-μgm^{-3}). The variation of $PM_{2.5}$ value is highest on SKP, which suggests that the contribution from different sources is varying to a great extent. The PM_{10} concentration value was always greater than the 24-hour NAAQS (150-$\mu g/m^3$ in industrial and commercial areas and 100-$\mu g/m^3$ in residential areas). The study shows that, at Okhla, $PM_{2.5}$ contribution to PM_{10} is 47%; it increases to 59% at VV, further increases to 69% at IIT, and achieves its maximum (77%) at SKP. The highest contribution of $PM_{2.5}/PM_{10}$ strongly shows that industrial sources are predominant in the SKP area. This shows that at Okhla RSP contribution to TSP is 39%; it increases to 50% at VV and decreases to 41% at IIT and again achieves its highest of 53% at SKP. The above discussion put forward that the secondary formation and diesel exhaust, which are prime important mechanisms of the formation of $PM_{2.5}$, are predominant at IIT but the primary particulate sources also seem like prime important reasons for the formation of RSP. In all of Delhi, 24-hour SO_2 concentration varies from 8.45-21.70-μgm^{-3} and 24-hour NO_2 concentration varies from 35.12-104.47-μgm^{-3}. The variation in Delhi shows that sources liberating NOx are also releasing SOx. The SOx levels were always lesser than the 24-hour NAAQS of SOx i.e. 80-$\mu g/m^3$ in the residential area and 120-$\mu g/m^3$ in the industrial area. The NOx levels were lesser than the 24-hour NAAQS of NOx i.e. 80-$\mu g/m^3$ in the residential area and 120-$\mu g/m^3$ in the industrial area. The SO_2 and NO_x data do not show any weekend or holiday effect. The variations observed are most likely due to differences in the air mass, emission characteristics, and the meteorology. The category of air quality based on AQI for SOx and NOx at all sites was always good for the public. The category of air quality based on AQI for PM at all sites was always unhealthy for the public. It can be concluded that PM is the critical parameter for the deterioration of ambient air quality in Delhi.

5. Acknowledgement

The authors are thankful to Envirotech Instruments Private Limited, Delhi for their sponsorship of the study. Thanks were due to all the persons, organizations, institutions, and companies who provided a secure location for the sampling equipment used in this study, as well as those who provided supplemental information. Authors were thankful to Mr. Ronald Zallocco for correcting the English.

6. References

Borja-Aburto, V., Loomis, D., Bangdiwala, S., Shy, C., Rascon-Pacheco, R. (1997). Ozone, Suspended Particulates, and Daily Mortality in Mexico City. *American Journal of Epidemiology,* 145, pp. (250-257).

Burton, R., Suh, H., Koutrakis, P. (1996). Spatial Variation in Particulate Concentrations within Metropolitan Philadelphia. *Environmental Science and Technology,* 30, pp. (392-399).

Cheng, S., Lam, K.C. (2000). Synoptic Typing And Its Application To The Assessment Of Climatic Impact On Concentrations Of Sulfur Dioxide And Nitrogen Oxides In Hong Kong, *Atmospheric Environment,* 34, pp. (585-594).

Crowther, G., Finlay, H., Raj P.A., Wheeler, T. (1990). India: *A travel survival kit,* (IV ed.), Lonely Planet Publications, pp. (140-169).

Das, D.B., Gupta, A.B., Bhargava, A., Kushwaha, Pandit, M.K. (1998). Quarrying Induced Particulate Air Pollution In Jaipur City, Western India. *Journal Of Nepal Geological Society,* 18, pp. (369-378).

Down To Earth. (2000). *Killing Pollutants,* Vol 9, No.10, pp. (23).

E.P.A, 1999, *Guideline for reporting daily air quality,* EPA-454/R-99-010 July 1999

Gajraj V., Jain R., Gupta T.P., Gupta, A.B. (2000). A study of air quality of Vishwakarma Industrial Area, Jaipur, *Proc. IPC 2000,* National seminar on industrial pollution and its control, Feb. 18-19, 2000, BHUIT, Varanasi, India, pp. (10-19).

Garg, A., Shukla, P.R., Bhattacharya, S., Dadhwal, V.K. (2001). Sub-Region (District) And Sector Level SO2 And NOx Emissions For India: Assessment Of Inventories And Mitigation Flexibility, *Atmospheric Environment,* 35, pp. (703-713).

Guenther, A., Geron, C., Pierce, T., Lamb, B., Harley, P., Fall, R. (2000). Natural Emissions Of Non-Methane Volatile Organic Compounds, Carbon Monoxide, And Oxides Of Nitrogen From North America. *Atmospheric Environment,* 34, pp. (2205-2230).

Hargreaves, P.R., Leidi, L.A., Grubb, H.J., Howe, M.I., Muggle Stone, M.A. (2000). Local & Seasonal Variations In Atmospheric Nitrogen Dioxide Levels At Both Amsted, UK & Relations With Metrological Conditions. *Atmospheric Environment,* 843-853.

http://www.envfor.nic.in/, Date of access: September, 9, 2010.

http://www.envfor.nic.in/divisions/cpoll/delpoll.html, Date of access: July, 28, 2009.

Jain, R. (2001). Dispersion Modeling of Gaseous Pollutants in Jaipur City and Assessment of Their Health Effects on Residents Ph.D. Thesis, Malviya Regional Engineering College, University of Rajasthan.

Kamyotra & Ashish. (2001). Characterization of different methods used for analysis of sulfur-di-oxide, M. Tech. Thesis, Indraprastha College, Delhi University, New Delhi.

Kumar, A., Kadiyala, A., Somuri, D., Shandilya, K.K., Velagapudi, S., Nerella, V.K.V., (2011). Characterization of Emissions and Indoor Air Quality of Public Transport Buses Using Alternative Fuels. In: *Biodiesel: Blends, Properties and Applications,* Editors: Marchetti, J.M., Fang, Z., In Press, Nova Science Publishers, Inc., ISBN 978-1-61324-660-3.

Mondal, R., Sen, G.K., Chatterjee, M., Sen, B.K., Sen, S. (2000). Short Communication: Ground-Level Concentration Of Nitrogen Oxides (NOx) At Some Traffic Intersection Points In Calcutta. *Atmospheric Environment*, 34, pp. (629-633).

Monn, C. (2001). Exposure Assessment of Air Pollutants: A Review on Spatial Heterogeneity and Indoor/Outdoor/Personal Exposure to Suspended Particulate Matter, Nitrogen Dioxide and Ozone. *Atmospheric Environment*, 35, pp. (1-32).

Peavy, H.S., Rowe, D.R., Tchobanoglous, G. (1985). Environment Engineering, International Ed., McGraw-Hill Book Company.

Pekkanen, J., Timonen, K.L., Ruuskanen, J., Reponen, A., Mirme, A. (1997). Effects of ultrafine and fine particles in urban air on peak expiratory flow among children with asthmatic symptoms. *Environmental Research*, 74, pp. (24-33).

Perkins, H.C. (1974). *Air Pollution*, McGraw -Hill Kogakusha Ltd., Tokyo.

Rao, M.N. & Rao, H.V.N. (1989). *Air Pollution*, Tata–McGraw Hill publishing Company Limited.

Sawyer, R.F., Hardely, R.A., Cadle, S.H., Norbeck, J.M., Slott, R., Bravo, H.A. (1998). Mobile Sources Critical Review: 1998 NARSTO Assessment, *Atmospheric Environment*, 34, pp. (2161-2181).

Shandilya, K.K., Khare, M., Gupta, A.B. (2006). Suspended Particulate Matter concentrations in Commercial East Delhi and South Delhi, India. *Indian Journal of Environmental Protection*, 26(8), pp. (705-717).

Shandilya, K.K., Khare, M., Gupta, A.B. (2007). Suspended Particulate Matter concentrations in Rural-Industrial Satna and in Urban-Industrial South Delhi. *Environmental monitoring and assessment*, 128(1-3), pp. (431-45).

Shandilya, K.K., Gupta, A.B., Khare, M. (2009). Defining Aerosols by physical and chemical characteristics, *Indian Journal of Air Pollution Control*, Vol. IX, No. 1, pp. (107-126).

Shandilya, K.K., Kumar, A. (2009). Analysis of Research Studies on Exhaust Emission from the Heavy Duty Diesel Engine Fueled by Biodiesel. In: *Handbook of Environmental Research*, Editors: Aurel Edelstein and Dagmar Bär, pp. (101-151), Nova Science Publishers, Inc., ISBN: 978-1-60741-492-6, NewYork..

Shandilya, K.K., Kumar, A. (2010a). Qualitative Evaluation of Particulate Matter inside Public Transit Buses Operated by Biodiesel. *The Open Environmental Engineering Journal*, 3, pp. (13-20).

Shandilya, KK., Kumar, A., (2010b). Morphology of Single Inhalable Particle inside Public Transit Biodiesel Fueled Bus. *Journal of Environmental Sciences*, 22(2), pp. (263–270).

Shandilya, K.K., Kumar, A. (2011). Physical Characterization of Fine Particulate Matter inside the Public Transit Buses Fueled by Biodiesel in Toledo, Ohio. *Journal of Hazardous Materials*, Volume 190, Issues 1-3, pp. (508-514).

Shandilya, K.K., Khare, M., Gupta, A.B. (2011). Particulate Matter Concentrations in Delhi before Changing to Compressed Natural Gas (CNG). *Indian Journal of Air Pollution Control*, Paper in Press.

Shandilya, K.K., Gupta, A.B., Khare, M. (2012). Formation of Atmospheric Nitrate under High Particulate Matter Concentration. *World Review Of Science, Technology And Sustainable Development, (WRSTSD), Special Issue: Waste And Environment Management*, Paper in Press.

Shandilya, K.K., Khare, M. (2012). Particulate Matter: Sources, Emission Rates and Health Effects, In: Advances in Environmental Research Volume 23, Editor: Daniels, J.A., In Press, NOVA publishers, ISBN: 978-1-62100-837-8.

Sharma, M., Kiran, Y.N.V.M., Shandilya, K.K. (2003). Investigations into Formation of Atmospheric Sulfate under High PM10 Concentration, *Atmospheric Environment*, 37, pp. (2005-2013).

Sharma, M., Kiran, Y.N.V.M., Shandilya, K.K. (2004). Subset Statistical Modeling of Atmospheric Sulfate. *Clean Air and Environmental Quality*, 38, No. 4, pp. (37-41).

Streets, D.G., Tsai, N.Y., Akimoto, H., Oka, K. (2000). Sulfur Dioxide Emissions In Asia In The Period 1985-1997, *Atmospheric Environment*, 34, pp. (4413-4424).

Thurston, G.D., Ito, K., Lippmann, M., Hayes, C. (1989). Reexamination of London, England, mortality in relation to exposure to acidic aerosols during 1963-1972 winters. *Environmental Health Perspectives*, 79, pp. (73-82).

Tuch, T., Brand, P., Wichmann, H.E., Heyder, J. (1997). Variation of Particle Number and Mass Concentration in Various Size Ranges of Ambient Aerosols in Eastern Germany, *Atmospheric Environment*, 31, pp. (4193- 4197).

WHO. (1979). *Fine Particulate Pollution: A Report of the United Nations Economic Commission for Europe (UNECE)*, Pargamon Press sec. I, A, 2.

Wilson, W.E., Suh, H.H. (1997). Fine particles and coarse particles: Concentration relationships relevant to epidemiologic studies. *Journal of the Air & Waste Management Association*, 47, pp. (1238-1249).

4

Environmental Parameter Effects on the Fate of a Chemical Slick

Stéphane Le Floch[1], Mélanie Fuhrer[1,2], Pierre Slangen[2] and Laurent Aprin[2]
[1]*CEDRE - Centre de Documentation, de Recherche et D'Expérimentations sur les Pollutions Accidentelles des Eaux, Brest*
[2]*ISR, LGEI, Ecole des Mines d'Alès, Ales*
France

1. Introduction

Recent data highlights the growing trend in the transport of dangerous substances and the consequent evolution in legislation concerning such substances, whether on a European level (evolution of the Standard European Behaviour Classification, or SEBC code) or on a worldwide scale (new MARPOL-Annex II classification which entered into force on 1st January 2007, the OPRC-HNS Protocol, as well as the HNS Convention). This evolution is however based exclusively on data from the literature which all too often cannot be applied to spills, as it does not take into account the influence of factors in the marine environment on the physico-chemical characteristics of the product spilt.

Yet, following a spill at sea of a bulk liquid chemical, the response authorities must immediately take measures in order to reduce the risks of exposure of the surrounding population as well as to prevent contamination of the marine ecosystem as a whole and, more widely, to protect all life forms. To succeed in this effort that forms a part of emergency response, it is vital to know which compartment of the ecosystem will be affected (atmosphere, water surface, water column, sediments…) and to monitor the evolution of concentrations in order to define the environmental risks.

The accident of the chemical tanker *Ievoli Sun* which sank in the English Channel in 2000 releasing approximately 1,000 tonnes of styrene into the sea (Law et al., 2003) has raised concerns about the fate of this compound at sea. In the aquatic environment, styrene is often reported to be of low risk for chronic effects due to its volatility and low bioaccumulation potential. Hence, the potential for styrene to produce long term adverse environmental impacts appears to be negligible. However, most of these statements are based on tests performed in fresh water environments whilst very few data exist concerning the marine environment. Also, the value of some of these tests can on some occasions be questionable as they are reliant on a static mode of exposure and measurements of the actual exposure concentrations are not performed during the tests. Hence, although styrene's volatility and biodegradability would in most cases significantly reduce the levels in water, the behaviour and bioavailability of styrene in saline water is not very well known. Thereby, a better evaluation of the actual toxic effects of this compound in marine aquatic organisms is needed. In the recent example of the *Ievoli Sun* vessel, a relatively large amount of styrene was discharged in a relatively short time. Yet, studies showed that only low levels of styrene contamination were measured in edible tissues of crabs caught in the immediate vicinity of the wreck site (Law et al., 2003). Bioaccumulation does not necessarily represent the real uptake of a contaminant as many aquatic organisms including fish and crustaceans are able to efficiently biotransform the mother compounds into more polar metabolites, thereby considerably reducing the bioaccumulation factor of the mother compound. Even at low concentrations, it is known that some substances can disturb important biochemical pathways which can impair biochemical functioning of cells and lead to deleterious effects in tissue, like cancer development, in the longer term (Alexander, 1997; Vaghef & Hellman, 1998; Moller et al., 2000). Furthermore, the styrene released from the wreck lying on the floor of the Channel, whose density is lower than that of seawater, rose through the water column to form slicks at the sea surface. These slicks drifted under the combined influence of the wind and marine currents, without ever reaching the coast. Under the effect of the wind and sunshine, the styrene evaporated to form an invisible cloud whose fate was then dependent, both in terms of drift and dilution on aerial currents alone. Following this succession of physico-chemical events, the response authorities had to deal with contamination of the atmosphere, in addition to water pollution, by a product recognised as being particularly neurotoxic (Laffon et al., 2002). The French Customs operators conducting aerial surveillance above the spill area were not prepared for this new risk and were exposed to styrene vapours, resulting in disabling headaches. Just as concerning, on the inhabited island of Aurigny, naval firefighters sporadically yet repeatedly detected the presence of styrene in the air. The inability to predict, i.e. to anticipate, arrivals of styrene gas on the coast rendered contingency plans ineffective for the protection of the population. While no consequences on the health of neighbouring populations were reported in this incident, as the concentrations always remained below the ADI for styrene, things could have been quite different with other more toxic substances.

While classifications can be used to predict the theoretical fate of a product in the event of a spill in the environment, they do not take into account the specificities of this environment and should therefore be used cautiously. For instance, wind and sunshine can significantly affect chemicals' evaporation. A similar observation can be made for the salinity and temperature of seawater and the chemicals' hydrosolubility. Furthermore, while the prediction of a substance's fate in the event of a spill has been facilitated by the emergence of forecast software (CLARA, CHEMMAP…), these predictions must nevertheless be validated by field measurements. Forecasting predicts behaviour, i.e. the fate of products in the environment, from the physico-chemical characteristics of products as determined in laboratory conditions, to which mathematical equations are applied which often underestimate certain environmental parameters such as the influence of surface agitation on the spreading of the slick or sunshine on evaporation. This software also has difficulty in predicting the interaction of chemicals with the sea water matrix in the case of a leaking wreck, due to a lack of information on the effects of pressure, temperature and salinity on dissolution processes (Le Floch et al., 2007; 2009).

This work aimed to obtain experimental data on the behaviour of 3 liquid chemicals when released at sea and to characterise them through *in situ* experiments. For this purpose, experimental open cells were set up in Brest Bay, each of them made of a 9 square metre surface surrounded with a 3 metre high nylon skirt set on a metal structure. As the cell sides were flexible, the contained water was subject to the influence of swell and waves. As the cells were not covered, the sea surface was subject to atmospheric influences (wind, sunshine and rain). After releasing the products at the surface, weathering processes were studied by measuring the viscosity of the slick, solubility in the water column and gas cloud formation.

2. Fate of chemicals in the environment after accidental release

Several international, regional and national authorities, aware of the risk connected with the transport of chemicals by sea, have published operational guides to describe the possible response options. For instance, IMO has published several manuals (IMO, 1987, 1992, 2006) and REMPEC has done the same for the Mediterranean area (REMPEC, 1996, 1999, 2004). We can also mention the Helsinki Baltic Sea Convention manuals (HELCOM, 1991, 2002, 2007) and the North Sea Bonn Agreement manuals (Bonn Agreement, 1985, 1994). At a national level in France, *Cedre* has produced response guides, each of which is specifically dedicated to a given chemical.

Alongside this information, which ranges from general (i.e. in the case of a guide on response at sea to hazardous substances) to specific (i.e. guides that characterize the behaviour and identify the risks related to a given product), the Standard European Behaviour Classification system or SEBC code determines the short term behaviour of any chemical as long as its physico-chemical characteristics are known. This tool is of the utmost importance for operational staff in charge of response, as it can be used in an emergency to indicate the main components of the ecosystem that will be affected. However, experience has made it clear that this code can be tricky to use as it can be difficult to understand how environmental factors (wind, sunshine, water and air temperature, etc.) will affect the chemicals' categorization. Indeed these parameters have a very significant influence on the short-term fate of chemicals in the marine environment, in particular on their evaporation kinetics (transfer from the water surface to the atmosphere), their dissolution kinetics (transfer from the water surface to the water column) and their transport and transformation processes in the aqueous phase.

2.1 Fate of substances with low reactivity

In the case of a spill, response usually follows three generally accepted scenarios: i) response is not possible because the spill occurred in a geographical environment that is incompatible with reasonable response times; ii) response is not possible due to the reactivity of the substances (major, imminent danger); iii) response is possible.

If, in the latter scenario, the substance spilled is not particularly reactive, it is important to bear in mind that the slick formed will evolve. Immediately after its accidental release into the marine environment, the substance will tend to move into different components of the ecosystem, i.e. the atmosphere, the sea surface, the water column and the seafloor. These different forms of distribution are gathered under the umbrella term of "fate", for which we generally distinguish short term behaviour (a few hours) and longer term behaviour (up to several years).

2.2 Short term fate

Short term behaviour covers a duration ranging from a few minutes to a few hours, and it must be known in order to implement response measures and techniques. It was in light of this observation that the Standard European Behaviour Classification, or SEBC code, was developed (Cedre, 1988; Bonn Agreement, 1994; GESAMP, 2002).

This code classifies the behaviour of chemical substances according to their state and some of their physical properties, i.e. density, vapour pressure and solubility. The state of the substance refers to whether it is in the form of a gas, a liquid or a solid at 20°C. The density, defined as the mass per unit volume compared to that of seawater (1.03 g.cm^{-3}, at 20°C), determines whether or not a substance will float. The vapour pressure is defined as the partial vapour pressure of a compound in equilibrium with its pure condensed phase (liquid or solid). It is generally accepted that a floating substance will not evaporate if its vapour pressure is lower than 0.3 kPa and it will evaporate rapidly if its vapour pressure is higher than 3 kPa. For dissolved substances, evaporation occurs when the vapour pressure is greater than 10 kPa. Solubility is defined as the maximum abundance of a compound per unit volume in the aqueous phase when the solution is in equilibrium with the pure compound in its initial state. It is usually measured at 20°C, at a pressure of 1 atmosphere, in distilled water (zero salinity). The criteria used are different according to the physical state of the substance. A substance is considered insoluble if its solubility is less than 0.1% for liquids and less than 10% for solids. At the other end of the scale, the phenomenon of dilution will take place if the solubility is greater than 5% for liquids and greater than 100% for solids.

These limits are presented in Figure 1 and can be used to identify 12 types of behaviour, which are summarized in Table 1.

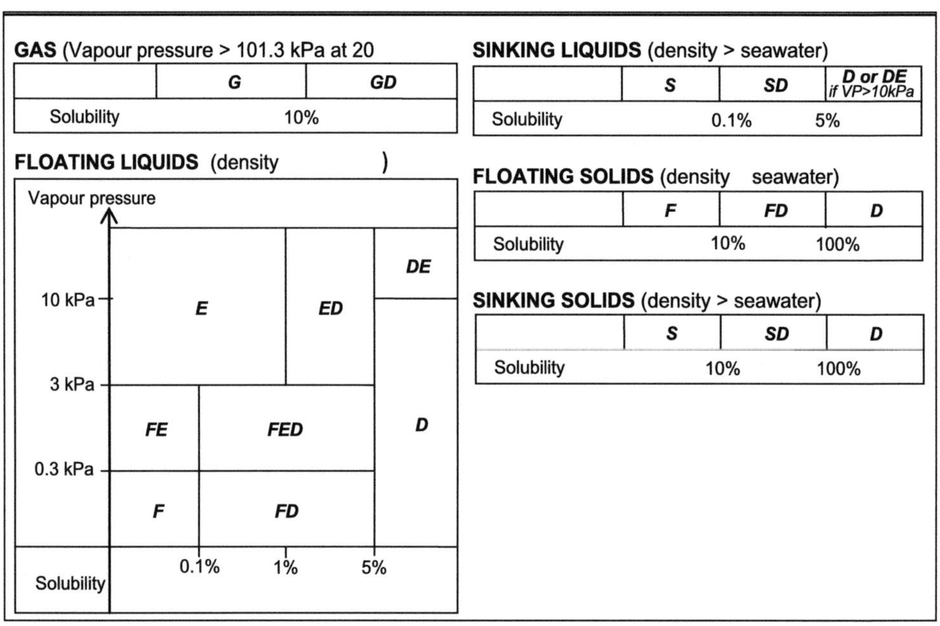

Fig. 1. SEBC Classification of Chemical Products according to their Physical State and Physical Properties

Main group		Subgroup	
G	Gas	GD	Gas that dissolves
E	Evaporator	ED	Evaporator that dissolves
F	Floater	FE	Floater that evaporates
		FD	Floater that dissolves
		FED	Floater that evaporates and dissolves
D	Dissolver	DE	Dissolver that evaporates
S	Sinker	SD	Sinker that dissolves

Table 1. The 12 Behaviour Groups according to the Standard European Behaviour Classification System

2.3 Long term fate

In terms of response, it is also useful to be aware of longer term behaviour as it provides information on the fate of a pollutant that cannot be removed from the environment through response operations. The residual chemical will spread through the atmosphere and/or the water column where it may become associated with sediment particles in suspension before settling. In water and air, the ultimate fate of chemicals will depend on the processes of dispersion, breakdown and other physical, chemical or biological transformations. These processes can extend over periods ranging from a few months to several years when the substance remains in the environment.

In terms of fate, all these phenomena will contribute, in the best case scenario, to the disappearance of the product and, in the worst case scenario, to its accumulation or even its transfer into another part of the ecosystem.

2.4 Influence of environmental factors on fate

Spill response fundamentally requires rapid identification of the compartment(s) of the ecosystem in which the pollutant will accumulate so as to implement the appropriate means to reduce, or even eliminate, its impact. While the SEBC code provides initial elements of response, the operational personnel in charge of response must interpret this classification critically so as to assess whether or not the specificities of the environment in which the incident has occurred will alter the results. This critical interpretation may go as far as altering the categorization of a substance as in the case of the industrial incident at the Jilin plant in China.

On 13 November 2005, an explosion occurred in a petrochemical plant resulting in a spill of around one hundred tonnes of various substances into the Songhua River. Analyses conducted by the Chinese Research Academy of Environmental Science showed that the water contained benzene and nitrobenzene concentrations 30 to 100 times higher than the accepted standards, of 0.01 $mg.L^{-1}$ and 0.017 $mg.L^{-1}$ respectively (Ambrose, 2006; UNEP, 2006). The pollution was carried by the current, and concentrations 34 times higher than standard levels were found in Harbin, 380 km downstream of Jilin, and 10 times higher than standard levels in Jiamusi, 550 km downstream. This decrease can be explained both by the processes of dilution and adsorption of the chemicals by organic matter present in the water, either in suspension, or on the banks (Levshina et al., 2009). These unexpectedly high concentrations can be explained by the particularly harsh weather conditions on site at this time of year (negative temperatures [< -10 °C] and river partially frozen) which restricted evaporation, promoting dilution.

With this as a background, *Cedre* developed an experimental system designed to study the fate of chemicals when released into the marine environment in field conditions, i.e. as close as possible to those encountered in the event of a spill.

3. Materials and methods

The aim of this experimental investigation was to characterise the fate of liquid chemicals released at the sea surface.

3.1 Floating cell enclosures

Six floating cells were constructed by *Cedre* and moored in Brest Bay, France (48°22'32N and 4°29'32W) to study the influence of natural conditions (i.e., prevailing air and sea temperatures, wind, radiant energy) on the weathering of products spilled onto the sea surface (Figure 2). Each cell consisted of a 3 x 3 m rigid aluminium framework surrounded by a nylon skirt which extended approximately 2 m below and 0.6 m above the sea surface. As cells were not covered and were open to the sea at the base, evaporation and dissolution processes were unimpaired but lateral dispersion at the sea surface was restricted. Observations have shown that while wind-induced capillary waves were reduced by the presence of the skirt, the majority of the mixing action of the sea passed into the cell almost unimpeded. The volume of seawater inside the skirt of each floating cell was approximately 18 m^3. As the floating cell enclosures were independent of each other, replica experiments could be conducted simultaneously. During this research programme only 3 of the 6 cells were used.

Fig. 2. Floating cell enclosure

3.2 Chemical products tested

Following an investigation into the chemicals transported most frequently and in the greatest quantities, three products were selected: xylene, methyl methacrylate and methyl ethyl ketone.

Xylene is a volatile aromatic hydrocarbon produced through petrochemistry. It is in liquid form at room temperature and is often a mixture of isomers (ortho, para and meta - often the most dominant) containing variable proportions of ethylbenzene, although always below <15%. In France, xylene is the most commonly used hydrocarbonated solvent, with 34,000 tonnes in 2004, in particular in the paint, varnish, glue and printing ink industries. The main physico-chemical characteristics of this substance indicate that it will, in theory, behave as a floater/evaporator with very low solubility (Vp = 8.9hPa at 20°C, density of 0.88g.cm^{-3}, and solubility between 175 and 200 mg per litre of seawater; Cedre, 2007 and Lyman et al., 1996).

Methyl methacrylate (MMA) is described as a transparent volatile liquid with a characteristic smell that can be detected at very low concentrations (< 1 ppm in air). It is mainly used in the polymer and copolymer industry (plastic sheets such as Plexiglas®, Perspex® and Lucite...). Its vapour pressure is 3.9 kPa at 20°C, its solubility 15 g.L^{-1} and its density 0.944 g.cm^{-3} (Cedre, 2008).

Methyl ethyl ketone (MEK) is a sweet-smelling colourless liquid. It is a volatile organic compound, that is readily flammable and is of little harm to the environment. MEK is an eye and respiratory irritant as well as a central nervous system depressant in humans. It is mainly used as a solvent in various coatings including vinyl, nitrocellulose and acrylic coatings. It is also used as an extraction agent in certain oils as well as in the processing of products and food ingredients. Its vapour pressure is 10.5 kPa at 20°C, its solubility 158 g.L^{-1} and its density 0.805 g.cm^{-3}. It is therefore a highly evaporative and soluble product, with particularly low persistence at the water surface (Cedre, 2009).

3.3 Experimental protocol

The test, replicated for each product, required 25 litres to be released at the sea surface of each floating cell. To monitor physical oceanographic conditions that may influence the behavioural processes of products, weather conditions, prevailing light, wind speed and temperature conditions were recorded, for both air and sea.

3.3.1 Slick sampling

Slicks were sampled daily, with a funnel equipped with a tap, in order to characterise the emulsification processes. Samples were transferred into 0.5 L amber bottles and brought back to the laboratory for viscosity measurement and water content determination.

Before the release of xylene, the internal standard pentacosane, a non-evaporating and insoluble substance, was introduced. By measuring the pentacosane/xylene ratio throughout the trial, it was possible to monitor the disappearance kinetics of xylene at the water surface.

3.3.2 Water column sampling

The natural dispersion of the products in the water column was monitored by performing *in situ* fluorescence measurements at 2 positions in each floating cell at 3 depths (0.5, 1 and 1.50 m). In addition, for each *in situ* fluorescence measurement (at a given position and a given depth), a sample of seawater (1 L) was taken. All samples were placed in an amber bottle and were stored in a refrigerator before analysis, performed the following day. These samples, which were analysed by GC-MS after liquid-liquid extraction, were used to correct data obtained by fluorimetry measurements (Katz, 1987).

3.3.3 Air monitoring

During the tests, evaporation was monitored by measuring the Volatile Organic Carbon using two Photo Ionization Detectors (PID MiniRAE 2000, RAE systems). A stationary PID was placed in a corner of the floating cell, while the second was placed on a wind vane so as to monitor evaporation independently of the wind direction.

During these trials, an infrared hyperspectral SPIM camera was deployed to assess its capacity to monitor the formation and movement of a gas cloud formed from slicks of MMA.

4. Results and discussion

4.1 Xylene

To study the fate of xylene, two 25 litre releases were performed.

4.1.1 Field observations

Photo 1, taken at T_0, illustrates the chemical release. One hour after the release, the slick began to emulsify and, two hours later, the emulsion covered the entire surface of the floating cell. After 5 hours, the slick began to fragment and several small slicks scattered across the entire cell appeared (Photo 2).

4.1.2 Prevailing sea and weather conditions

The week of the trials was marked by relatively harsh weather conditions, with a slight sea according to the Douglas sea scale (sea 3), a north-easterly wind at force 3 on the Beaufort scale (12 to 19 km.h^{-1}) and low temperatures (between 3 and 5°C). Several snow showers occurred during this week.

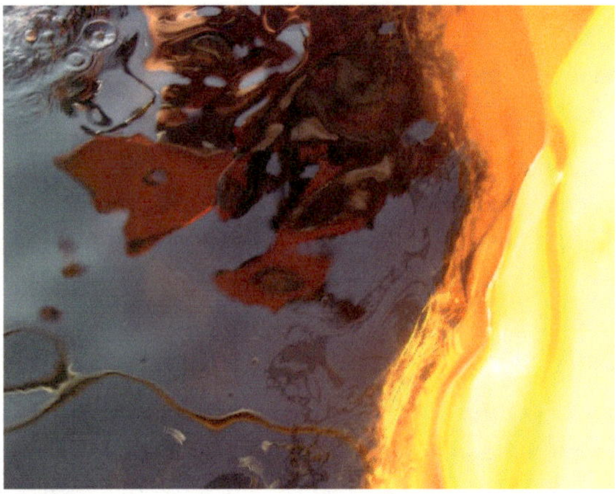

Photo 1. Xylene release (T_0)

Photo 2. Xylene slick at T_{+5h}

4.1.3 Slick persistence

Figure 3 shows the persistence of the xylene slicks at the water surface. The results are expressed as a percentage of the quantity spilt at T_0. After 2 hours, over 50% of the slick had disappeared, and after 5 hours, less than 30% of the initial quantity remained. The quantities lost were either transferred from the surface to the air (evaporation) or from the surface to the water column (natural dispersion). After day 2, no traces of the slick remained visible at the surface. The time during which the slick was present was estimated at 34 hours.

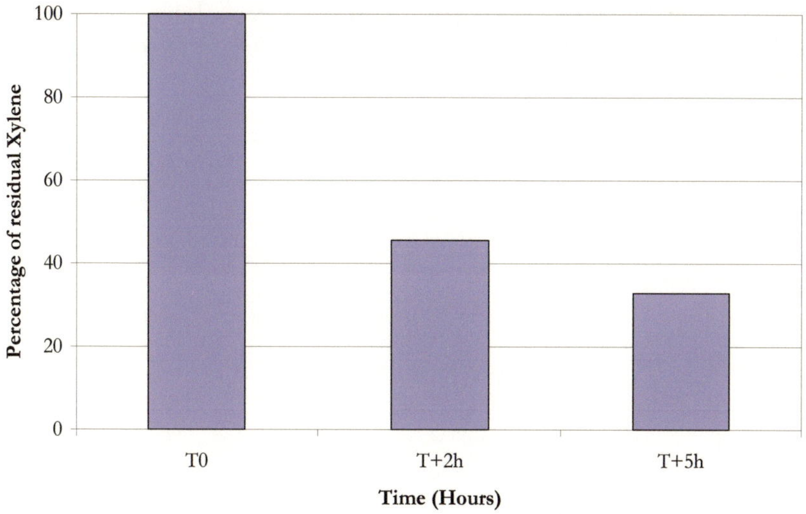

Fig. 3. Evolution over time of the slicks remaining at the surface (average of results obtained using samples taken from cells 1 and 2).

4.1.4 Monitoring of emulsification

The emulsification kinetics could only be monitored during the first day of the experiment (Figure 4).

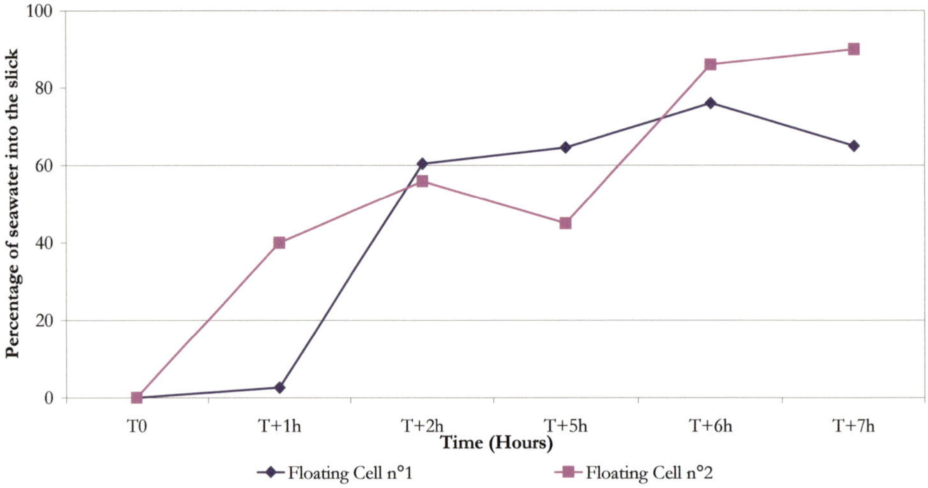

Fig. 4. Emulsification kinetics of xylene slicks within cells 1 and 2.

Xylenes have a high water sorption capacity and rapidly form emulsions. At T_{+7h} the water content of slicks in cells 1 and 2 were respectively 65% and 90%. However, these emulsions are unstable as they are non-persistent: from day 2, it was no longer possible to sample the surface slick due to insufficient quantities, and on day 3, no pollution remained visible.

4.1.5 Monitoring of dispersion in the water column

Figures 5 and 6 show the xylene concentrations in the water column, respectively in cells 1 and 2, at the three sampled depths (0.5, 1 and 1.5 m) over time. After 29 hours, concentrations of the substance were no longer detected.

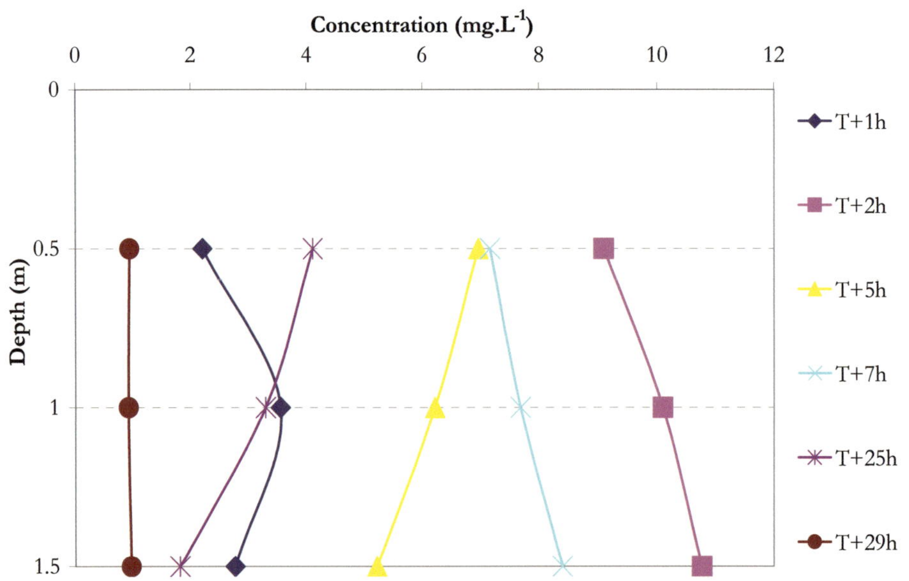

Fig. 5. Evolution over time of xylene concentrations in the water column in cell 1.

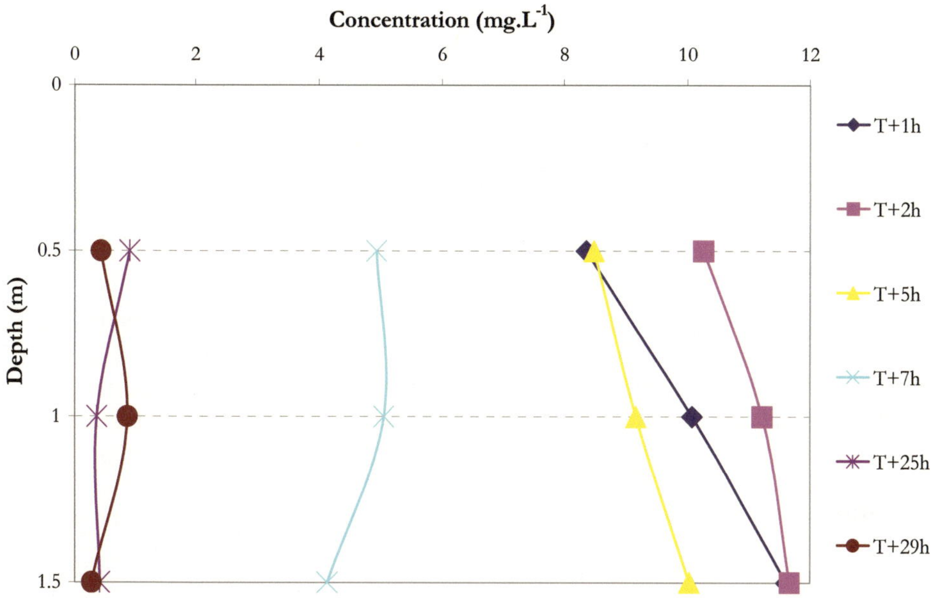

Fig. 6. Evolution over time of xylene concentrations in the water column in cell 2.

In the two cells, the maximum concentrations were obtained at T_{+2h} (10.8 and 11.7 $mg.L^{-1}$, respectively in cells 1 and 2), and represented a quantity of around 200 g of dispersed xylenes. These concentrations then gradually decreased over time, and dropped below 1 $mg.L^{-1}$ at T_{+29h}. Beyond this sample time, traces of xylene were no longer detected in the water column.

During these *in situ* trials, the natural dispersion of xylenes in the water column (dissolution + emulsification) showed rapid kinetics, and the concentrations measured were higher than expected: 100 times higher than the theoretical solubility found in the literature.

4.1.6 Fate of xylenes

These trials demonstrated that the persistence of a slick of xylenes at the water surface is relatively short-lived (around 2 days in these trial conditions, not very conducive to evaporation) and that this disappearance could be explained by dissolution processes which were higher than expected (in the literature this chemical is identified as having low solubility). In terms of the transfer of the chemical into the atmosphere, evaporation kinetics could not be measured due to the particularly harsh weather conditions (the snow interfered with the photoionisator). However, by comparing it to styrene (this chemical was previously tested and these two products have an equivalent vapour pressure and are classified as FE according to the SEBC code), it can be presumed that xylenes evaporate less quickly than theory predicts: after 5 hours, the entire styrene slick had disappeared, while this took 34 hours with xylenes. These results can be explained by emulsification processes: the xylene slicks rapidly emulsified to a great extent during the first day (90% at T_{+7h}), a process which does not take place with styrene. This raises the question of surface tension which is not taken into account in the SEBC classification; yet it is this factor that defines a product's capacity to emulsify, and emulsification promotes natural dispersion in the water column.

The results obtained *in situ* therefore directly contradict the literature which mainly describes the behaviour of a xylene slick in terms of transfer into the atmosphere. It is clear that the slick will initially emulsify, causing evaporation kinetics to slow down and promoting transfer into the water column by natural dispersion processes (dissolution and emulsification). It is important, however, to remember that the sea and weather conditions on site should be taken into account when considering these results, as, in these trials, the low temperatures curbed evaporation processes and surface agitation promoted emulsification. This fate is in accordance with what was observed during the benzene spill in Jilin (higher dissolution than predicted and low evaporation processes).

In terms of response, this means that all the issues related to pollution by soluble products (contamination of the water column and problem of water intakes, impact on benthic flora and fauna…) must be taken into account even if the product is classified as a floater and evaporator. Nevertheless, the gas cloud must not be overlooked, as it presents a risk for responders both in terms of intoxication and as an explosion hazard. In the case of xylene, vapours are heavier than air and will move around just above the water surface, pushed by the wind. This danger is even greater if the spill occurs in a coastal area, or even in a port or harbour, where the population density is liable to be high.

4.2 Methyl methacrylate

Methyl methacrylate (or MMA) is classed ED according to the SEBC classification, and its main behaviour is evaporation according to its Henry's law constant.

4.2.1 Observations

The MMA spread out across the water surface forming a heterogeneous slick: various sized clusters of floating droplets were observed (Photo 3). This phenomenon was only temporary, as in less than two hours all of the substance had disappeared.

Photo 3. Appearance of the MMA slick immediately after the spill

4.2.2 Prevailing sea and weather conditions

The prevailing weather conditions onsite at the time of the trial are presented in Table 2. The key point is the presence of wind.

Date	25/06	25/06	25/06
Time	11:00-12:00	12:00-13:00	13:00-14:00
Air T (°C)	16.4	16.7	17.1
Wind speed (m/s)	5.5	5.6	5.6
Solar intensity (mW/cm²)	61.5	71.2	83.4

Table 2. Evolution of weather conditions during the release of MMA

4.2.3 Evaporation

The MMA evaporated very rapidly: it was released at 11:18 and the PIDs began to detect significant values from 11:25 (1 ppm at 11:24, then 53 ppm at 11:25). A peak in concentration was reached at 11:27 with 130 ppm, the time at which a local maximum was recorded for light intensity (130 mW/cm²), while the wind varied little throughout the duration of measurements (5 to 6 m/s on average with peaks at 7 m/s and minima at 4 m/s). These results, illustrated in Figure 6, show the importance in evaporation processes of sunshine, which had a greater influence than the wind: positive correlation between solar peaks and evaporation peaks (T_{+9}, T_{+14}, T_{+35}, T_{+49}, T_{+56}, $T_{+74\ min}$).

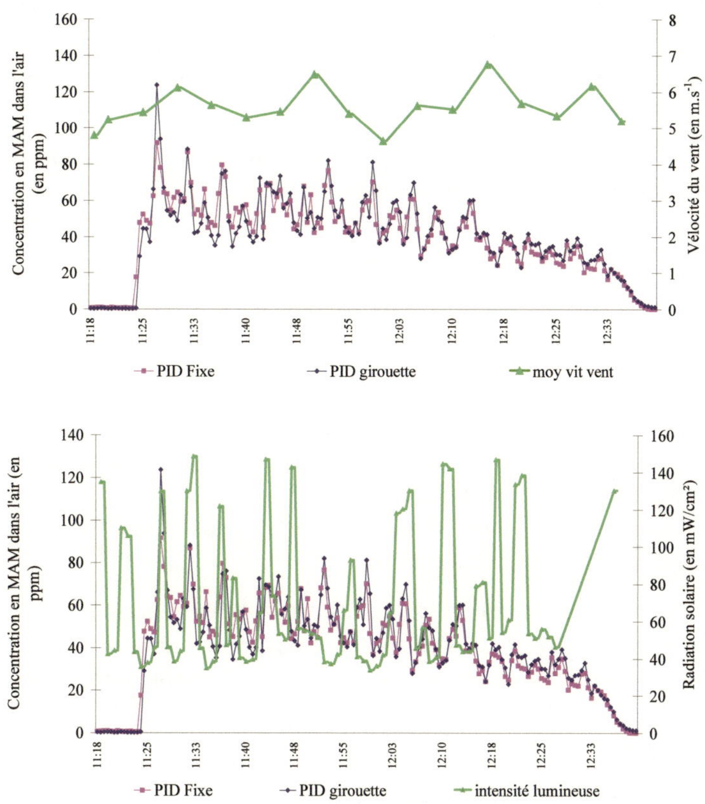

Fig. 7. MMA concentrations in the air measured using PIDs (stationary and on the wind vane) and variations in wind and sun intensity throughout the experiment.

The advantage of using the wind vane to monitor the concentration of the product in the atmosphere can be observed in Figure 7: while the shapes of the two curves (stationary PID and PID on wind vane) match, the concentration peaks are more intense for the PID on the wind vane and, in general, its curve shows slightly higher values, indicating better quantification of the vapours emitted by the slick. This result was obtained despite a relatively constant wind direction throughout the experiment (250 - 270°), therefore suggesting that the gain would be even greater in the case of quantification with a swirling wind.

4.2.4 Dissolution

Despite its solubility (10 $g.L^{-1}$), the MMA did not dissolve at all in these experimental conditions: the chemical was not detected at any of the three depths sampled (0.5, 1 and 1.5 m).

4.2.5 Fate of methyl methacrylate

MMA is listed as ED according to the SEBC code, i.e. it is first and foremost an evaporator, but also dissolves. In the prevailing weather conditions during the trial, this product only evaporated, and it did so with rapid kinetics: in less than 90 minutes, no traces of MMA were visible at the water surface or detected by the PIDs. Given its negligible dissolution, an atmospheric transfer coefficient can be estimated at around 4.1 litres per square metre per hour (4.1 $L.m^{-2}.h^{-1}$), in these experimental conditions.

In the event of a spill, response should therefore focus on the gas cloud as it will be explosive (MMA can form flammable vapours with air) and toxic (TEEL[1] of 400 ppm). It is therefore essential to have operational tools to visualise the formation and movement of a gas cloud as well as a computer forecast model.

Due to its marked evaporation, MMA was selected to test the SPIM camera. Using this camera, three periods were identified:

- Beginning of detection at T_{+30min} and until T_{+40min} the cloud has spread little and appears denser around the outside (Photo)
- Between T_{+40min} and T_{+60min}, the cloud spreads and the edges become more diffuse, while remaining detectable by the SPIM
- Between T_{+60min} and T_{+75min}, the cloud no longer spreads but the wind promotes its dilution so that it completely disappears.

These results, although partial, are very promising in order to monitor the formation and movement of an invisible gas cloud. This technology, already employed for the detection of gas leaks at industrial facilities, could be put to good use in the field of pollution response.

4.3 Methyl ethyl ketone

Methyl ethly ketone (MEK), like MMA, is classed ED.

---

[1] TEEL = TEmporary Exposure Limit

Photo 5. Infrared image of the MMA gas cloud near the floating cells

4.3.1 Observations

Immediately after the spill, the MEK spread across the entire water surface within the floating cell to form a homogeneous, colourless slick that was difficult to detect. This observation is consistent with the results obtained in the laboratory on MEK-water interfacial tension which predict complete spreading, or even infinite spreading in the absence of containment. This high degree of spreading promoted the transfer of the product into the atmosphere and the water column, resulting in complete disappearance of the slick in less than 1 hour and the end of sampling and PID measurements at 12:00 (no signal from PID).

4.3.2 Prevailing sea and weather conditions

The prevailing weather conditions on site at the time of the experiment are presented in Table 3, and can be summarised as alternating sunny spells and threatening clouds due to the constant wind.

Date	25/06	25/06	25/06
Time	10:25-11:00	11:00-11:30	11:30 -12:00
Air T (°C)	15.6	15.7	15.7
Wind speed (m/s)	5.1	5.1	5.2
Solar intensity (mW/cm^2)	93.1	106.1	104.6

Table 3. Evolution of weather conditions during the MEK release

Note that the mean solar intensity on this day was higher than that measured during the MMA trial.

4.3.3 Evaporation

The MEK evaporated very swiftly and to a great extent: less than 5 minutes after the spill, a peak in concentration was measured both by the stationary PID and the PID on the wind vane (350 ppm). Concentrations then gradually decreased and levelled out at around 20 ppm at $T_{+30\,min}$ before disappearing completely at $T_{+90\,min}$ (Figure 8). The evaporation of MEK was so intense that it was unaffected by the wind or sun; no peaks could be correlated with a gust of wind or spell of sunshine.

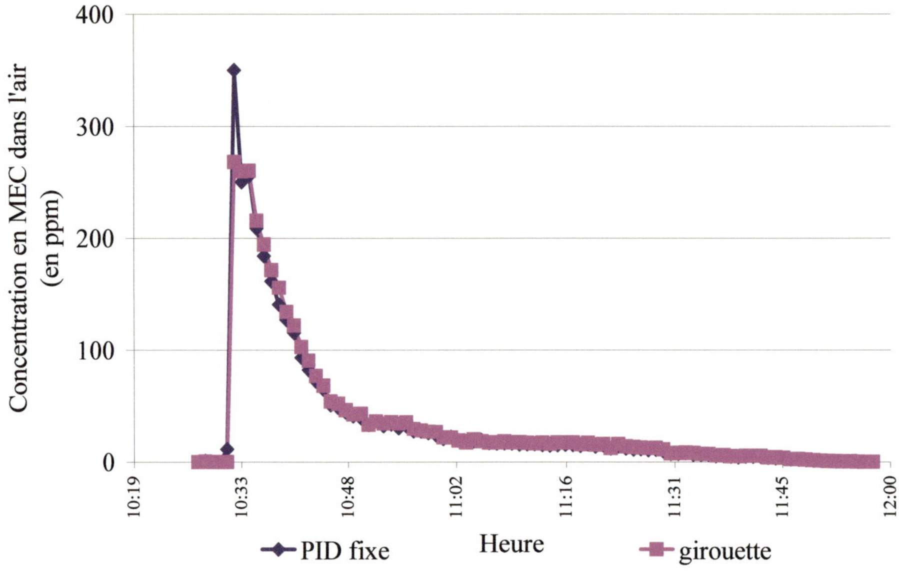

Fig. 8. MEK concentrations in the air measured by the PIDs (stationary and wind vane) during the trial.

4.3.4 Dissolution

Despite its high solubility (≈ 300 $g.L^{-1}$ i.e. 30 times higher than MMA), MEK did not dissolve at all in the conditions of this experiment: the chemical was not detected at any of the three depths sampled (0.5, 1 and 1.5 m).

4.3.5 Fate of methyl ethyl ketone

MEK is listed as ED according to the SEBC code, i.e. it is first and foremost an evaporator, but also dissolves. In the weather conditions during this trial, the product only evaporated, and did so with intense, rapid kinetics: in less than 5 minutes, an evaporation peak was observed and thereafter concentrations in the air continually decreased. This behaviour was similar to MMA with, however, even faster evaporation kinetics, unaffected by the wind or sun. This trend can be explained by MEK's capacity to spread widely at the water surface and to form a homogeneous slick. The coefficient is the same as for MMA, i.e. 4.1 $L.m^{-2}.h^{-1}$, bearing in mind that for MEK it is not constant throughout evaporation. It is estimated that 80% of the slick evaporated in 15 minutes, giving a coefficient of 8.9 $L.m^{-2}.h^{-1}$, while the remaining 20% took around 75 minutes, giving a considerably lower transfer coefficient of around 0.45 $L.m^{-2}.h^{-1}$.

In the event of a spill, response should therefore focus on the gas cloud which will form rapidly and will be flammable and toxic.

5. Conclusion

In the event of a chemical spill at sea, it is crucial to be aware of the behaviour of the substances involved to determine the most appropriate response strategy. While classifications exist and are valuable tools, they are based on laboratory data for which the influence of environmental parameters is not taken into consideration, i.e. the effects of wind and sunshine on evaporation and of surface agitation on dissolution are not taken into account. Moreover, they consider the physico-chemical properties of substances independently of each other, while in the event of a spill, dissolution and evaporation processes will occur simultaneously. Therefore, if evaporation kinetics are rapid, dissolution processes will not have time to occur and a product, even if soluble, could evaporate completely into the atmosphere. With this in mind, *Cedre* developed an experimental tool, floating cells, which can be used to characterise the main processes governing a substance's fate after its release at sea.

This paper presents the results obtained during a series of trials in these floating cells, which aimed to characterise the behaviour of three chemicals: xylene, methyl methacrylate and methyl ethyl ketone.

These trials provided two vital pieces of information on these chemicals, to add to the information provided by the SEBC classification. The first relates to xylene which is listed as FE. It became apparent that, in certain weather conditions, xylene will indeed behave as a Floater/Evaporator, however if the temperature drops and especially if the surface agitation and wind are sufficient (wind speed greater than 3 $m.s^{-1}$), it will form an emulsion and will mainly dissolve, thus exhibiting FDE behaviour. The two other products (MMA and MEK) are classed ED while, in these experimental conditions, they behaved only as Evaporators.

These results highlight the need to study *in situ* the behaviour of chemicals to add to the theoretical classification based only on laboratory data, especially as it is clear that factors such as surface agitation, wind, sunshine and outside temperature significantly affect the transfer of products from the surface to another environmental compartment. The floating cells, developed by *Cedre,* adequately fulfil this purpose, and, what's more, are an innovative tool which, in addition to helping to define a response strategy in the event of a spill, can be used routinely as realistic input for forecast software.

Concretely, such experiments on the behaviour of HNS releases in sea water provide:

i. further information in addition to the chemical emergency response guides. For example, on the one hand, xylene, which is considered as a Floater/Evaporator, can behave as a FDE; in the case of emergency response in the same weather conditions as the trials, the attention of the intervention team must also be focused on the water column. On the other hand for both other substances (MMA and MEK), the atmospheric compartment has to be especially considered by the intervention team which has, firstly, to protect itself against the gas cloud.

ii. experimental data for the implementation of existing software. Forecast software is a useful tool but few programs are validated by field experiments. Here, results are used to validate the Clara software which is fully dedicated to predicting the behaviour of HNS at sea.

iii. the Authorities in charge of the response with operational information. They require accurate data in order to organize the most appropriate response.

6. Acknowledgments

This study was financially supported by the French Navy which is in charge of the response at sea following an incident and the French Ministère de l'Ecologie, du Développement durable, des Transports et du Logement (MEEDDM).

7. References

Alexander, M., 1997. Environmental fate and effects of styrene. Critical Reviews in Environ. Sci. Technol. 27, 383-410.

Ambrose, P. (2006). China Involved in Massive Chemical Spill". In *Marine Pollution Bulletin,* 52, news.

Bonn Agreement. (1985). *Bonn Agreement: Counter-Pollution Manual,* Bonn Agreement, London

Bonn Agreement. (1994). European classification system. Bonn Agreement: *Counter-Pollution Manual.* London.

Cedre. (1988). Système européen de classification pour les produits chimiques déversés en mer". *18^{E} réunion du groupe de travail de l'Accord de Bonn sur les questions opérationnelles, techniques et scientifiques concernant les activités de lutte contre la pollution.* BAWG OTOPSA 18/13/3-F, 3-6 mai 1988, 8p

Cedre. (2007). Xylènes. *Guide d'intervention chimique.* ISBN 978-2-87893-085-6. ISSN 1950-0556. p 69.

Cedre. (2008). Méthacrylate de méthyle stabilisé. *Guide d'intervention chimique.* ISBN 978-2-87893-091-7. ISSN 1950-0556. p 72.

Cedre. (2009). Méthyléthylcétone. *Guide d'intervention chimique.* ISBN 978-2-87893-095-5. ISSN 1950-0556. p 70.

Gesamp. (2002). (IMO/FAO/UNESCO-IOC/WMO/WHO/IAEA/UN/UNEP Joint Group of Experts on the Scientific Aspects of Marine Environmental Protection). *Revised GESAMP Hazard Evaluation Procedure for Chemical Substances Carried by Ships.* Rep. Stud. Gesamp No. 64, IMO, London, 126pp.

HELCOM. (1991). Manual on Co-operation in Combating Marine Pollution within the framework of the Convention on the Protection of the Marine Environment of the

Baltic Sea Area, 1974 (Helsinki Convention), vol. III Response to incidents involving chemicals

HELCOM. (2002). HELCOM Response Manual, Volume 2, ANNEX 3, A3- Case histories of marine chemical accidents

HELCOM. (2007). Report on shipping accidents in the Baltic Sea area for the year 2007. http://www.helcom.fi/stc/files/shipping/shipping_accidents_2007.pdf

IMO. (1987). *Manual on Chemical Pollution,* Section 1: Evaluation and intervention, IMO, London, p 82.

IMO. (1992). Manual on Chemical Pollution, Section 2: Search and Recovery of Merchandise and Packages Lost at Sea, IMO, London, p 41.

IMO. (2006). International Convention for the Prevention of Pollution from Ships, 1973, as modified by the Protocol of 1978 relating thereto (MARPOL). London.

Katz, E. (1987). Quantitative analysis using chromatographic techniques. *Separation science series.* John Wiley & Sons, Chichester, New York, Brisbane, 427p.

Laffon, B., Pasaro, E., Mendez, J., 2002. Evaluation of genotoxic effects in a group of workers exposed to low levels of styrene. Toxicology. 171, 175-186.

Law, R.J., Kelly, C., Matthiessen, P., Aldridge, J., 2003. The loss of the chemical tanker Ievoli Sun in the English Channel, October 2000. Mar. Pollut. Bull. 46, 254-257.

Le Floch S., Dumont J., Jaffrennou C., Olier R. (2007). Accidental spills: the C.E.C., a new device to test initial dissolution rate of chemicals in seawater column. In *7th International Symposium on Maritime Safety, Security and Environmental Protection,* 20-21 September 2007, Athènes, Grèce.

Le Floch S., Benbouzid H., Olier R. (2009). Operational Device and Procedure to Test the Initial Dissolution Rate of Chemicals after Ship Accidents: The Cedre Experimental Column. In *The Open Environmental Pollution & Toxicology Journal.* Vol. 1, 1-10.

Levshina S.L., N.N. Efimov and V.N. Bazarkin. "Assessment of the Amur River Ecosystem Pollution with Benzene and its Derivatives Caused by an Accident at the Chemical Plant in Jilin City, China". *Bull Environ Contam Toxicol,* 83: 776-779, 2009.

Lyman, W.J., Reehl, W.F., Rosenblatt, D.H. (1996). Handbook of chemical property estimation method: Environmental behavior of organic compounds. *American Chemical Society,* 4ème Ed., Whashington.

Moller, P., Knudsen, L.E., Loft, S., Wallin, H., 2000. The comet assay as a rapid test in biomonitoring occupational exposure to DNA-damaging agents and effect of confounding factors. Cancer Epidem. Biomar. 9, 1005-1015.

REMPEC. (1996). Compendium of notes on preparedness and response to maritime pollution emergencies involving hazardous substances. REMPEC, Malta

REMPEC. (1999). Practical Guide for Marine Chemical Spills. REMPEC, Malta

REMPEC. (2004). Oil Accidents Report. REMPEC, Malta

UNEP. (2006). The Songhua River spill. China. in *Field Mission Report.* UNEP, 26p., December

Vaghef, H., Hellman, B., 1998. Detection of styrene and styrene oxide-induced DNA damage in various organs of mice using the comet assay. Pharmacol. Toxicol. 83, 69-74.

5

Indoor Air Quality and Thermal Comfort in Naturally Ventilated Low-Energy Residential Houses

M. Krzaczek and J. Tejchman
Gdańsk University of Technology,
Faculty of Civil and Environmental Engineering, Gdańsk,
Poland

1. Introduction

The indoor environmental quality (IEQ) and occupant comfort are closely related. The current indoor environmental assessment includes four aspects, namely thermal comfort (TC), indoor air quality (IAQ), visual comfort (VC) and aural comfort (AC). IAQ, as the nature of air in an indoor environment with relation to the occupant health and comfort is not an easily defined concept. In a broad context, it is the result of complex interactions between building, building systems and people. Comparative risk studies performed by the United States Environmental Protection Agency (USEPA) ranked IAQ as one of the 5 top environmental risks to the public health. Over the past decades, exposure to indoor air pollutants increased due to a variety of factors including: construction of tightly sealed buildings, reduction of ventilation rates (for energy saving) and use of synthetic building materials and furnishings as well as chemically formulated personal care products, pesticides and household cleaners. The effect of chemical pollutants on the perceived IAQ was investigated in several studies. The volatile organic compounds (VOCs) were suspected to cause "sick-building" symptoms, like headache, eye and mucous membrane irritation, fatigue and asthmatic symptoms (Redlich et al. 1997). The WHO air quality guidelines exist for major ambient air pollutants such as nitrogen dioxide and ozone, a few organic pollutants including mainly chlorinated and aromatic hydrocarbons (World Health Organization, 2000). The International Agency of Cancer Research recently upgraded formaldehyde to the group 1, known human carcinogen (IARC, 2004). However, there are still inadequate data about health effects of other VOCs. The total amount of VOCs and TVOC was not proven to correlate with symptoms. Investigations of all types of indoor air pollutants for the general air quality monitoring and assessment are complicated. In many studies, it was suggested that the measurement and analysis of the indoor carbon dioxide (CO2) concentration could be useful for understanding the Indoor Air Quality (IAQ) and ventilation effectiveness. Healthy people can tolerate the CO_2 level up to 10,000 ppm without serious health effects. An acceptable indoor CO2 level should be kept below 1000 ppm or 650 ppm above the ambient level in order to prevent any accumulation of associated human body odors. The indoor carbon dioxide is relatively easy to measure and its low level in the indoor air usually corresponds to a low level of VOCs.

The chapter focuses on the concentration of CO_2 in the indoor air. The term comfort is not commonly used in relation to the indoor air quality and it is mainly linked with the lack of the discomfort due to odor and sensory irritations. The acceptable air quality is defined as "air in which there are not known contaminants at harmful concentrations as determined by cognizant authorities and with which, a substantial majority (80% or more) of people do not express a dissatisfaction". Consequently most standards including requirements for the indoor air quality define conditions by providing the minimum percentage of persons who are dissatisfied with the air quality. They are mainly based on the discomfort and annoyance caused for visitors in indoor spaces. Recently, some standards also deal with requirements for occupants.

A problem of the thermal and indoor air quality rises in low-energy buildings. The low-energy buildings are tight and impermeable. A relative contribution of the heat consumption due to the ventilation in the total heat consumption in low-energy buildings is very high. Thus, the ventilation system significantly affects the thermal comfort and indoor air quality in low-energy buildings. However, the mechanical ventilation and air-conditioning of buildings are responsible for a lot of the non-renewable fossil-based energy consumption in the world. Therefore, the natural ventilation can be used as a cheap air-exchange system for buildings. The natural ventilation effectiveness depends on many factors such as: size and location of openings (Hummelgaard et al. 2007, Stavrakakis et al. 2010), location of inlet gaps, ambient climate conditions, etc. During the natural ventilation, the air exchange is caused by temperature gradients, but can be enforced by fans located in the exhaust duct inlets (outlets of the indoor zone). The air quality and thermal comfort in buildings equipped with the natural ventilation depend strongly on air flow patterns which are influenced by a type of the space heating system.

In the chapter, the indoor air quality and thermal comfort in the naturally ventilated low-energy buildings are discussed. The investigations are limited to residential houses equipped with the natural ventilation system with and without exhaust ventilators only. Two locations of the fresh air inlets are considered: above and below the window. The CFD simulations were carried out for two heating systems: a radiator heating system and a floor heating system. The results of numerical investigations are compared with the corresponding experimental measurements.

2. Low-energy buildings

The term 'low-energy building' is not well defined. The definition is different in several countries and in national and international standards and regulations. However, modern residential buildings can be characterized by the following features with respect to ventilation systems (Meyer 1993, Reinmuth 1994):

a. relative contribution of the heat consumption due to ventilation in the total heat consumption can be even higher than 50% (thus, the influence of the mechanical ventilation on the heat consumption is very high),
b. buildings are tight and impermeable; however, they must be ventilated due to the minimum air exchange rate related to hygienic requirements in order to reach the appropriate air quality in rooms and good comfort perception of residents; thus, the type of ventilation system has a great meaning for a comfort perception,

c. buildings with a low heat consumption consume heat by 1/3 less than conventional houses,
d. buildings equipped with ventilation systems do not reach a theoretically calculated level of the heat consumption,
e. consumption of the electric energy by ventilators and water pumps increases the consumption of the electric energy in buildings up to 50%,
f. when installing a ventilation system, the demand for the ventilation heat per year can be reduced by 5%, and in the case of heat exchangers even by 20%.

Low-energy houses that use very small amounts of energy for heating purposes do not need any conventional heating system. If windows have low transmission losses, the risk of drafts is minimized and the heating appliances (i.e. radiators) are redundant (Karlsson & Moshfegh 2006). When a house can be designed to require less than 10 W/m^2 of the heating capacity to maintain 20°C by −10 °C ambient conditions, a conventional heating system (i.e. a gas fired furnace, circulation pipes and radiators) can be omitted and the total energy consumption drops to a small fraction of normal levels. The result is a drastic reduction in both operating costs and environmental impacts. Hastings (2004) studied five building projects as example of successful solutions. He reported that the estimated space heating demand varied from 12.6 kWh/(m^2 a) up to 22.4 kWh/(m^2 a), while the thickness of the insulation layers varied between 30-50 cm. In apartment projects, the ventilation system typically ran 24 h/day and 365 days/year. He concluded that high performance housing depended on keeping the heat in. To achieve the required *U*-values, the walls had to be too thick. The superior glazing (U=0.5 W/m^2 K) together with highly insulating frame constructions are already on the market today, but they are still too expensive and are one of several reasons for a lager cost of the high performance housing. The ventilation is ambiguously related to the energy saving rationale originating from the global warming problem and non-renewable energy sources saving. Since it makes up for about half of the energy consumption in well-insulated buildings, it is an attractive target for energy saving measures.

However, simply reducing ventilation rates has unwanted repercussions on the indoor air quality. Advances in several disciplines of the knowledge such as the growing understanding of the global warming (IPCC, 2007) and its effects on our environment, the increasing evidence of the limited nature of our major energy supply and large costs, both economical and human, of the air pollution related illnesses are dramatically altering the goals of innovations in the building technology. The focus is shifted towards low-energy, 'green' or sustainable buildings, seeking concepts that allow to maintain or even further increase the comfort level that we are accustomed to, while significantly to reduce the associated energy used in every aspect of the human life. Consequently, this field represents a massive gross energy saving potential. Simply reducing ventilation rates, however, deteriorate the indoor air quality and therefore cause unwanted effects such as an increase in the incidence of the respiratory illness and loss of productivity. The purpose of each ventilation system is to provide an acceptable micro-climate (in this context, the micro-climate refers to the thermal environment and air quality). The process to create an acceptable interior micro-climate can be conducted basically by two sides: ventilation to achieve a good air quality and heating or cooling inside air to have a thermal comfort. It is relatively easy to control mechanical ventilation systems to maintain the best air exchange rate (Wong |& Huang 2004, Ho et al. 2011, Paul et al. 2010, Laverge et al. 2011). However, the mechanical ventilation and air-conditioning of buildings are responsible for a lot of non-renewable fossil-based energy consumptions in the world. Therefore, the natural ventilation can be used as a cheap air-exchange system for buildings.

The comfort and health factors are the major elements for designing proper indoor environments. However, they often act opposite to each other (Behne 1999). This opposing behavior of design elements is often not taken into account. In contrast, modern designers make an imaginative use of glass and space to create well-lit and attractive interiors. These buildings are perfectly isolated from outdoors, and often create unusual conditions leading to a violation of traditional "rules of thumb" for the thermal comfort and hygiene (Hastings 2004). Poorly designed naturally ventilated buildings are uncomfortable to live and work in and lead to a reduced quality of the life and loss of productivity. The natural ventilation is an energy-efficient alternative for reducing the energy use in buildings and, with a proper design, it is able to create the thermal comfort and healthy indoor conditions. Typically, the energy cost of naturally ventilated building is 40% less than that of air-conditioned buildings (Energy Consumption Guide 1993)

3. Reference measurements

3.1 Reference buildings

To investigate the indoor air quality and thermal comfort in the naturally ventilated low-energy buildings, the colony of identical residential houses located in Germany by Leipzig was chosen for experiments to obtain reliable experimental results (Maier et al. 2009) and to compare them to the results of our CFD simulations. From the total number of 149 houses, 22 houses standing in one row were chosen for measurements (Fig.1). They were situated in the central part of the entire complex. The residential area of the medium size house was 114 m^2, where 102 m^2 was the heating area (the ratio of the area to the volume was A/V=0.54). Each measured object had two external walls exposed to the air influence. The residential rooms with a large window area were oriented towards the south. The transmission heat coefficient U was 0.35 W/(m^2 K) (ground floor), 0.22 W/(m^2 K) (external walls), 1.30 W/(m^2 K) (windows) and 0.22 W/(m^2 K) (roof). The theoretical heat demand per year in the houses was 36.8 kWh/(m^2 a). For further investigations, only two residential houses were considered:

a. equipped with natural gravitational ventilation (Fig.2),
b. equipped with natural ventilation enforced by single exhaust ventilators (Fig.3).

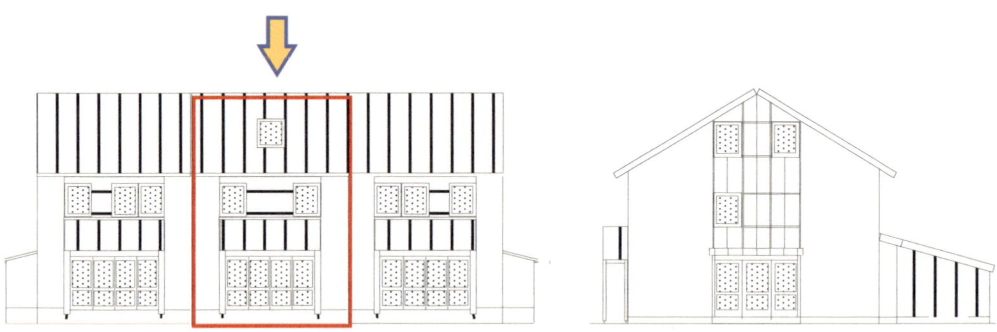

Fig. 1. The reference house for experiments (Maier et al. 2009)

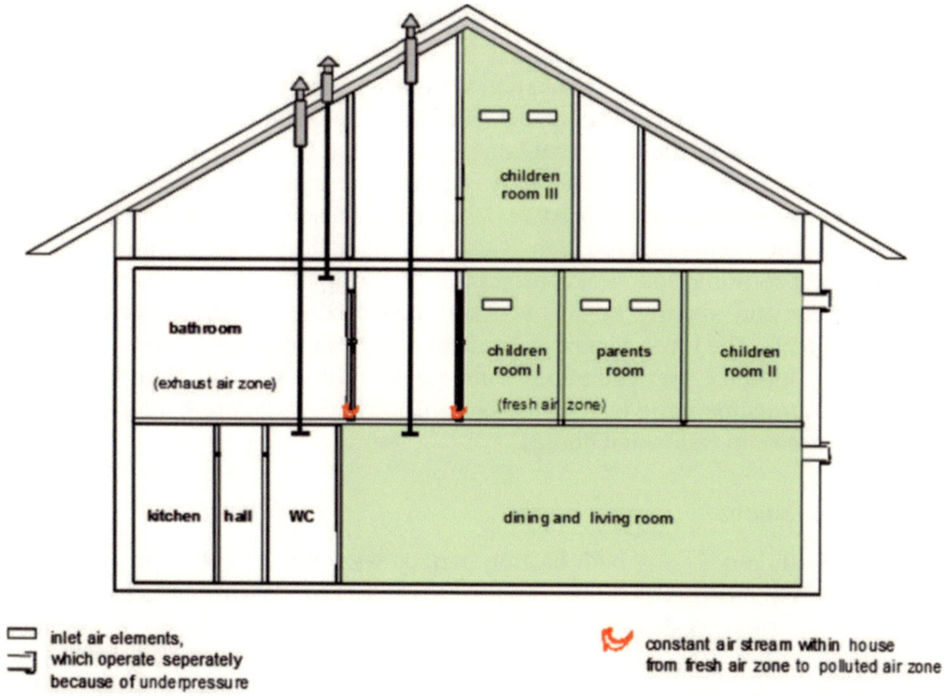

Fig. 2. Natural ventilation without mechanical exhaust ventilators (Maier et al. 2009)

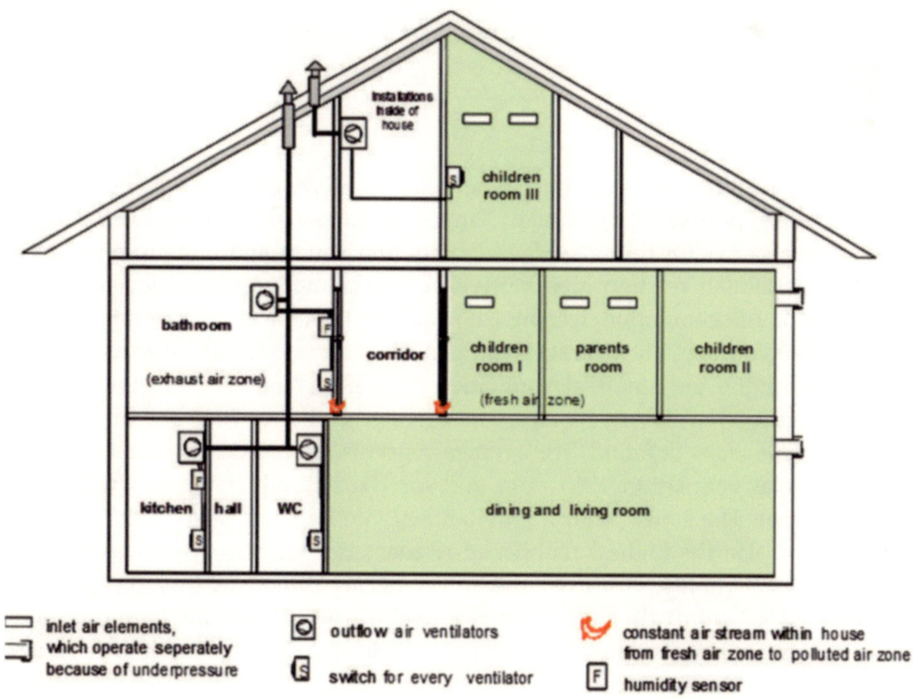

Fig. 3. Natural ventilation with mechanical exhaust ventilators (Maier et al. 2009)

In the case of the natural ventilation with exhaust ventilators, the area of the air supply included children rooms, living rooms and bedrooms. However, a bathroom, toilet and rooms with the internal building installation were associated to the area of the air outlet. The single air ventilators were used, and each of them operated separately. The output of the stream in the mode of the basic ventilation was around 0.32 Wh/m^3. The value of 0.25 Wh/m^3 for the ventilation output friendly for users was unfortunately exceeded. The following parameters were monitored in houses by sensors: air humidity, air quality (expressed by the emission of both CO_2 and malodors), temperature, electricity consumption, gas consumption, heat consumption, working time of the window ventilation (windows totally and slightly open), working time of the mechanical ventilation and number of residents. The low-temperature radiation was not measured due to high costs of transducers. In addition, sun radiation, wind power, wind direction, air humidity, air quality and the air temperature were registered outside houses. Figure 4 shows the location of measuring sensors in residential houses.

3.2 Weather conditions

The weather conditions during both heating periods were similar (Maier et al. 2009). The time interval between consecutive measurements was 1 hour. According to the occurrence frequency concept (Stavrakakis et al. 2010), the prevailing climatic data were determined as recorded-range weighted averages for the time period between 8:00 and 18:00, i.e. the weights used were the occurrence frequencies of the weather data ranges. Consequently, the mean values of temperature, total solar radiation and wind speed were 4.0 °C, 20.0 W/m^2 and 1.6 m/s respectively. Referring to the associated wind direction, south-west (SW) wind directions were dominant, as they corresponded to the occurrence frequency of approximately 39%.

3.3 Experimental results

The experimental results covered a long time period of 2 years. The weather conditions during both heating periods were similar. Thus, the relationships were established only between the mean values for the sake of readability (maximum and minimum values were not shown). To establish relationships between experimental results, a linear regression method was used. All ventilation systems showed the lower heat consumption than the theoretically calculated one. The average year heat consumption for all systems (22 houses) was 3063 kWh: for the gravitational ventilation 3168 kWh and natural ventilation with exhaust ventilators 3586 kWh. Any relation between the ventilation time and the parameters of the indoor climate was not found. The average concentration of CO_2 for each system was: for the gravitational ventilation 1266 ppm and for the natural ventilation with exhaust ventilators 647 ppm. The house with a gravitational ventilation system had the worst air quality, but it had also the highest number of residents (5 persons) what was the second reason for a poor air quality. However, the air quality in all houses was in a permissible comfort range (<1500 ppm) (ISO 2008). In the case of the relative air humidity and air quality (Fig.5), the number of residents realistically depicted a distribution of their values.

A strong relationship between the concentration of CO_2 per resident and opening time of windows is visible in Fig.6 (including 22 houses with different ventilation systems).

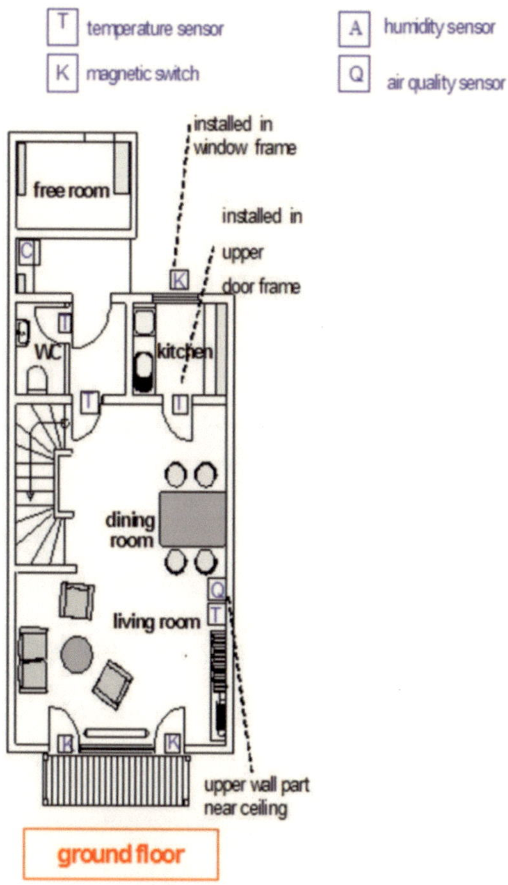

Fig. 4. Location of measuring sensors in residential houses (Maier et al. 2009)

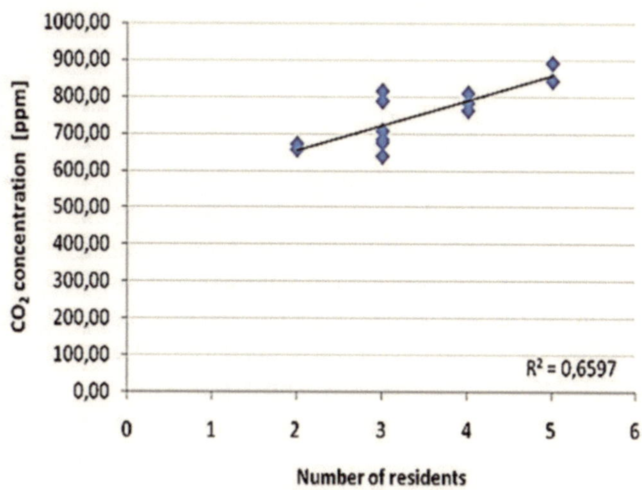

Fig. 5. The relationship between number of residents and air quality (Maier et al. 2009)

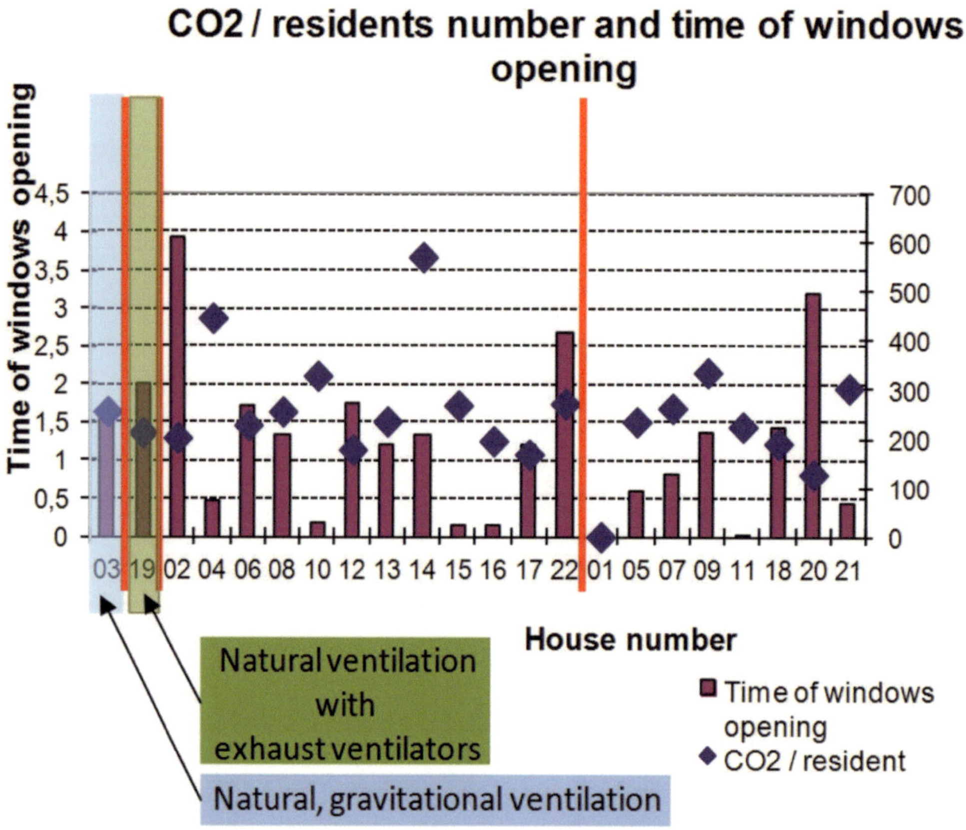

Fig. 6. Results of experimental measurements: time of windows opening and CO_2 concentration/residents number (Maier et al. 2009)

3.4 Analysis of questionnaire results

After the measurements program was finished, a detailed questionnaire program was performed (Maier et al. 2009). The results obtained from questionnaires performed in 21 houses equipped with different ventilation systems were compared (Fig.7). In the case of the natural ventilation with exhaust ventilators, the new designed in-take ventilator elements did not disappoint expectations. No problems took place in reference to the air draught. The operational elements were estimated positively. Considering the air quality in rooms and installation regulation, the residents saw the necessity for introducing several improvements (however, from the technical point of view it was impossible). Any negative opinion of the comfort perception or ventilation system was not found. The air quality and comfort perception obtained the notes "good" and "very good" for all ventilation systems and no evident differences of the level of the residents' satisfaction with the ventilation system were detected. In spite of "good" and "very good" notes in questionnaires, in all cases occupants adapted the thermal environment to improve their comfort perception by opening operable windows. It was an unexpected observation because notes in questionnaires were high and the measured air quality was in the comfort range.

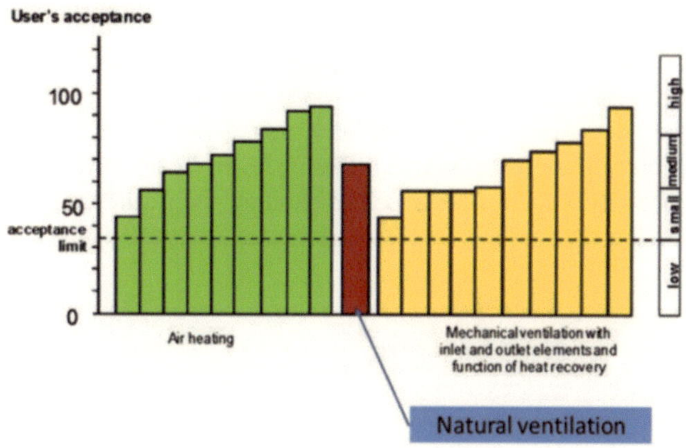

Fig. 7. Acceptance level of separate ventilation systems in residential houses (on the basis of the questionnaire) (Maier et al. 2009)

4. Physical model of ventilated zone

To investigate the fluid flow patterns and to track CO_2 concentration in the indoor air for two inlet gap configurations and two natural ventilation systems, the reference house was considered as one indoor zone which represented the dining and living room (Figs.2 and 3). The indoor zone was modeled together with exhaust ducts and created one fluid flow domain (Fig.8). The exhaust duct was added to the model to make it more realistic, especially in the case of the natural ventilation without fans, induced by the mixture temperature differences.

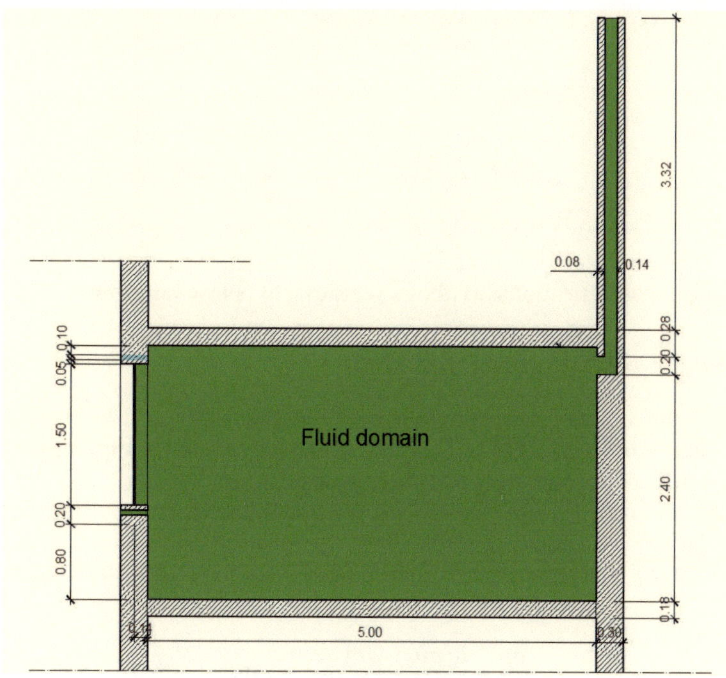

Fig. 8. Geometry of ventilated zone and fluid domain

Two different locations of the fresh air inlet are the most common in the engineering practice. Hence, the two inlet gap configurations were considered: above the window (Fig.9a) and below the window (Fig.9b). Two space heating systems, the most common in engineering practice, were considered: radiator space heating system and floor space heating system. The natural ventilation was either induced by air temperature differences or enforced by a fan located in the outlet gap (the exhaust duct inlet). The complete physical model is presented in Fig.10.

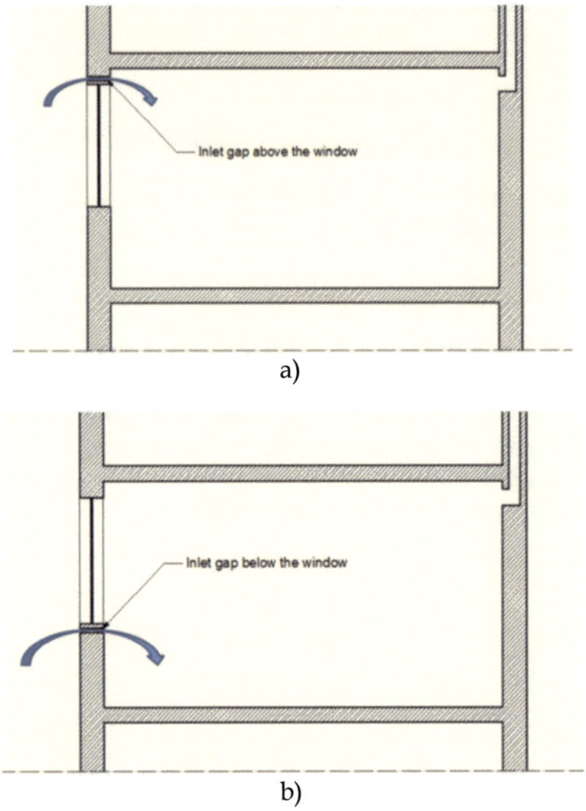

Fig. 9. The inlet gap configurations: a) above window, b) below window

5. Mathematical model of ventilated zone

In the ventilated zone (Fig.10), the fluid velocity magnitude can reach 1.6 m/s. It corresponds with the Reynolds number of 21000. Thus, turbulent flow of the compressible fluid is to be considered. Considerations for the governing equations are: steady state, two-dimensional state, Newtonian and compressible fluid flow in turbulent regime. The fluid flow problem is defined by the law of the conservation of mass, momentum and energy. It is assumed that there is only one phase. From the law of mass conservation, the continuity equation is derived:

$$\frac{\partial p}{\partial t} + \frac{\partial(\rho v_x)}{\partial x} + \frac{\partial(\rho v_y)}{\partial y} = 0 \quad (1)$$

where v_x and v_y are components of the velocity vector in the x and y direction, respectively, ρ is density (kg/m^3) and t is time.

For the Newtonian fluid, the momentum equations including the viscous loss terms T_x and T_y, are as follows:

$$T_x = \frac{\partial}{\partial x}\left(\mu \frac{\partial v_x}{\partial x}\right) + \frac{\partial}{\partial y}\left(\mu \frac{\partial v_y}{\partial x}\right) \tag{2}$$

$$T_y = \frac{\partial}{\partial x}\left(\mu \frac{\partial v_x}{\partial y}\right) + \frac{\partial}{\partial y}\left(\mu \frac{\partial v_y}{\partial y}\right). \tag{3}$$

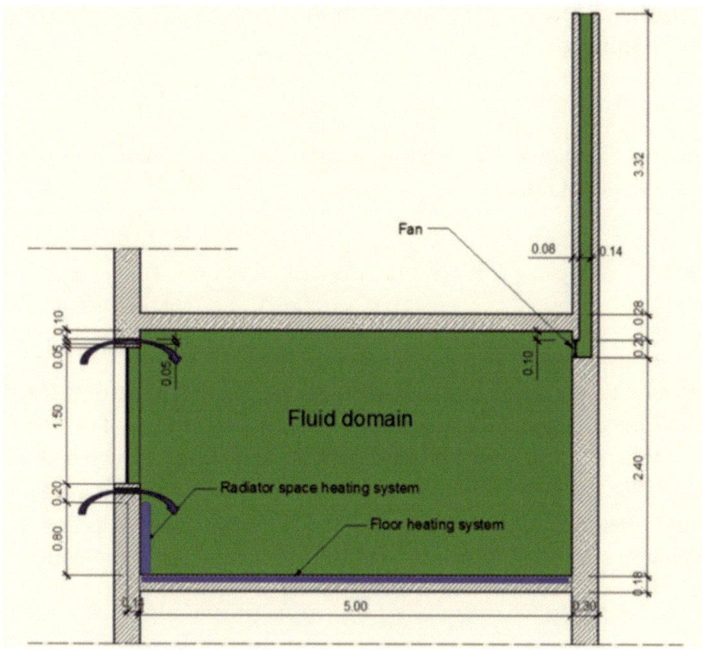

Fig. 10. The complete physical model of ventilated zone

The complete energy equation is solved in the compressible case with the heat transfer. In terms of the total (or stagnation) temperature, the energy equation is:

$$\frac{\partial}{\partial t}(\rho c_p T_0) + \frac{\partial}{\partial x}(\rho v_x c_p T_0) + \frac{\partial}{\partial y}(\rho v_y c_p T_0) = \frac{\partial}{\partial x}\left(\lambda \frac{\partial T_0}{\partial x}\right) + \frac{\partial}{\partial y}\left(\lambda \frac{\partial T_0}{\partial y}\right) + W^V + E^k + Q_V + \Phi + \frac{\partial P}{\partial t}, \tag{4}$$

where: c_p - specific heat [J/(kg K)], T_o - total (or stagnation) temperature [°C], λ - thermal conductivity [W/(m K)], W^v - viscous work term, Q_v - volumetric heat source, Φ - viscous heat generation term and E^k - kinetic energy.

In our study, the standard *k*-ε turbulence model is applied (Launder & Spalding 1974). The turbulence model is modified to model buoyancy (Viollet 1987), however, it is still only a fit for fully developed turbulent flow. Two different components (air and CO_2) are tracked. A single momentum equation (Eqs.2 and 3) is solved for the flow field. The properties for this equation are calculated from the species fluids and their mass fractions for density, viscosity and conductivity. The governing equation for the air transport is

$$\frac{\partial(\rho Y_{air})}{\partial t}+\nabla\cdot(\rho Y_{air}v)-\nabla\cdot(\rho D_{m\,air}\nabla Y_{air})=0 \tag{5}$$

where Y_{air} is the mass fraction for the air, ρ denotes the bulk density [kg/m^3], v is the velocity vector [m/s] and $D_{m\,air}$ denotes the mass diffusion coefficient [m^2/s]. The equation for CO2 is not solved directly. The mass fraction Y_{CO2} for CO_2 is calculated at each node from the identity condition:

$$Y_{CO2}=1-Y_{air} \tag{6}$$

At each node the gas density is calculated as a function of the mass fractions and molecular weights of gases (air and CO2):

$$\rho=\frac{P}{RT\sum_{1}^{2}\frac{Y_i}{M_i}}, \tag{7}$$

where R is the universal gas constant and M_i is the molecular weight of the *ith* species, P is pressure degree of freedom and T is the absolute temperature [K]. The opaque of the fluid domain is in a contact with a solid body of external and internal walls of the indoor zone and exhaust duct (Fig.8). Two locations of the air-CO_2 mixture inlet gap are considered: above the window opening (Fig.9a) and below the window opening (Fig.9b). The boundary conditions for velocity components at the mixture inlet gap during the windy weather are

$$v_x=v_{inlet},\quad v_y=0 \tag{8}$$

and during the calm weather are

$$\frac{\partial v}{\partial n}=0, \tag{9}$$

where n is the vector normal to the flow direction.

The boundary conditions for velocity components at the mixture outlet in the exhaust duct are

$$\frac{\partial v}{\partial n}=0 \tag{10}$$

The boundary conditions for the temperature are: at the mixture inlet gap

$$T = T_{inlet} \tag{11}$$

and at the mixture outlet in the exhaust duct

$$\frac{\partial T}{\partial n} = 0 \tag{12}$$

The mixture velocity inlet profile is applied following the power-law equation[3]:

$$v_{inlet} = v_{ref} \left(\frac{z}{h}\right)^{a} \tag{13}$$

where v_{inlet} is the wind speed at an arbitrary height z from the ground level, h is the weather station sensor height, 5 m, where the reference prevailing velocity v_{ref} is recorded in the climate database, and a - the exponential coefficient is taken as 0.2 (Stavrakakis et al. 2010), the value corresponding to a rough-rural terrain. Following the assumptions mentioned above, the wind speed equal to 3.0 m/s and south-west west (SW-W) wind direction, the reference value of the wind speed is vref=1.13 m/s for the wall faced south and consequently v_{inlet}=0.98 m/s (Eq.13). In the case of the calm weather, v_{inlet}=0.0 m/s. The solar radiation reflecting the external wall surface is modeled by the equivalent convection heat flux. The equivalent value increases the convection heat exchange rate on the surface of the external wall up to a sum of the convection and radiation rate. The convection and radiation heat flux is characterized by the sol-air temperature $T_{e\text{-}sol}$ defined as follows (ISO 2007, Duffie & Beckman 1991):

$$T_{e-sol} = T_e + \frac{aI}{h_e}, \tag{14}$$

where: T_e - the outdoor air temperature [°C], a - the solar absorptivity of outdoor surface of the wall [-], and I - the incident total solar radiation [W/m^2].

It is assumed that the absorptivity coefficient, 0.65, is equivalent to the grey painting. Besides the sol-air temperature, the wind driven changes of the convective heat coefficient h_e influence the heat exchange rate on the external wall surface. Thus the convective heat coefficient is defined by the empirical formula (Duffie & Beckman 1991):

$$h_e(t) = \max\left[5, \ \frac{8.6\, v(t)^{0.6}}{l^{0.4}}\right] \tag{15}$$

where $v(t)$ is the wind speed [m/s] and l is the cubical root of the building volume [m]. Following the above assumptions, the inlet temperature T_{inlet} is 5.3°C and for the calm weather T_{inlet} is 5.6°C. For the natural ventilation with the exhaust fan, the boundary conditions for the velocity components at the inlet to the exhaust duct are

$$v_x = v_{fan}, \ v_y = 0 \tag{16}$$

and for the temperature are $\frac{\partial T}{\partial n} = 0$. The $'x'$ component of the velocity at the inlet to the exhaust duct is computed according to the assumption of the air exchange rate in the ventilated zone of 0.5 1/h. Hence, v_{fan} is 0.23 m/s. The boundary conditions for the velocity components on the surfaces in a contact with internal walls, floor, ceiling and the exhaust duct are

$$v_x = 0 \quad \text{and} \quad v_y = 0 \tag{17}$$

On the boundary with the external wall and window, the equivalent convection is assumed (Fig.11):

$$q = h_{e\,eqv}(T_{Fi} - T_e) \tag{18}$$

where T_{Fi} is the temperature of the internal surface of the wall or window [°C], T_e is the ambient air temperature [°C] and $h_{e\,eqv}$ is the equivalent convective heat transfer coefficient [W/(m K)]. The convection heat flux q is characterized by the equivalent convective heat transfer coefficient $h_{e\,eqv}$. For steady-state heat transfer problems, the convective heat transfer coefficient $h_{e\,eqv}$ is independent of ambient and indoor climate conditions. The conduction heat flux q_{cond} in a steady-state heat transfer for the external wall and window is:

$$q_{cond} = U(T_i - T_e) \tag{19}$$

and the equivalent convection heat flux $q_{e\,eqv}$ is

$$q_{e\,eqv} = h_{e\,eqv}(T_{Fi} - T_e), \tag{20}$$

where U is the coefficient of the heat transmission [W/(m^2 K)], T_e is the ambient air temperature [°C] and T_i is the indoor AIR-CO_2 mixture temperature [°C].

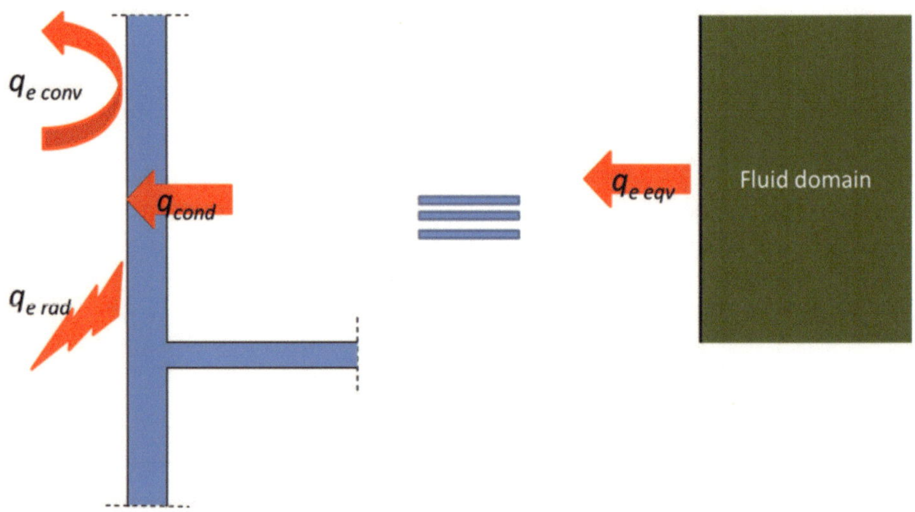

Fig. 11. The equivalent convection heat flux at boundary with external wall and window

According to the assumption q_{cond} equals to $q_{e\,eqv}$, the convective heat transfer coefficient $h_{e\,eqv}$ is

$$h_{e\,eqv} = \frac{U(T_i - T_e)}{\left(T_i - U(T_i - T_e)\frac{1}{h_e} - T_e\right)} \tag{21}$$

where h_e is the convective heat transfer coefficient (ISO 2007) on the internal surface of the external wall or window. For the above assumptions, the convective heat transfer coefficient $h_{e\,eqv}$=0.222 W/m^2 K and 1.370 W/ m^2 K, for the external wall and window, respectively. Opposite to surfaces in a contact with the external wall and window, other surfaces of the fluid domain are considered to be adiabatic. To simplify an analysis of flow patterns in the indoor zone, it is assumed that the only heat source in the indoor zone is a resident. The human body is concentrated in one heat source which is distributed over the edges of the rectangle (0.3 m × 1.6 m). Figure 12 presents the location of the heat source in the fluid domain. Murakami el at. (2000) studied a problem of airflow, thermal radiation and moisture transport for predicting a heat release from a human body. They reported that a heat release from a human body (in a standing position) to the surrounding environment by convection was 29.14 W/m^2 and the mean value of the convective heat transfer coefficient was 4.95 W/m^2 K. Following the above conclusions and considering the skin surface area of 1.5696 m^2 (Murakami et al. 2000), the constant heat flux of 13.26 W/m^2 along the human body's rectangle edges is assumed.

In a composite gas analysis, the species concentration depends strongly on boundary and initial conditions. The boundary conditions are defined by the CO_2 concentration over the fluid domain boundaries and the initial conditions are defined by the initial CO_2 distribution over the fluid domain. According to the research results of Li et al. (2010), it is assumed that the ambient CO_2 concentration is approximately 700 mg/m^3 (389 ppm). Consequently, the CO_2 concentration at the inlet gap C_{inl} =389 ppm. The same assumption is taken for initial conditions over the fluid domain. For any other boundaries of the fluid domain, one took:

$$\frac{\partial C}{\partial n} = 0 \tag{22}$$

In the indoor zone model, the main CO_2 sources are residents. To track the CO_2 concentration, two residents occupying the indoor zone are assumed. The respiration process of human beings was investigated by several researchers (Lawrence & Braun 2007, Hyun & Kleinstreuer 2001). They concluded that the frequency of respiration under light physical work is 17 times per minute with the time-mean rate of 8.4 l/min. Following their conclusions, a steady inhalation process is assumed. Consequently, it is chosen that a resident exhales CO_2 into the indoor zone with a rate $\dot{S}$ of 0.14×10^{-3} m^3/s. According to research results of Yanes et al., the average value of the CO_2 concentration in exhaled air by human beings is 55.100 ppm. Hence, the CO_2 source is modeled as the source generating the mixture of 55,100 ppm CO_2 concentration with the velocity $v_x = v_{exh}$ and $v_y = 0$ m/s.

6. Numerical input data

The Ansys FLOTRAN software package (2007) is used to solve Eqs.1-5. In numerical simulations, the two fluid domain configurations are considered: the inlet gap located above the window (Fig.13a) and the inlet gap located below the window (Fig.13b). A fine mesh generation requires the mesh to be well suited to the boundary conditions, heat and CO_2 sources, etc. Thus, the fluid domain is simulated as a multi-area region and a variable mesh density is applied (Fig.14). To compute the IAQ indices, the average breathing zone is modeled according to the international guideline EN ISO (2007) (Fig.15).

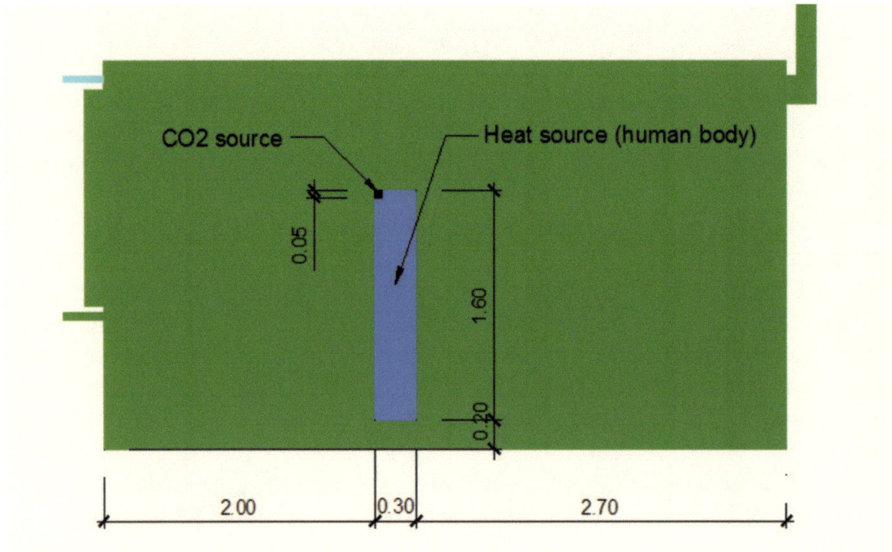

Fig. 12. Location of heat source and CO_2 source in fluid domain

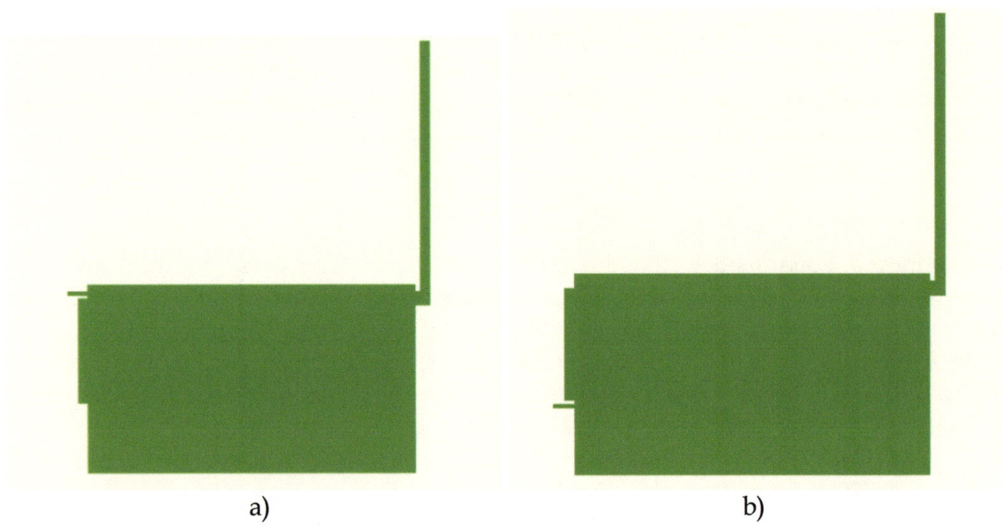

Fig. 13. Fluid domain configurations: a) inlet gap located above window, b) inlet gap located below window

7. Thermal comfort and IAQ indices

The comfort indices used in this study are PMV and PD. The calculation procedure is performed by means of a special post-processing algorithm, which predicts the indices in each node and element of the FE mesh. The PMV nodal value is computed according to the algorithm defined by ISO (2005):

$$\mathrm{PMV}_i = \left[0.303e^{(-0.036M)} + 0.028\right]TL \quad (23)$$

where M is the initial standard value of the prescribed activity level and TL is the thermal load on the body. For PMV, the metabolic rate for a standing position and light activity is considered. The local indices values of PMV are averaged within the area of the occupied zone as follows:

$$\mathrm{PMV}_{\mathrm{avg}} = \frac{\sum_{k=1}^{n} \mathrm{PMV}_k \cdot A_k}{\sum_{k=1}^{n} A_k} \tag{24}$$

where PMV_k is averaged over the k-th element, A_k is the area of the k-th element and n is the number of finite elements. If the floor is too warm or too cool, the occupants feel uncomfortable owing to thermal sensation of their feet.

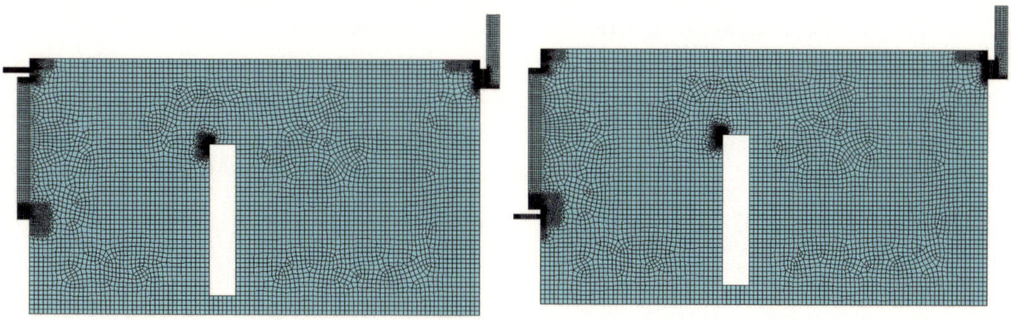

a) b)

Fig. 14. The FE mesh: a) inlet gap located above window, b) inlet gap located below window

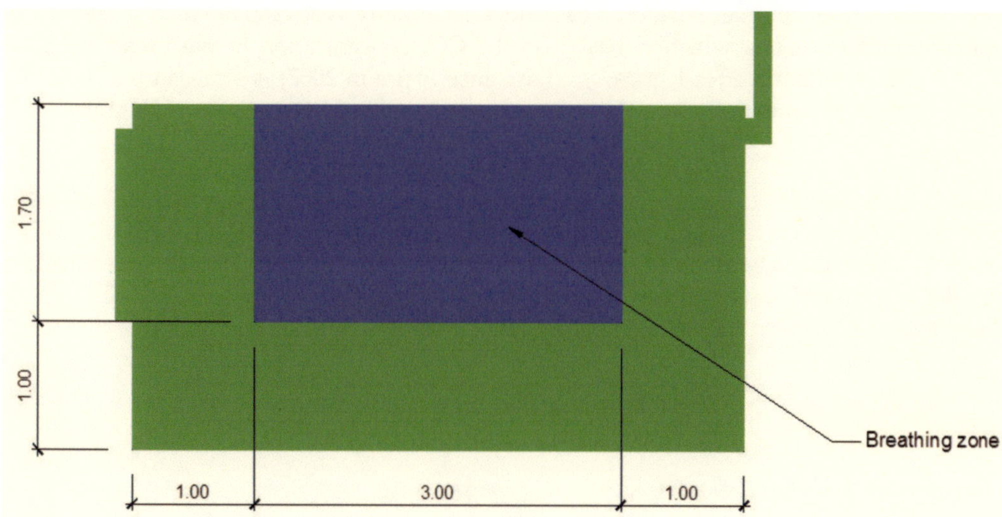

Fig. 15. The breathing zone location in fluid domain

The percentage dissatisfied PD as a function of the floor temperature [%], is computed as follows (ISO 2005):

$$PD = 100 - 94 \cdot \exp(-1.387 + 0.118 \cdot T_f - 0.0025 \cdot T_f^2), \tag{25}$$

where T_f is the average floor temperature in [°C]. The overall ventilation effectiveness for the temperature distribution $\bar{\varepsilon}_t$ provides a quantitative index related to the way in which the heat is distributed inside the indoor zone. The higher value of $\bar{\varepsilon}_t$ the more is the homogeneous temperature distribution. Following the expression by Awbi (2003), the overall ventilation effectiveness for the temperature distribution $\bar{\varepsilon}_t$ is defined as

$$\bar{\varepsilon}_t = \frac{T_{outlet} - T_{inlet}}{T_{average} - T_{inlet}} \quad (26)$$

where T_{outlet} is the average temperature for the air-CO_2 mixture at the outlet, $T_{average}$ denotes the average temperature for the air-CO_2 mixture all over the breathing zone and T_{inlet} is the average temperature for the air-CO_2 mixture at the inlet. The temperatures $T_{average}$, T_{outlet} and T_{inlet} are computed by the following equations:

$$T_{average} = \frac{\int_{A_z} T\,dA}{A_z} \quad (27)$$

$$T_{outlet} = \frac{\int_{S_r} T\,dS}{S_r} \quad (28)$$

$$T_{inlet} = \frac{\int_{S_s} T\,dS}{S_s} \quad (29)$$

where A_z is the area of the breathing zone, T is temperature in point of (x, y), S_r is the outlet surface and S_s is the inlet surface. The indoor air quality is investigated in terms of the ventilation effectiveness, which is based on the CO_2 concentration in the breathing zone (Fig.11). The ventilation effectiveness η_V (Lawrence & Braun 2007) is a measure of how the supply airflow mixes with the breathing zone for a removal of CO_2 or other pollutants:

$$\eta_V = \frac{C_r - C_s}{C_z - C_s}, \quad (30)$$

where C_r - average CO_2 return concentration [ppm], C_s - average CO_2 supply concentration [ppm], C_z - average CO_2 concentration in the breathing zone [ppm].
The average CO_2 concentrations C_r, C_s and C_z are computed from:

$$C_z = \frac{\int_{A_z} C\,dA}{A_z}, \quad (31)$$

$$C_r = \frac{\int_{S_r} C\,dS}{S_r} \quad (32)$$

$$C_s = \frac{\int_{S_s} C\,dS}{S_s} \quad (33)$$

where C is the CO_2 concentration at the point of (x, y).

8. Results and discussion

The simulation results are divided into two main groups, called:

House 1: residential house equipped with natural, gravitational ventilation (cases 1 - 12),
House 2: residential house equipped with natural ventilation with exhaust ventilators (cases 13 - 24).

Next, the main 2 groups (houses) are divided into the groups which differ in the space heating system and ventilation type:

Group (a): gravitational ventilation and radiator heating system (cases 1 - 6),
Group (b): gravitational ventilation and floor heating system (cases 13 - 18),
Group (c): with exhaust ventilators and radiator heating system (cases 7 - 12),
Group (d): with exhaust ventilators and floor heating system (cases 19 - 24).

Finally, the subgroups differ in the air inlet location:

Subgroup (i): air inlet located above window,
Subgroup (ii): air inlet located below window.

For each subgroup, three ambient climate and operating conditions are considered: windy weather and windows closed, calm weather and windows closed, windy weather and windows opened. The simulation cases are presented in Tab.1.

The health and toxicity implications of CO_2 on the human health were reviewed by Hodgson from University of Connecticut Health Sciences Center. His investigations indicated that there was no evidence that CO_2 influenced the normal functions of human beings, if the concentration of CO_2 remained lower than 8500 ppm. However, for normal living conditions, the measured values of CO_2 above 1000 ppm resulted in people dissatisfaction. Hence, all figures presenting the investigations results of the CO_2 concentration are limited up to 8500 ppm. Alike CO_2 concentration, the figures presenting PMV index are limited to the seven point-scale defined by ISO (2004). All results of the simulations are presented in Tab.2.

8.1 Influence of air-CO_2 mixture flow characteristics on local air quality

The average CO_2 concentration in the breathing zone is a main index which assesses the indoor air quality. However, the resident's body can disturb the mixture flow pattern in the indoor zone and change local values of the index. Thus, the following analysis focuses on the local indoor quality in the zone located at the front of the resident's head. The figures show the distribution of the air velocity for different air inlet locations and different space heating systems supported or not by the mechanical exhaust ventilators. The plotted colors show the velocity and the CO_2 concentration distribution. In the case of the windy weather (case 1) and radiator heating system installed without exhaust ventilators, the upward flow is dominated by the forced convection from the imposed inlet velocity. Next, main flow bends down to the resident's head with the velocity of about 0.36 m/s. The air mixture, enforced by the breathing affect, flows down along a body of the resident reaching the flow velocity of 0.7 m/s (Fig.16a). At the bottom of the indoor zone, the main flow turns back to the external wall and circulates in the closed cell. The cell is bounded by the floor, ceiling, human's body and the external wall. The cell height is about 1.8 m. This flow pattern results in a fresh air zone located at the front of the resident's face (Fig.16b). The average CO_2 concentration in a cellular zone (diameter of 1.0 m) at the front of the resident's head is 735 ppm, while over the breathing zone 1015 ppm.

Case name		House	Group	Subgroup	Inlet location	Weather	Open window	Exhaust fan	Heating system
Gravitational	Case 1	1	a	i	UP	Windy	NO	NO	Radiator
	Case 2				UP	Calm	NO	NO	Radiator
	Case 3				UP	Windy	YES	NO	Radiator
	Case 4			ii	DOWN	Windy	NO	NO	Radiator
	Case 5				DOWN	Calm	NO	NO	Radiator
	Case 6				DOWN	Windy	YES	NO	Radiator
	Case 7		b	i	UP	Windy	NO	NO	Floor
	Case 8				UP	Calm	NO	NO	Floor
	Case 9				UP	Windy	YES	NO	Floor
	Case 10			ii	DOWN	Windy	NO	NO	Floor
	Case 11				DOWN	Calm	NO	NO	Floor
	Case 12				DOWN	Windy	YES	NO	Floor
Mechanical exhaust	Case 13	2	c	i	UP	Windy	NO	YES	Radiator
	Case 14				UP	Calm	NO	YES	Radiator
	Case 15				UP	Windy	YES	YES	Radiator
	Case 16			ii	DOWN	Windy	NO	YES	Radiator
	Case 17				DOWN	Calm	NO	YES	Radiator
	Case 18				DOWN	Windy	YES	YES	Radiator
	Case 19		d	i	UP	Windy	NO	YES	Floor
	Case 20				UP	Calm	NO	YES	Floor
	Case 21				UP	Windy	YES	YES	Floor
	Case 22			ii	DOWN	Windy	NO	YES	Floor
	Case 23				DOWN	Calm	NO	YES	Floor
	Case 24				DOWN	Windy	YES	YES	Floor

Table 1. Specification of the simulation cases

Case	C_z [ppm]	PMV_{avg} [-]	η_V [-]	PD [%]	$\bar{\varepsilon}_t$ [-]
Case 1	1015	0,81	0,1917	6,0	1,0139
Case 2	1007	1,22	0,3300	6,1	0,8265
Case 3	729	-1,88	0,0500	10,6	2,3943
Case 4	1243	0,92	0,3244	6,4	0,9408
Case 5	991	1,07	0,3455	6,0	0,8278
Case 6	1776	-0,44	0,2206	8,0	1,6427
Case 7	1016	0,77	0,1914	13,8	0,9668
Case 8	1008	1,15	0,3344	13,7	0,9068
Case 9	1266	-3.00	0,1676	13,5	1,7275
Case 10	1243	0,93	0,3244	13,9	0,9783
Case 11	991	1,18	0,3455	13,7	0,9058
Case 12	1776	-1,01	0,1846	13,5	1,6528
Case 13	743	0,77	0,2571	6,6	0,9581
Case 14	688	1,24	0,0010	7,0	0,8974
Case 15	645	-1,34	0,4805	8,5	1,9367
Case 16	904	1,11	0,2738	6,9	0,9330
Case 17	688	1,05	0,0010	7,0	0,9004
Case 18	967	-1,49	0,3443	7,7	1,6136
Case 19	743	0,52	0,2571	13,8	0,9666
Case 20	688	1,07	0,0010	13,9	0,8913
Case 21	957	-0,53	0,2165	13,7	1,5133
Case 22	904	-2,78	0,2738	13,9	0,9483
Case 23	688	1,06	0,0010	13,8	0,8913
Case 24	962	-1,30	0,3473	13,5	1,5942

Table 2. The CFD simulation results

Figure 17a shows the distributions of the air velocity for the air inlet located below the window (case 4). The indoor space is heated by radiators without exhaust ventilators. Unlikely the air inlet located above the window, in the case of windy weather, the downward flow is still dominated by the forced convection from the imposed inlet velocity. However, main flow moves forward, parallel to the floor, just to the resident body with the velocity of about 0.63 m/s (Fig.17a). The air mixture reaches the resident body at the height of 0.6 m and turns down along the body, slowing down to the velocity of about 0.33 m/s. At the bottom of the indoor zone, the main flow turns back to the external wall and circulates in the closed cell. The cell is bounded by the floor, horizontal surface located at the height of 0.6 m, human body and external wall. The height of the cell is about 0.6 m. This flow pattern results in a poor air zone located just above the floor (Fig.17b). Unlikely the air inlet located above window, the fresh air zone is moved up and out of the resident face. A resident is not in a contact with the fresh air zone.

In the case of the floor heating system (case 7), the flow pattern is alike the radiator heating system. However, main flow bends down to the resident head and reaches the velocity of about 0.51 m/s (Fig.18a). It is much higher than in the case of the radiator heating system (0.36 m/s). The air mixture, enforced by the breathing affect, flows down along a body of the resident reaching the flow velocity of 0.57 m/s and is smaller than in the case of the radiator heating system (Fig.18b). Even though, the CO_2 concentrations in the breathing zone for the radiator and floor heating system are almost the same and equal to 1015 ppm and 1016 ppm, respectively, the fresh air zone at the front of the resident is bigger for the radiator heating system. The average CO2 concentration in a cellular zone (diameter of 1.0 m) at the front of the resident head is 628 ppm.

The exhaust ventilators installation does not change the flow pattern significantly. In case of the radiator heating system applied and the air inlet located above the window, for the windy weather, the main flow bends down to the resident head with a slightly higher velocity of 0.43 m/s (Fig.19a). The air-CO2 mixture flows down and back to the external wall, creating a cellular flow pattern in front of the resident. The average flow velocity in the cell is 0.24 m/s and the average CO_2 concentration in a cellular zone (diameter of 1.0 m) at the front of the resident head is 453 ppm (Fig.19b), while the average CO_2 concentration in the breathing zone is 743 ppm.

The calm weather is more unfavorable for the natural ventilation. In the case of the calm weather, when the gravitational ventilation and the radiator heating system are applied, for the air inlet located above the window there is no fresh air zone at the front of the resident (Fig.20a). There can be observed the mixture flow along the resident body downward to the floor with the velocity of 0.4 m/s and the mixture flow upward and along the radiator, and the window with the velocity of 0.082 m/s. These two mixture streams are not able to create a clear cellular flow of the fresh air. Thus, at the front of the resident head, the CO_2 concentration is high, about 1900 ppm (Fig.20b), while the average CO_2 concentration is 1002 ppm in the breathing zone.

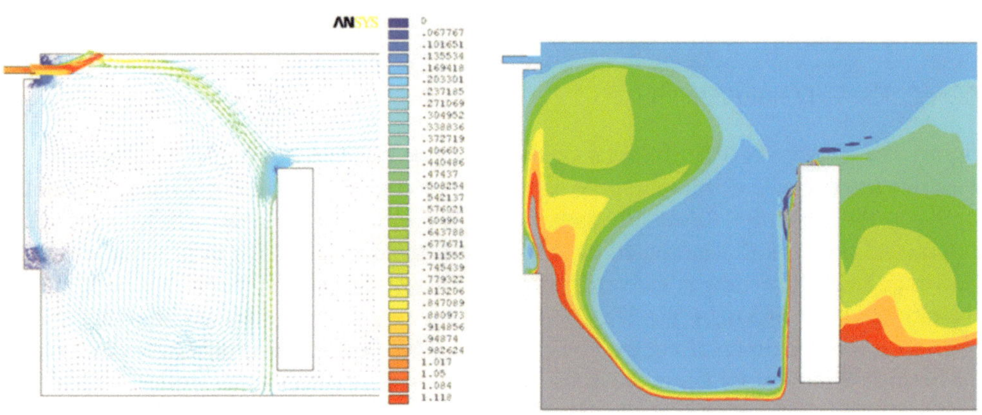

a) b)

Fig. 16. The air-CO_2 mixture local flow characteristics (case 1): a) velocity vector field, b) CO_2 concentration distribution

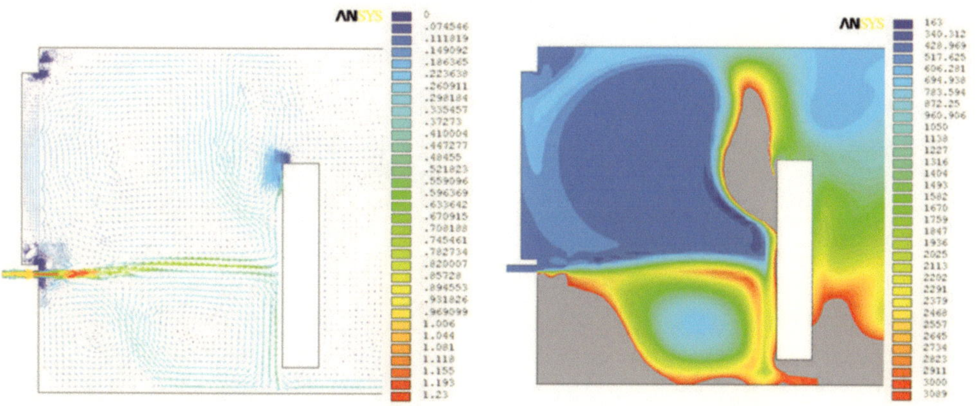

a) b)

Fig. 17. The air-CO_2 mixture local flow characteristics (case 4): a) velocity vector field, b) CO_2 concentration distribution

In the case of the calm weather, the location of the air inlet below the window is better. The air-CO_2 mixture flows along the resident body downward to the floor with the velocity of 0.27 m/s (Fig.21a). Next, it turns back to the external wall and slows down to the velocity of 0.13 m/s. This flow pattern and flow velocities are enough to create the fresh air zone at the front of the resident head. The CO_2 concentration of 1352 ppm in the fresh air zone is still high (Fig.21b), but significantly lower than at the location of the air inlet above the window. The average CO_2 concentration in the breathing zone is 991 ppm.

The exchange of the heating system into the floor heating system can improve a little the air quality in the fresh air zone located at the front of the resident's head during the calm weather. Opposite to the radiator heating system, the floor heating system decreases slightly the flow intensity, but surprisingly it results in a creation of the cellular "fresh" air zone at the front of the resident's head (Fig.22a). Even though the "fresh" air zone is created, the average CO_2 concentration is high and equals to 1425 ppm (Fig.22b), while the average CO_2 concentration in the breathing zone is 1008 ppm. It can results in the resident dissatisfaction.

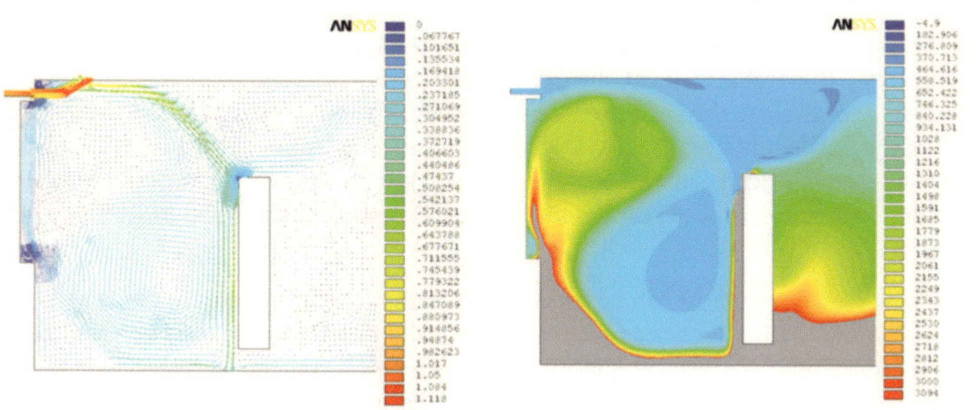

a) b)

Fig. 18. The air-CO_2 mixture local flow characteristics (case 7): a) velocity vector field, b) CO_2 concentration distribution

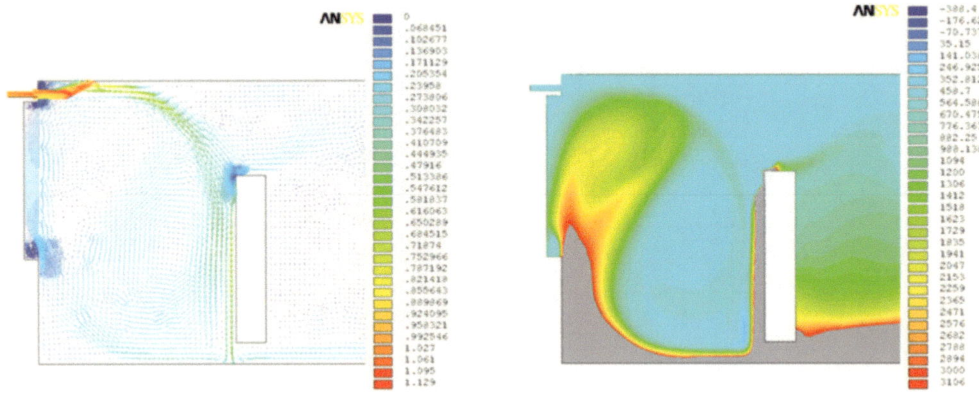

a) b)

Fig. 19. The air-CO_2 mixture local flow characteristics (case 13): a) velocity vector field, b) CO_2 concentration distribution

Moving down the air inlet location below the window, the air-CO_2 mixture flow pattern is not changed (Fig.23a) and the average value of the CO_2 concentration in the "fresh" air zone is reduced down to 1390 ppm (Fig.23b), while the average CO_2 concentration in the breathing zone is 991 ppm. However, this still can result in the resident's dissatisfaction.

In the case of the calm weather, the exhaust ventilators slightly improve the indoor air quality. In the following discussion, only the fresh air inlet located above the window is considered and two simulation cases are discussed: radiator heating system and floor heating system. For the radiator heating system installed, there is not a clear cellular fresh air zone (Fig.24a). However, the average CO_2 concentration at the front of the resident's head is 986 ppm (Fig.24b) and the value is still in the comfort range, while the average CO_2 concentration in the breathing zone is 688 ppm.

The heating system exchange into the floor heating system improves significantly the air quality near the resident head, however there is not a clear cellular fresh air zone (Fig.25a). In contrast to the radiator heating system, when the floor system is installed, the zone of the fresh air is significantly larger and the average value of the CO_2 concentration near the resident's head is about 583 ppm (Fig.25b). It means that the air quality near the resident head is excellent, while the average CO_2 concentration in the breathing zone is 622 ppm.

Analyzing the FE results, it can be concluded that the local air quality in the zone located at the front of the resident head can be significantly different than the one predicted by indices concerning the breathing zone. It can result in the unpredicted resident dissatisfaction. This conclusion is confirmed by experiments (Section 3.3). The momentary dissatisfaction of the local air quality results in actions aiming at the comfort level recovery (e.g. windows opening), while the overall indoor air quality indices indicate IAQ in a comfort range.

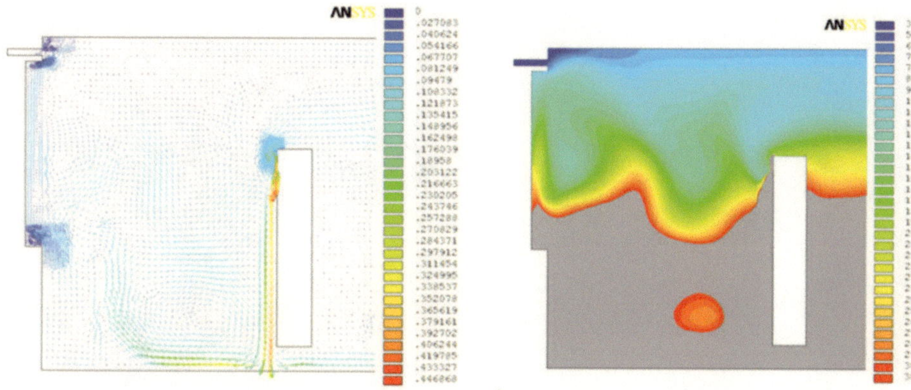

a) b)

Fig. 20. The air-CO_2 mixture local flow characteristics (case 2): a) velocity vector field, b) CO_2 concentration distribution

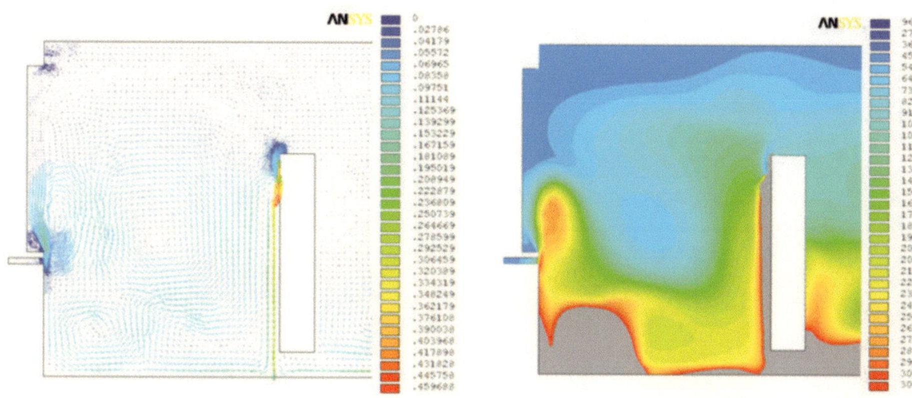

a) b)

Fig. 21. The air-CO_2 mixture local flow characteristics (case 5): a) velocity vector field, b) CO_2 concentration distribution

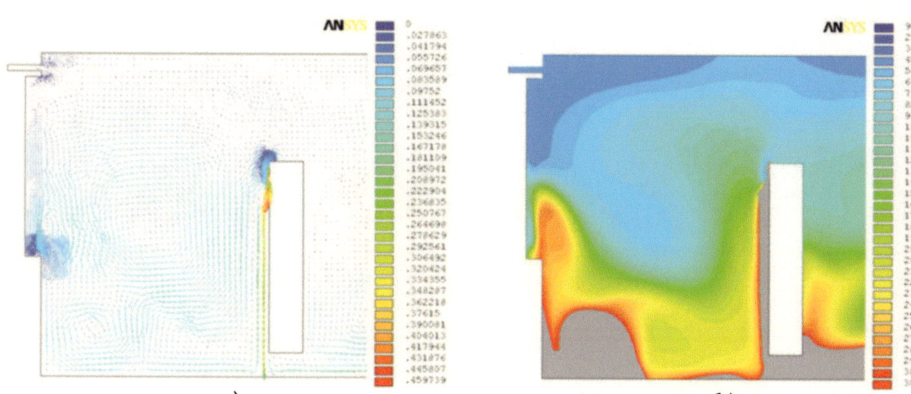

a) b)

Fig. 22. The air-CO_2 mixture local flow characteristics (case 8): a) velocity vector field, b) CO_2 concentration distribution

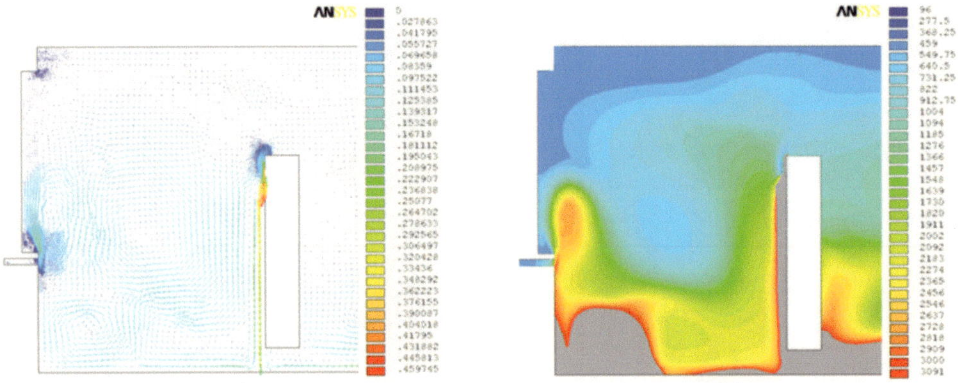

a) b)

Fig. 23. The air-CO_2 mixture local flow characteristics (case 11): a) velocity vector field, b) CO_2 concentration distribution

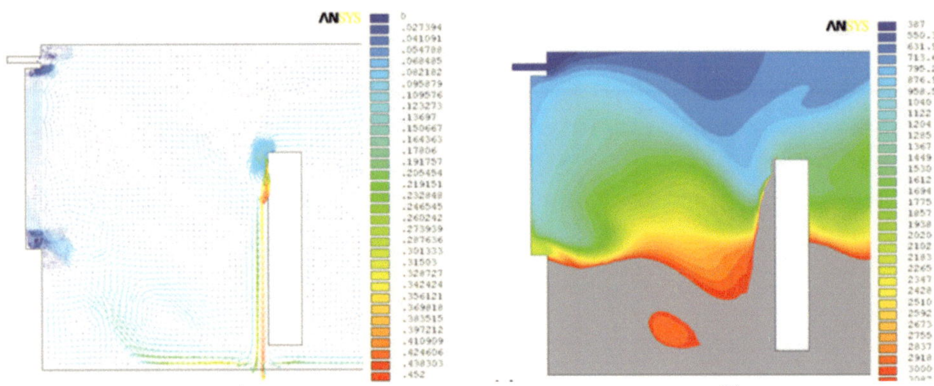

a) b)

Fig. 24. The air-CO_2 mixture local flow characteristics (case 14): a) velocity vector field, b) CO_2 concentration distribution

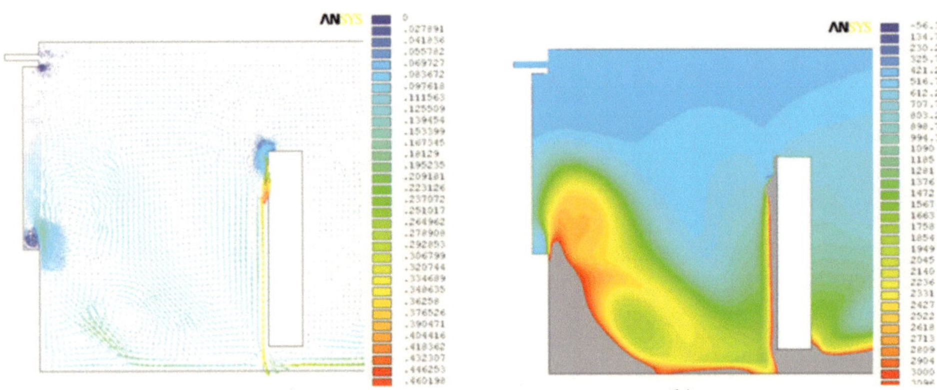

a) b)

Fig. 25. The air-CO_2 mixture local flow characteristics (case 20): a) velocity vector field, b) CO2 concentration distribution

8.2 Overall indoor air quality

The indoor air quality is usually assessed by indices averaged over the average breathing zone according to the international guideline EN ISO (2004): the average CO_2 concentration C_z and the ventilation effectiveness η_V (which is a measure of how well the supply airflow mixes with the breathing zone for a removal of CO_2). In this study, the average breathing zone is modeled as a separate area (Fig.15). The average CO_2 concentration over the breathing zone is computed according to Eq.31. The results of simulations for the house equipped with the natural ventilation without exhaust ventilators (gravitational) are presented in Fig.26.

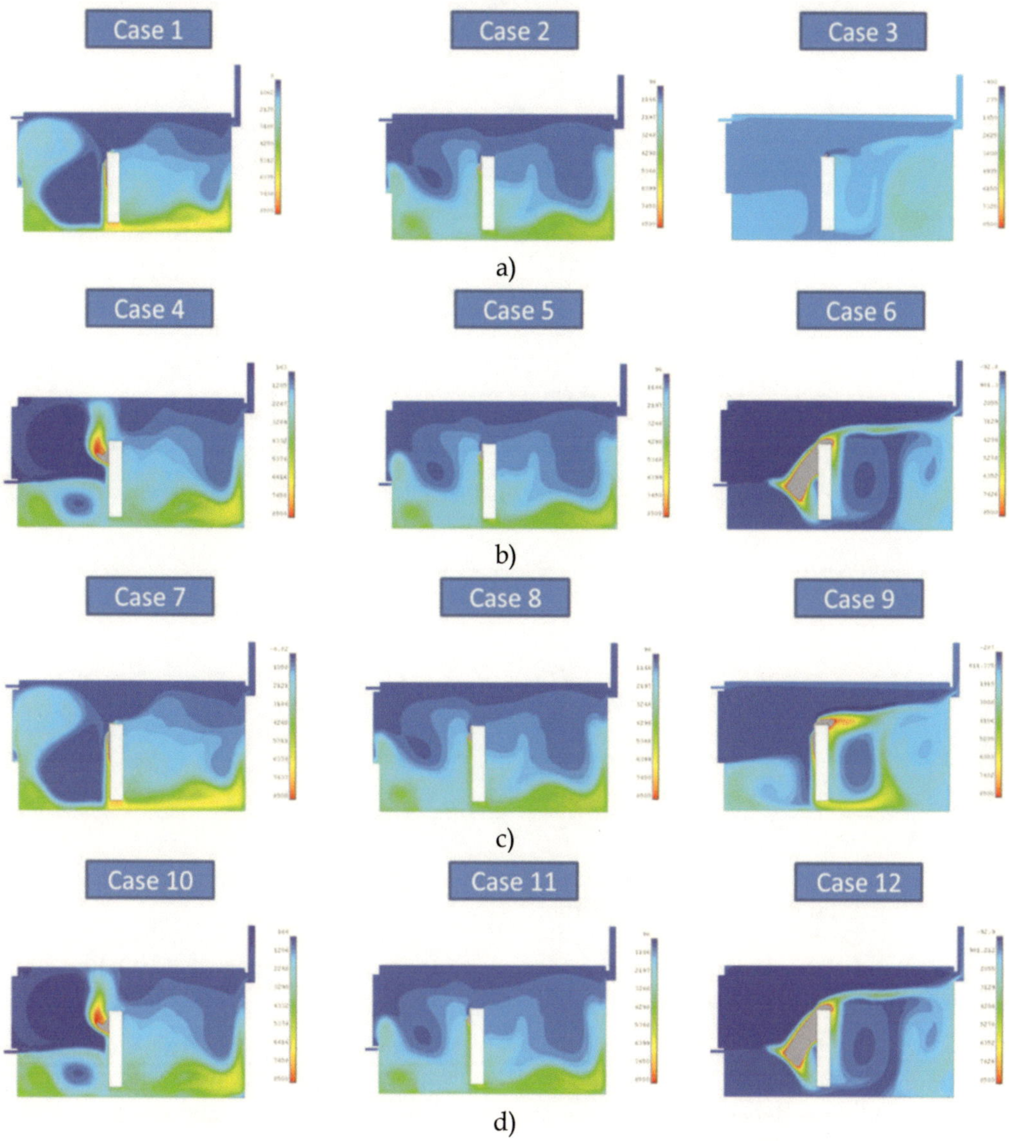

Fig. 26. The average CO_2 concentration C_z in breathing zone for house equipped with natural ventilation without exhaust ventilators

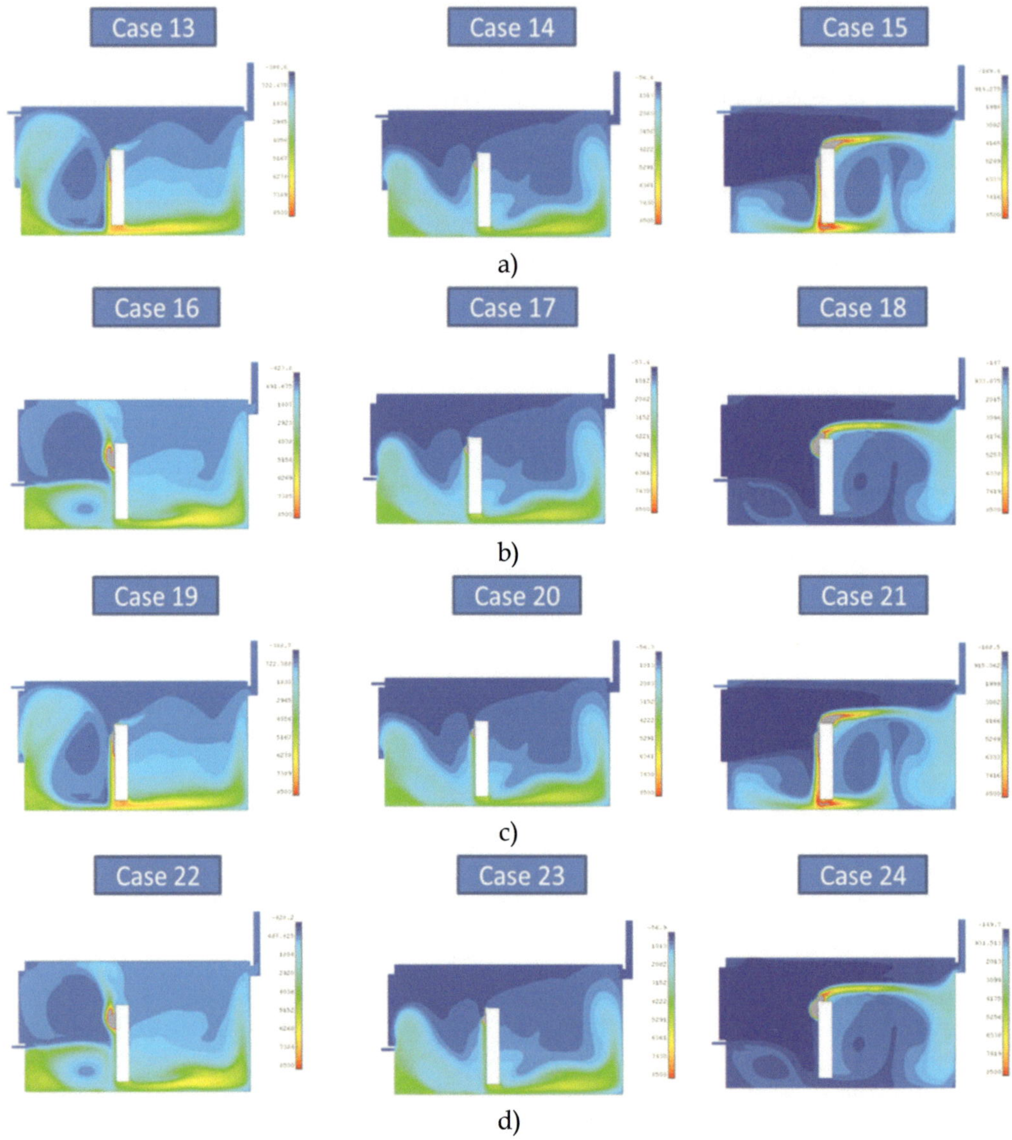

Fig. 27. The average CO_2 concentration C_z in breathing zone for house equipped with natural ventilation with exhaust ventilators

The results of the simulations for the house equipped with the natural ventilation with exhaust ventilators are presented in Fig.27.

In the house equipped with the natural ventilation system without exhaust ventilators (gravitational ventilation), the average CO_2 concentration varies significantly under the weather conditions depending upon the location of the fresh air inlet and the type of the heating system (Fig.28). Surprisingly, the very high CO2 concentration (1776 ppm) can be observed for the windy weather and the open window (cases 6 and 12). Studying Figs.26b and 26d, it can be noticed that a stream of the fresh air incoming through the open window is passing by and over the resident. Between the resident and the fresh air stream, there is a a relatively large zone of the poor quality air. The poor quality air zone is large enough to increase significantly the value of the average CO_2 concentration in the breathing zone. The best air quality (729 ppm) is reached by window opening in the case of the air inlet located above the window and the radiator heating system. The poorest air quality (2104 ppm) is obtained at the calm weather with the radiator heating system. The results of numerical simulations are confirmed by the experimental measurements (Section 3.3). The simulations show that the average CO_2 concentration in the breathing zone varies in the range of (729 - 2104) ppm with the mean value of 1352 ppm and is very close to the measured long term average value of 1266 ppm.

Contrary to the ventilation system without exhaust ventilators, the average CO_2 concentration C_z in the breathing zone varies very slightly in the range of (645 - 967) ppm (Fig.28). The worse results are obtained for the floor heating system rather than for the radiator heating system. The simulations for the radiator heating system show that the average CO_2 concentration in the breathing zone varies in the range of (645 - 967) ppm with the mean value of 772 ppm and is very close to the measured long term average value of 647 ppm. The indoor air quality depends strongly on the ventilation system effectiveness (Fig.29), which is expressed in terms of the η_V index (Eq. 30).

The simulation results show that the ventilation system effectiveness does not depend on the type of the heating system. Surprisingly, the higher effectiveness is obtained with the gravitational ventilation system without exhaust ventilators.

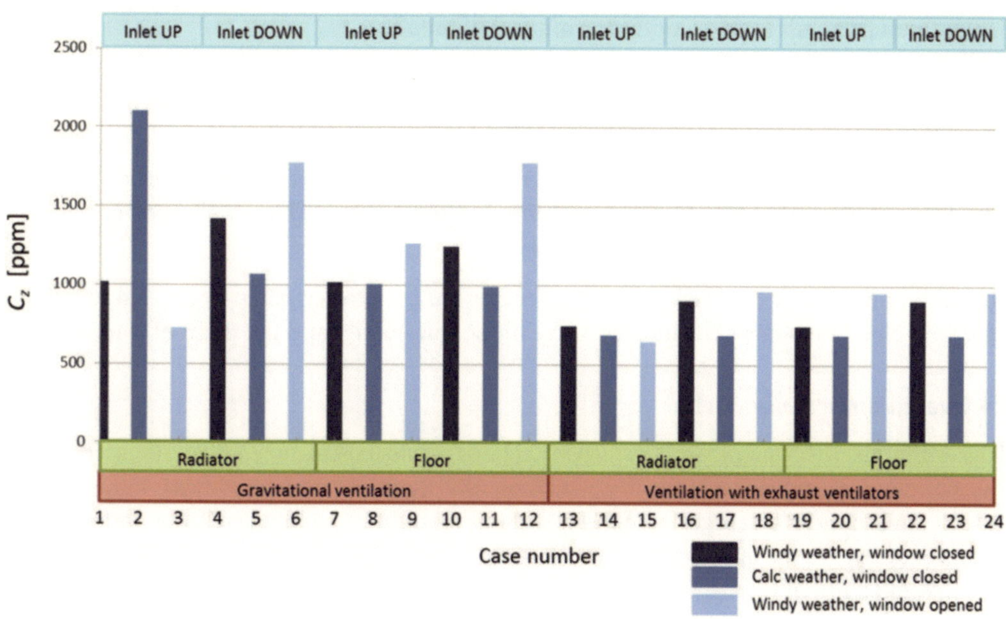

Fig. 28. The average CO_2 concentration C_z in the breathing zone

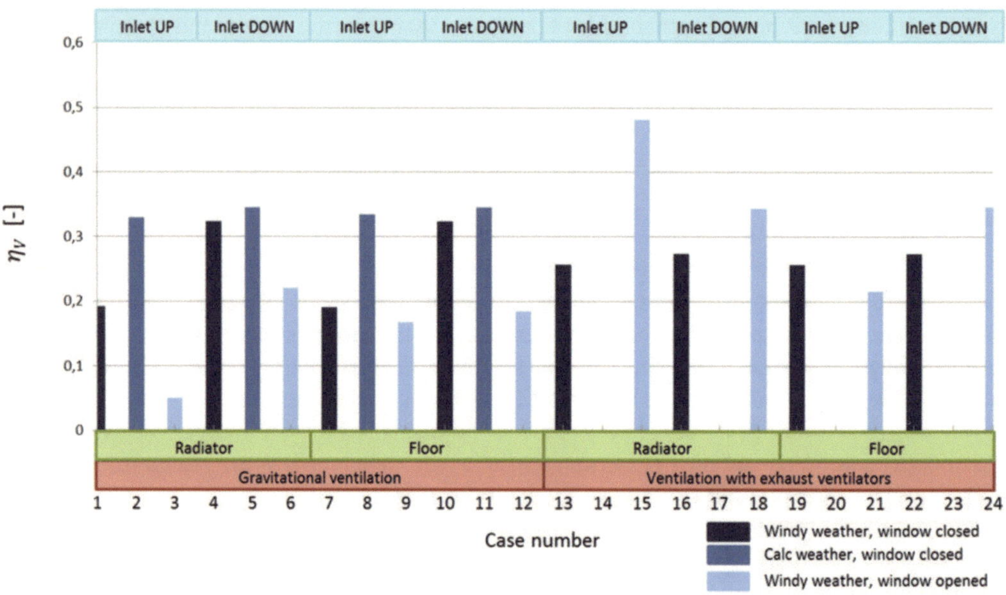

Fig. 29. The ventilation effectiveness in terms of η_V index

8.3 Optimal fresh air inlet velocity for natural ventilation

According to the simulations results, the indoor air quality in houses with natural ventilation systems depends strongly on the fresh air inlet velocity. To find the optimum inlet velocity, some simulations were carried out (Figs.30-33). The investigations were limited to the case of the natural ventilation system without exhaust ventilators, fresh air inlet location above the window and radiator heating system. The simulations considered inlet velocities in the range of (0.00 - 0.98) m/s.

The optimum range of the fresh air inlet velocities varies between 0.2 m/s and 0.4 m/s. This range of velocities satisfies the indices of the indoor air quality and the thermal comfort.

8.4 Overall thermal comfort

Concerning the problem of IAQ in residential houses it is reasonable to take into account the thermal comfort indices. As the IAQ indices, the thermal comfort indices influence the Indoor Environment Quality. All of them depend on the air-CO_2 mixture flow patterns and, in consequence, on the temperature distribution in the indoor zone. Thus, in this section the main overall thermal comfort indices are considered.

Figure 34 presents the average PMV value (Eq.24) for each simulation, where the gravitational ventilation is installed. In turn, Fig.35 presents the average PMV value (Eq.24) for each simulation with the gravitational ventilation and exhaust ventilators.

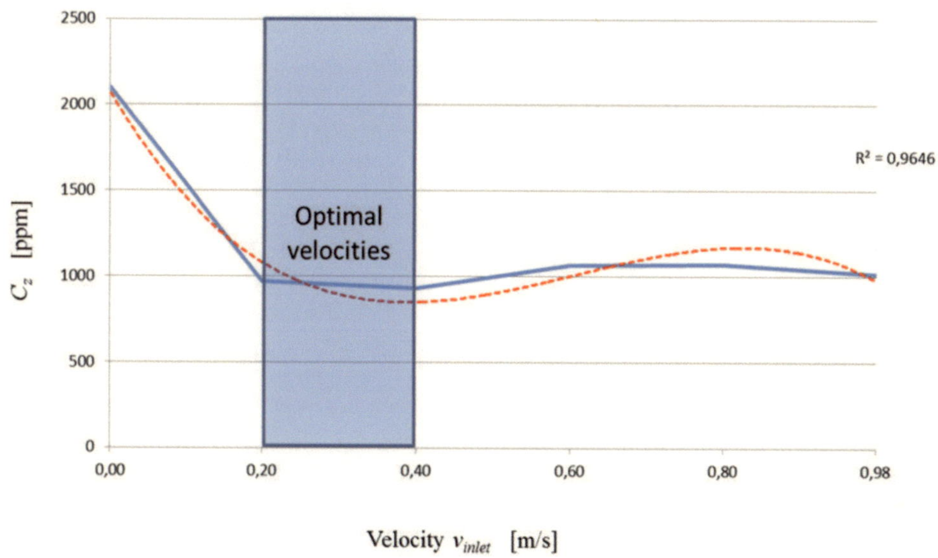

Fig. 30. The average CO2 concentration C_z in function of fresh air inlet velocity

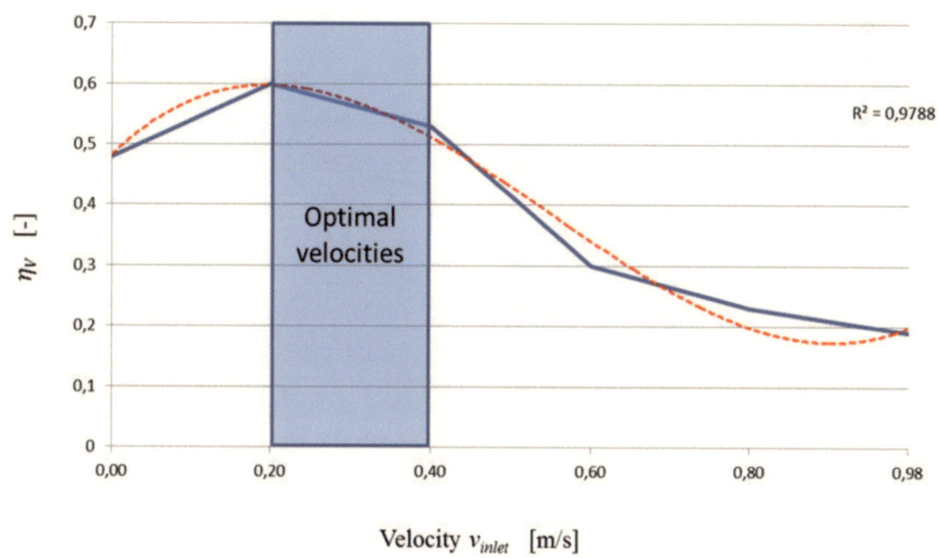

Fig. 31. The ventilation effectiveness η_V in function of fresh air inlet velocity

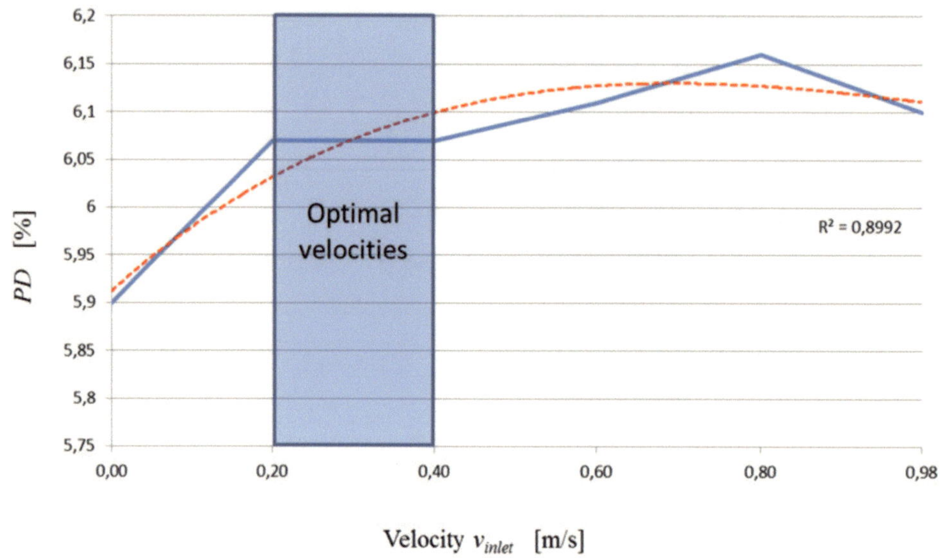

Fig. 32. The percentage dissatisfied *PD* in function of fresh air inlet velocity

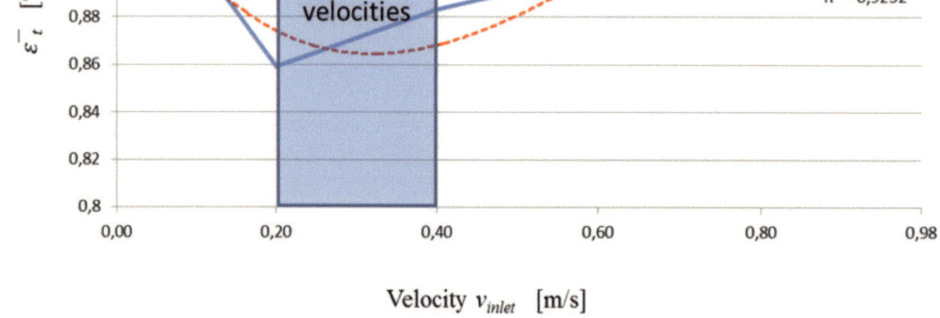

Fig. 33. The overall ventilation effectiveness for temperature distribution $\bar{\varepsilon}_t$ in function of fresh air inlet velocity

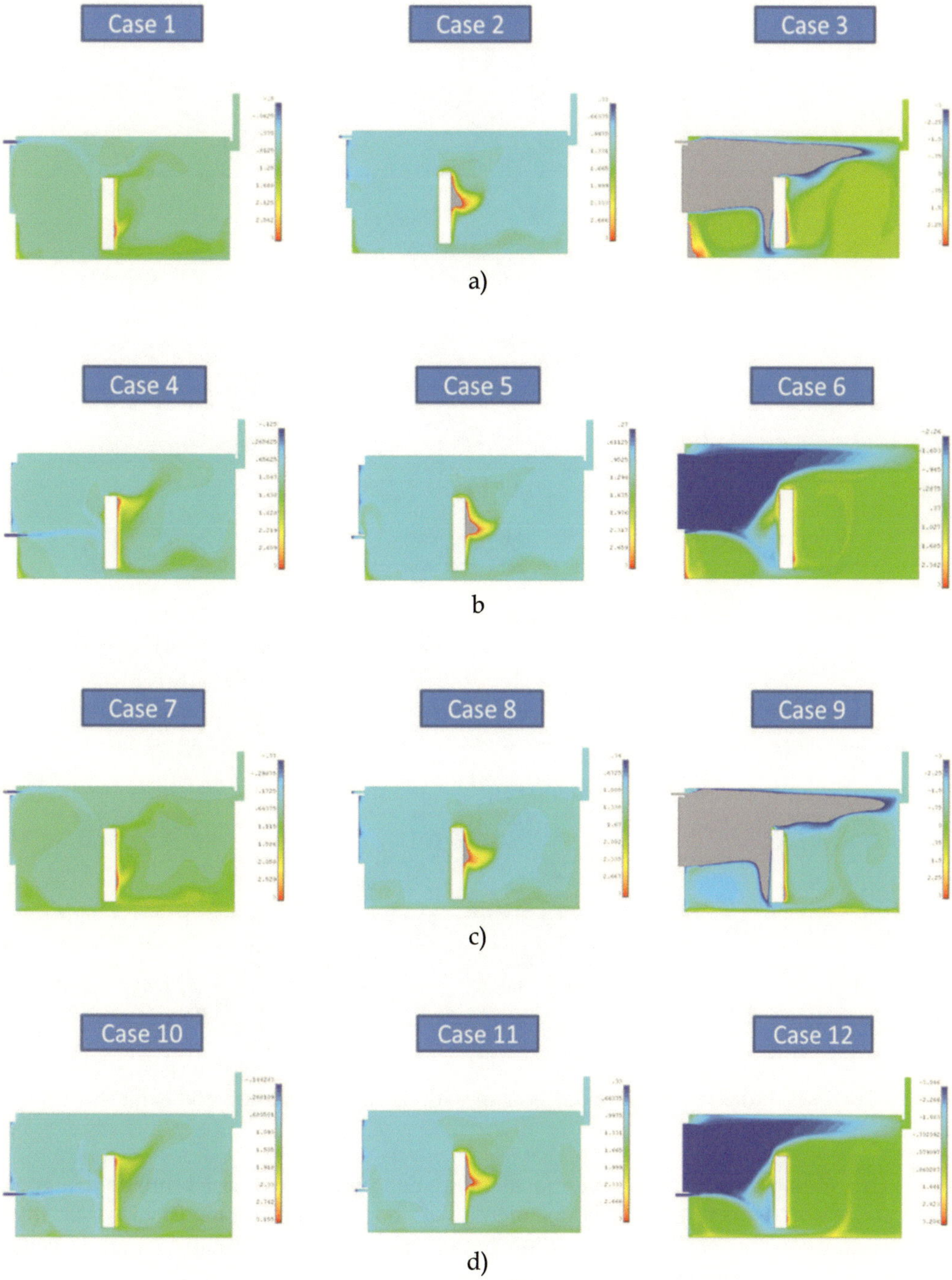

Fig. 34. The PMV index distribution for houses with natural ventilation and without exhaust ventilators

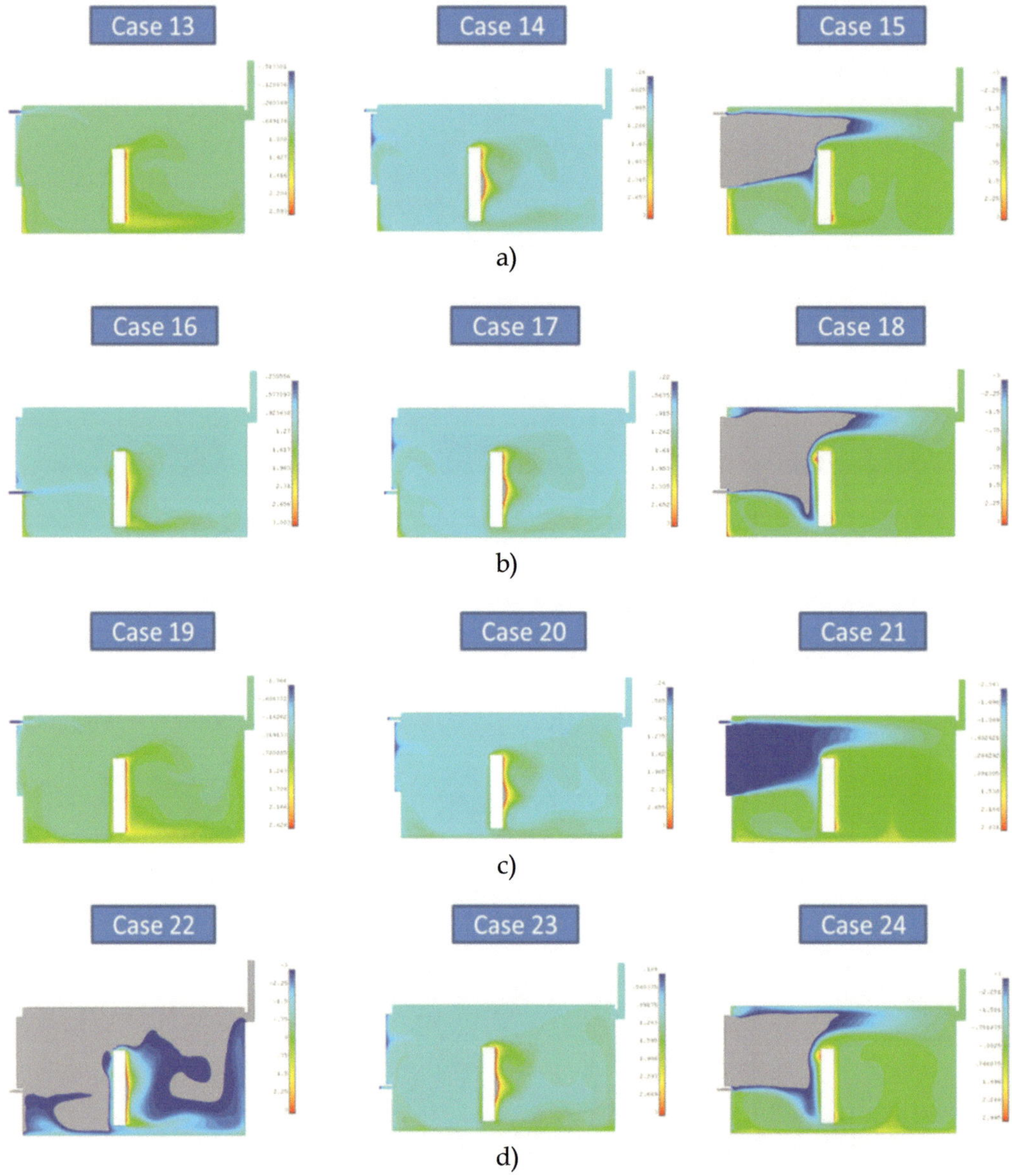

Fig. 35. The PMV index distribution for houses with natural ventilation and exhaust ventilators

In the case of houses with the gravitational ventilation without exhaust ventilators, a good thermal comfort is maintained independently of the heating system (Fig.36), air inlet location and weather conditions. It is not surprising that the thermal comfort is better during the calm weather (PMV_{avg}=1.22) than during the windy weather (PMV_{avg}=0.81). For houses with natural ventilation and exhaust ventilators, the results are significantly different (Fig.36). The house with the radiator heating system maintains the good thermal comfort level (PMV_{avg} is in range of (0.77-1.11)) independently of weather conditions and the air inlet location, while for the floor heating system, the thermal comfort strongly varies and depends on weather conditions and the air inlet location. Surprisingly, the best comfort level is reached in the house equipped with the radiator heating system and air inlet located below the window (PMV_{avg}=1.11), while the worst result is obtained in the house with the floor heating system and air inlet located below the window (PMV_{avg}=-2.78).

According to the PMV_{avg} index, it can be concluded that the best comfort level is observed in the house with the natural ventilation and without exhaust ventilators. More precise conclusions can be derived by studying the simulations results of the PD index (Fig. 37). It can be noted that the percentage of dissatisfied ones (PD index) depends only on the type of the heating system installed. The biggest values of the PD index of 13.9 % are obtained for the floor heating system.

The overall ventilation effectiveness for a temperature distribution is provided by the quantitative index $\bar{\varepsilon}_t$ related to the way in which the heat is distributed over the indoor zone (Fig.38). However, for all investigated cases, the $\bar{\varepsilon}_t$ index is almost the same (varies in the range of (0.8265 - 1.0139)). It means that a temperature homogeneity level is high enough. The only exception is observed for the cases with open windows; the $\bar{\varepsilon}_t$ index reaches 2.3943. However, the open window is an abnormal situation and the corresponding values of the $\bar{\varepsilon}_t$ index are artificial and thus can't be taken into account. Analyzing the above results, it can be seen that the natural ventilation without exhaust ventilators can maintain the satisfying thermal comfort level. It can be even higher by installing the radiator heating system. The location of the air inlet gap does not influence the thermal comfort when using the natural ventilation system with or without exhaust ventilators.

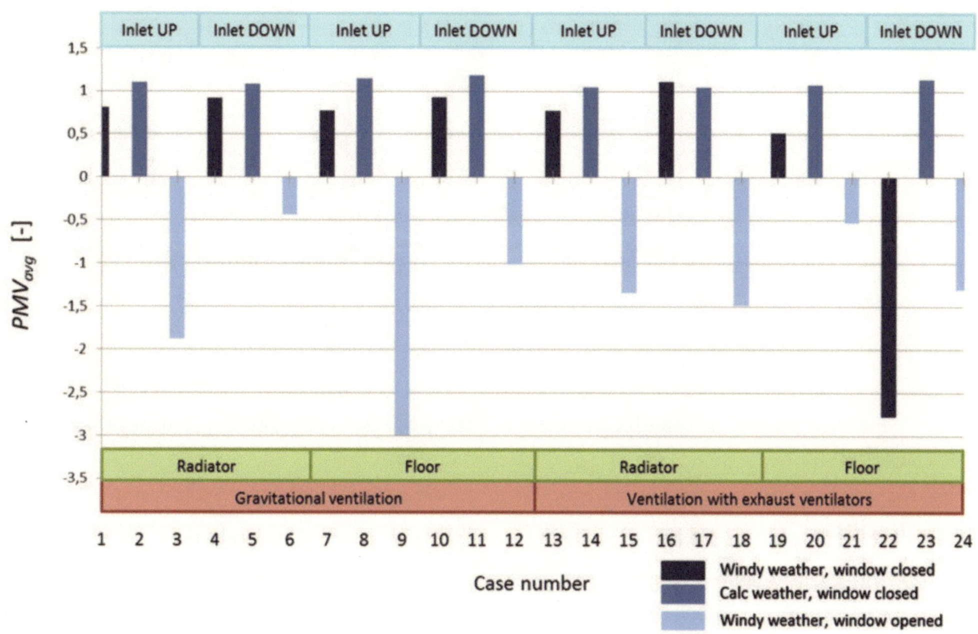

Fig. 36. The average PMV_{avg} index for houses with natural ventilation

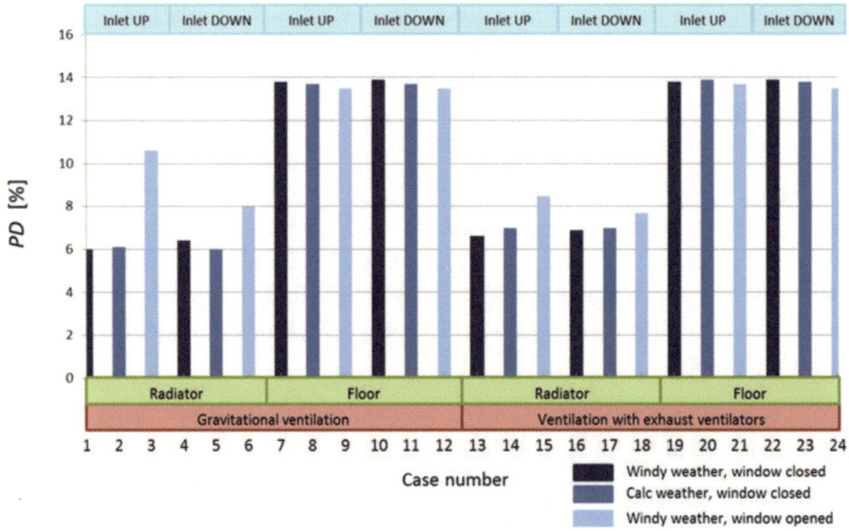

Fig. 37. The average *PD* index for houses with natural ventilation

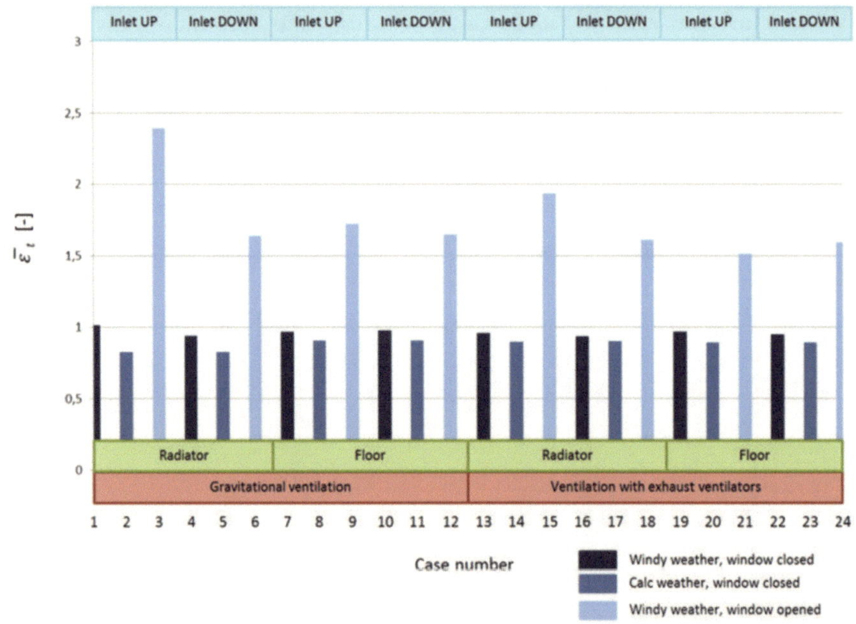

Fig. 38. The average ventilation effectiveness for temperature distribution $\bar{\varepsilon}_t$ in houses with natural ventilation

9. Conclusions

The CFD simulations were carried out to study the indoor air quality and the thermal comfort in residential houses with the natural ventilation and with or without exhaust ventilators. The average results were compared with the experimental measurements and a good correlation was found between them.

The fresh air inlet location above the window opening leads to a better indoor air quality, however decreases the thermal comfort.

A type of the heating system does not influence the indoor air quality, while strongly affects the thermal comfort. From the thermal comfort point of view, the radiator heating system is significantly better (based on the PD index) than the floor one.

The installation of exhaust ventilators significantly improves the indoor air quality, however, slightly deteriorates the thermal comfort. The exhaust ventilators installed in houses with the fresh air inlet located below the window deteriorate the thermal comfort even stronger.

The C_Z index of the average CO_2 concentration in the breathing zone as the assessment method is not well suited for the natural ventilation in residential houses located in cold-climate regions. The results indicate significant differences between the C_Z index and the average CO_2 concentration in the local fresh air zone located at the front of the resident's head. It can be a main reason of the dissatisfaction, leading to the unpredicted resident's actions aimed at recovering the satisfaction, e.g. by windows opening.

The natural ventilation without exhaust ventilators can maintain the good thermal comfort and acceptable indoor air quality, but the fresh air inlet has to be installed above the window and the radiator heating system should be applied. However, this type of ventilation is very sensitive to ambient conditions, thus the fresh air inlets should be controlled in order to maintain a constant optimum air velocity.

Our further research will focus on experimental investigations of the local fresh air zone and its influence on the subjective satisfaction sensation.

10. References

Ansys Inc., ANSYS Release 11.0 Documentation for ANSYS, 2007.

Awbi, H. (2003). Ventilation of building, E & FN Spon, 2003.

Behne, M. (1999). Indoor air quality in rooms with cooled ceilings. Mixing ventilation or rather displacement ventilation? *Energy and Buildings*, 30(2), pp.155–66

Duffie, J. A. & Beckman, W. A. (1991). Solar Engineering of Thermal Processes, New York: John Wiley and Sons

Energy Consumption Guide 19, Energy efficiency in offices, London: Energy Efficiency Office/HMSO, 1993.

Hastings, S. R. (2004). Breaking the "heating barrier". Learning from the first houses without conventional heating. *Energy and Buildings*, 26, pp.373-380

Ho, S. H., Rosario, L. & Rahman, R. R. (2011). Comparison of underfloor and overhead air distribution systems in an office environment. *Building and Environment*, 46, pp.1415-1427

Hummelgaard, J., Juhl, P., Sebjornsson, K. O., Clausen G., Toftum, J. & Langkilde, G. (2007). Indoor air quality and occupant satisfaction in five mechanically and four naturally ventilated open-plan office buildings. *Building and Environment*, 42, pp.4051–4058

Hyun, S. & Kleinstreuer, C. (2001). Numerical simulation of mixed convection heat and mass transfer in a human inhalation test chamber. *International Journal of Heat and Mass Transfer*, 44, pp.2247–60

ISO 16000-1, (2004). E, Indoor air - part 1: General aspects of sampling strategy

ISO 16814 (2008), Building environment design - Indoor air quality - Methods of expressing the quality of indoor air for human occupancy

ISO 6946:2007, Building components and building elements. Thermal resistance and thermal transmittance. Calculation method, 2007.

ISO 7730:2005, Ergonomics of the thermal environment. Analytical determination and interpretation of thermal comfort using calculation of the PMV and PPD indices and local thermal comfort criteria, Annex D, 2005.

Karlsson, J. F. & Moshfegh, B. (2006). Energy demand and indoor climate in a low energy building - changed control strategies and boundary condition. *Energy and Building*, 38, pp.315-326

Launder, B. E. & Spalding, D. B. (1974). The Numerical Computation of Turbulent Flows. *Computer Methods in Applied Mechanics and Engineering*, Vol. 3, pp 269-289

Laverge, J., Van Den Bossche, N., Heijmans, N. & Janssens, A. (2011). Energy saving potential and repercussions on indoor air quality of demand controlled residential ventilation strategies. *Building and Environment*, 46, pp.1497-1503

Lawrence, T. M. & Braun, J. E. (2007). A methodology for estimating occupant CO_2 source generation rates from measurements in small commercial buildings. *Building and Environment*, 42, pp.623-639

Li, J., Way, O. W. H., Li,Y. S., Zhan, J., Ho, Y. A., Li, J. & Lam, E. (2010). Effect of green roof on ambient CO_2 concentration. *Building and Environment*, 45, pp.2644-2651

Maier, T., Krzaczek, M. & Tejchman, J. (2009). Comparison of physical performances of the ventilation systems in low-energy residential houses. *Energy and Building*, 41, pp.337-353

Meyer, F. (1993). Niedrigenergiehäuser Heidenheim: Hauskonzepte und erste Meßergebnisse, Eggenstein-Leopoldshafen: Fachinformationszentrum, Bine Projekt Info-Service Nr. 9, Karlsruhe

Murakami, S., Kato, S. & Zeng J. (2000). Combined simulation of airflow, radiation and moisture transport for heat release from a human body. *Building and Environment*, 35, pp.489–500

Paul, T., Sree, D. & Aglan, H. (2010). Effect of mechanically induced ventilation on the indoor air quality of building envelopes. *Energy and Buildings*, 42, pp. 326–332

Redlich, C. A., Sparer, J. & Cullen M. R. (1997). *Sick-building syndrome*. Lancet, 349 (9057), pp.1013–6

Reinmuth, F. (1994). Energieeinsparung in der Gebäudetechnik: Baukörper und technische Systeme der Energieverwendung, Vogel Verlag, Würzburg

Stavrakakis, G. M., Zervas, P. L., Sarimveis, H. & Markatos, N. C. (2010) Development of a computational tool to quantify architectural design effects on thermal comfort in naturally ventilated rural houses. *Building and Environment*, 45(1), pp.65-80

Viollet, P. L. (1987). The Modelling of Turbulent Re-circulating flows for the purpose of reactor thermal-hydraulic analysis. *Nuclear Engineering and Design*, 99, pp.365-377

Wong, N. H. & Huang, B. (2004). Comparative study of the indoor air quality of naturally ventilated and air-conditioned bedrooms of residential buildings in Singapore. *Building and Environment*, 39, pp.1115-1123

6

Particles in the Indoor Environment

Hermann Fromme
Bavarian Health and Food Safety Authority,
Dep. of Chemical Safety and Toxicology
Germany

1. Introduction

As a result of a change in living and work habits, we now stay in industrial countries every day more than 90% of the time inside buildings. Against this backdrop findings about the exposure of users are relevant. Given their heterogeneity, very complex exposure patterns exist in the indoor environment in respect of which not only input from the outdoor air but also important sources inside the rooms themselves have to be taken into account. At any rate, different indoor environments have to be identified (e.g. living, bed, handicraft, leisure and cellar rooms, working rooms and workplaces in buildings, public buildings, restaurants and inns, community facilities such as schools and kindergartens as well as spaces in motor vehicles and other public transportation systems). Furthermore, it has been shown that the amount of airborne particle content in indoor environments can be highly variable in terms of space but also in terms of time. Apart from the conditions prevailing in the outdoor air close to the indoor environment (e.g. location close to a heavily trafficked street or in a rural region) and the current climatic conditions, the structural conditions of the building and the ventilation conditions are important. Furthermore, activities in indoor environments, such as the deposition and resuspension of house dust, cooking and cleaning activities or smoking can make a considerable contribution to the respective pollution situation.

Particulates (particulate matter, PM) which are dispersedly distributed in the air form colloidal systems with the gases which are also referred to as aerosols. Overall, the composition of aerosols strongly depend on the specific sources. The particles of the fine fraction develop primarily through transformation processes from gases or within the framework of combustion processes. They are typically composed of nitrates, sulphates, ammonium, elementary carbon, a large number of organic compounds and trace elements. By contrast, the particles in the coarse fraction develop largely mechanically following the disintegration of larger solid particles and consist typically of whirled up dust from industrial processes and biological material such as pollen and bacteria and their fragments.

PM in indoor environments consist of very different particles which are considerably varying in terms of size, form and chemical composition. Whereas the larger particles determine primarily the mass of the environmental aerosol, the particle number concentration (PNC) and the particle surface are dominated almost exclusively by the ultra fine particles (<100 nm).

Concerning the measurement of PM in air, different sampling conventions have established themselves, often using the aerodynamic diameter of the particles. In order to better reflect the human respiratory characteristics, conventions such as PM_{10} (Particulate Matter) or $PM_{2.5}$ were introduced by the US Environmental Protection Agency and European authorities. $PM_{2.5}$ is, for instance, the particle fraction which passes through a size-selective air inlet which has a separation efficiency of 50% for an aerodynamic diameter of 2.5 µm. Depending on the specific context, other definitions may be applied, for example in indoor working environments.

2. Behaviour, transport, and fate of particles in the indoor environment

The transport and fate of particles in indoor environments are fundamentally influenced by a series of physical and chemical processes (Fig. 1). This can lead to considerable changes in terms of their chemical composition, their physical characteristics, their distribution patterns and finally the measurable contents (Thatcher et al., 2001; Morawska & Salthammer, 2003; Nazaroff, 2004).

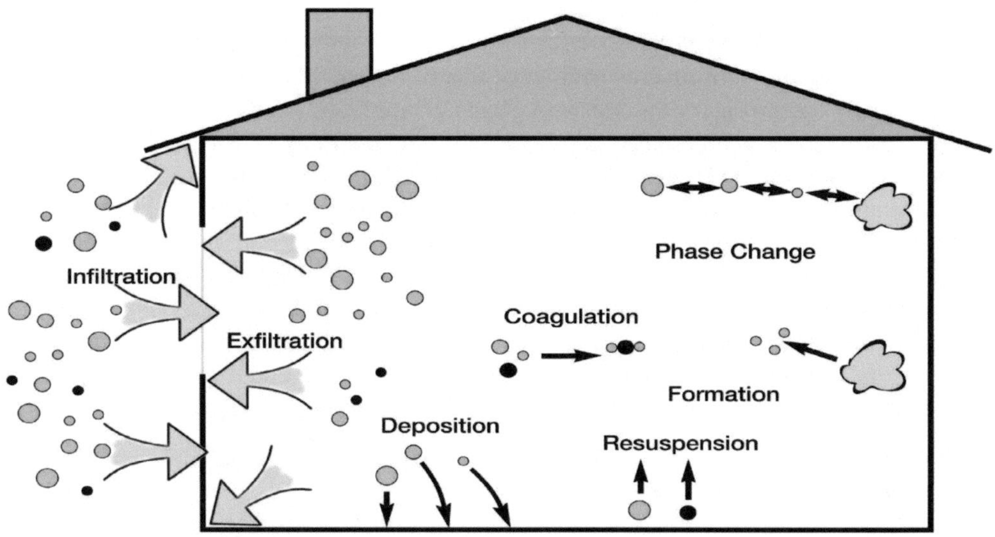

Fig. 1. Transport and transformation processes with impact on the indoor concentration of particulate matter (modified from Thatcher et al., 2001)

2.1 Infiltration/ penetration

The dimensionless penetration factor (P) is defined as the share of the particle fraction with a specific diameter which reaches the indoor environment through the inflow of outdoor air. In the scientific literature there are results of different studies which are based on the observation of the indoor to outdoor ratio of the particles, manipulations of the external building envelope, experimental simulations in the laboratory or mathematical modelling (e.g. Long et al., 2000; Vette et al., 2001; Riley et al., 2002; Riley et al., 2002; Liu & Nazaroff, 2003; Chen & Zhao, 2011). The results show that for different types of buildings and gap /

crack diameters and geometry, the largest penetration factor seems to exist for the particles with diameter between approximately >0.05 and < 1 µm (see Fig. 2).

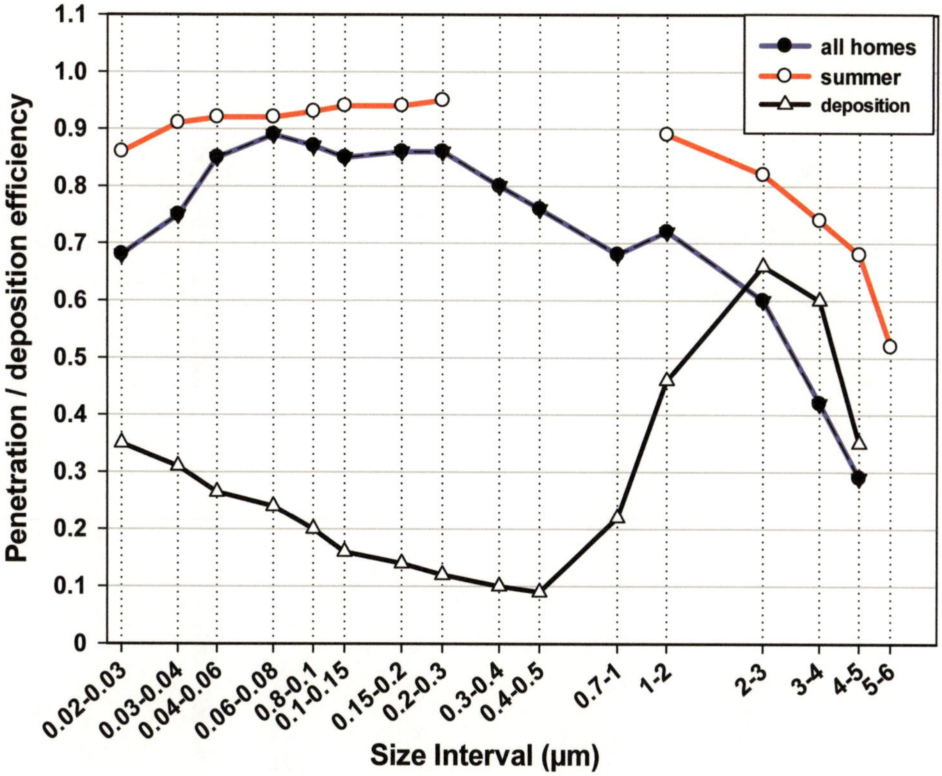

Fig. 2. Penetration efficiencies (P) and deposition rates (k) (all homes nightly averaged data from n=98-106 and in summer from n=8) (modified from Long et al., 2001)

2.2 Deposition

The deposition of particles on surfaces is based on different physical mechanisms such as gravitation and diffusion. Apart from the deposition speed, this process is described by the so-called deposition rate (k) (example, Fig. 2). This process is strongly dependent on the particle diameter and reaches a minimum for particles with an aerodynamic diameter of approximately 0.4 µm. However, there is a considerable variation range (Morawska & Salthammer, 2003; Miguel et al., 2005; Hussein et al., 2009). The particle deposition, in particular of coarse particles, increases with a rising draught in the room and an increasing room area and also varies depending on the degree of interior decoration.

2.3 Resuspension

Particles deposited on the surfaces of the room can become resuspended in the indoor air in particular through activities in the indoor environment (Thatcher & Layton, 1995; Hussein et al., 2006). Hu et al. (2005) state that essentially three parameters like mechanical vibration, aerodynamic as well as electrostatic forces can achieve a stronger effect than gravitation and hence influence the resuspension of particles. In different field studies it was shown that activities in the indoor environment (e.g. running, playing kids) resulted in a significant increase in PM contents, whereby essentially coarse particles were whirled up (Thatcher & Layton, 1995; Long et al., 2000; Miguel et al., 2005) (Fig. 3). Moreover it could be shown that the resuspension in rooms with wall to wall carpet was significantly higher compared to rooms with a smooth flooring (Long et al., 2000).

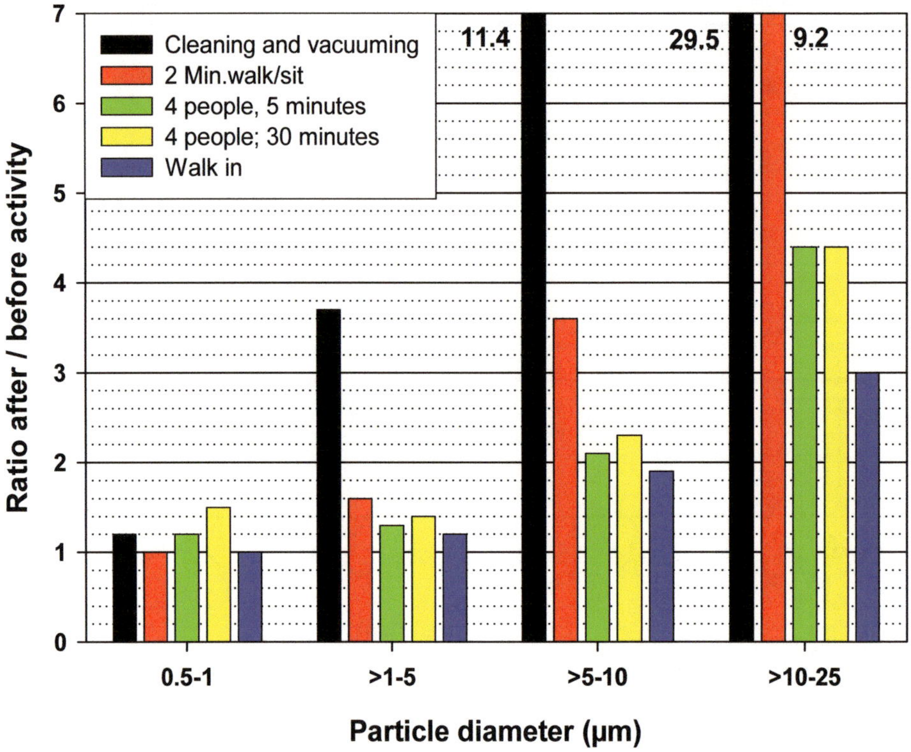

Fig. 3. The ratio of the suspended particle concentration after a resuspension activity to the indoor concentration before that activity, by particle size (modified from Thatcher & Layton, 1995)

2.4 Particle formation

Within the framework of chemical processes in the indoor environment particles can be newly formed or there can be a growth in particle size. The coagulation of particles in the indoor environment is based on the fact that e.g. depending on the particle number they come together with a certain probability and then tend to agglomerate. This process is, for instance, relevant for ultra fine particles in indoor environments, since the latter exist e.g. in high number concentrations when for example burning candles. They then agglomerate over time; this can be observed through a shift in the peak value of particle distribution (Dennekamp et al., 2001). The phenomenon of phase transition, too, describes an "ageing process" during which a growth of the particles is observed through the adsorption of organic substances or water.

3. Sources of particles indoors

3.1 Burning processes

Tobacco smoking constitutes an essential particle source in indoor environments which results in an increase in the particle mass as well as the ultrafine particles. In the Harvard Six City Study, for instance, the annual mean values in smoker households were higher by approximately factor 3 compared to non smoker households (Neas et al., 1994). Fig. 4 shows the increase in indoor pollution depending on the number of cigarettes smoked. In the same way the particle number increases considerably during cigarette smoking, partly to values up to 213,000 particles/cm³ (He et al., 2004; Afshari et al., 2005; Hussein et al., 2006).

When burning candles or oil lamps in indoor environments, an increase in ultrafine particles was likewise observed (Fine et al., 1999; Hussein et al., 2006; Wallace & Ott, 2011). This involved significantly higher concentrations when extinguishing candles compared to the burning itself (Hussein et al., 2006). During the burning of incense sticks it is also possible to detect high particle contents, in particular in the range from 0.06 to 2.5 µm, in indoor air (Chao et al., 1998; Jetter et al., 2002).

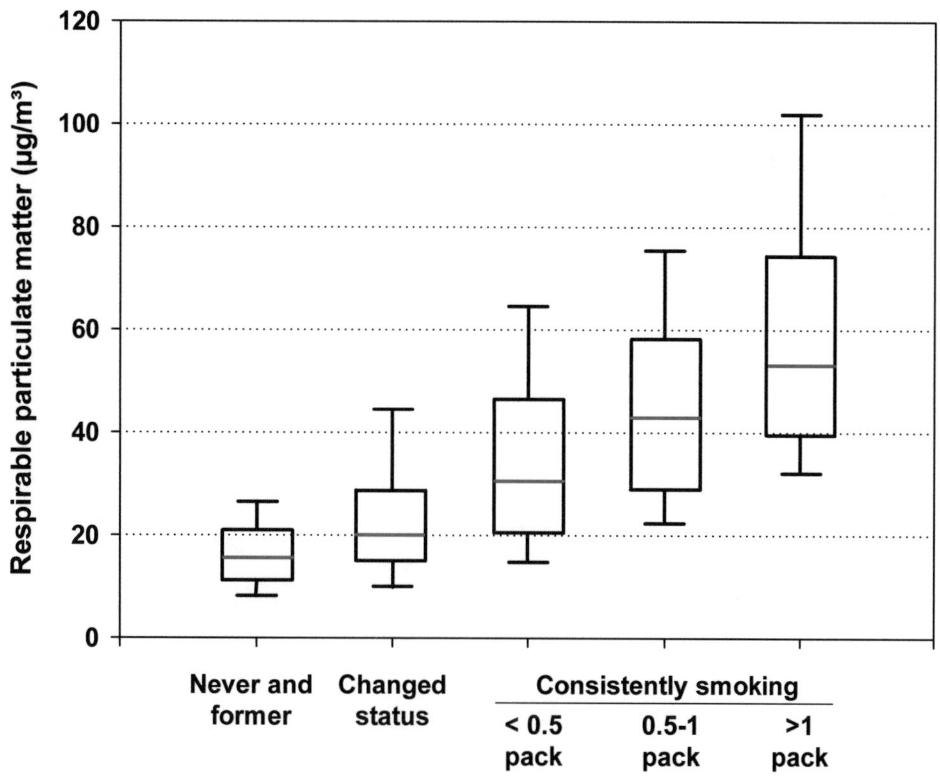

Fig. 4. Distribution percentiles for the annual average concentration of indoor particulate matter by household smoking status and the estimated number of cigarette packs smoked in the home (modified from Neas et al., 1994)

3.2 Cooking activities

During cooking, too, fine and ultrafine particles are released. Different working groups were able to detect very high peak pollutions during cooking with electric stoves and in particular gas stoves of 100,000 to 560,000 particles /cm³ (Morawska et al., 2003; Dennekamp et al., 2001; He et al., 2004; Afshari et al., 2005; Ogulei et al., 2006; Hussein et al., 2006). The large concentration range is attributable to the different cooking activities (eg baking, roasting, frying, toasting), the use of energy, the respective cooking goods, the ventilation conditions and the room geometry. Dennekamp et al. (2001) describe PNCs of up to 110,000 or 150,000 particles/cm³ when using four electric or gas rings. Peak values of up to 590,000 ultra fine particles/cm³ were reached at the frying of bacon on a gas stove. After a short period of time the particles grew up in the indoor air and a displacement towards larger diameters. (Abt et al., 2000; Dennekamp et al., 2001; Hussein et al., 2006). After the end of the cooking activity the concentration rapidly decreases (Fig. 5). Referred to the particle mass, these activities likewise constitute a certain source. In the American PTEAM Study it was determined by means of a regression model that cooking increased the basic load of PM_{10} in the indoor environment by approximately 12 - 26 µg/m³ ($PM_{2.5}$: approximately 13 µg/m³) (Wallace et al., 2003). Extremely high pollutions are to be expected when cooking on open fireplaces as, for instance, in third world countries (e.g. Naeher et al., 2000).

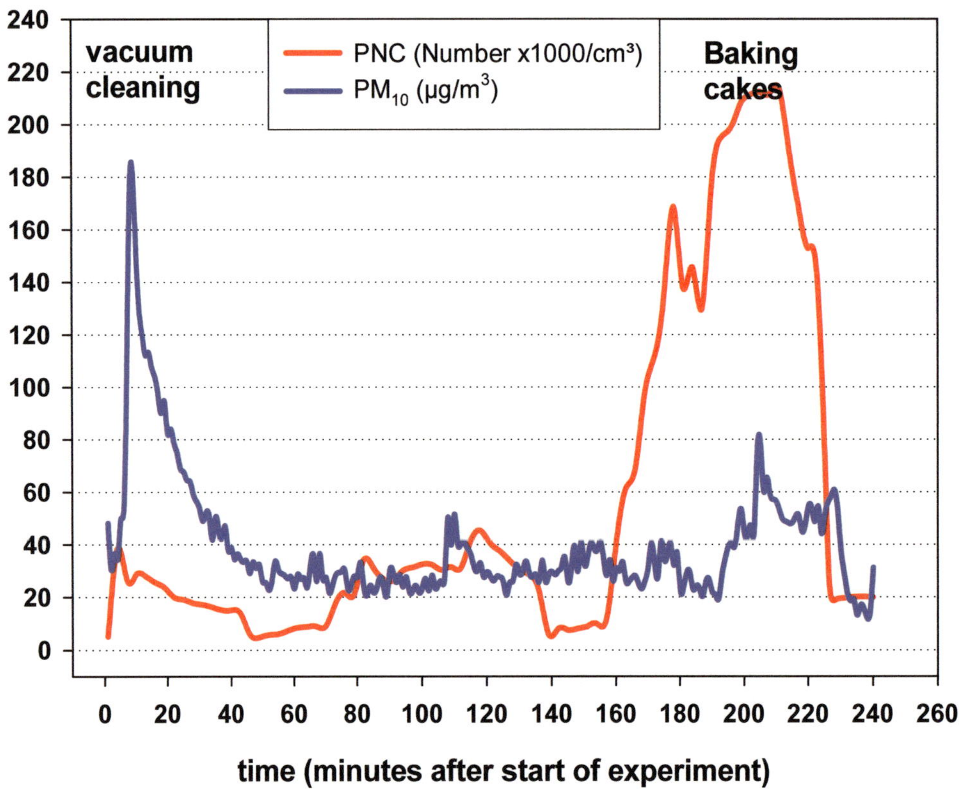

Fig. 5. Particle number concentration (PNC) and PM_{10} in the kitchen baking with an electric oven or vacuuming

3.3 Cleaning activities

During cleaning and in particular vacuum cleaning, an increase in coarse particles and hence in particular the particle mass is observed in indoor air (Abt et al., 2000). See also figure 5. In two US studies the contribution of cleaning activities to the $PM_{2.5}$ indoor pollution was estimated at 23-32 µg/m³ (Long et al., 2000; Ferro et al., 2004). Afshari et al. (2004) describe, by contrast, merely an insignificant minor increase for ultrafine particles.

Long et al. (2000) investigated the influence of the use of commercial cleaning agents on a pine oil basis on the exposure in a living room. During the activities the PNCs rose from initially 2,000 particles/cm³ to a maximum of 190,000 particles/cm³ and the $PM_{2.5}$ contents increased from 5 to 38 µg/m³. This phenomenon was explained by referring to the new particle formation and / or particle growth through oxidative processes in the indoor environment. Other working groups, too, were able to detect in test chambers in the presence of ozone and the simultaneous application of terpene-containing cleaning agents a significant increase in particle number concentrations and the particle mass (Sarwar et al., 2004; Singer et al., 2006; Destaillats et al., 2006).

3.4 Secondary organic aerosols (SOA)

Following chemical reactions of the gas and aerosol phase, so-called secondary organic aerosols (SOAs) are newly formed in indoor environments (Weschler et al. 2006). The formation of SOAs through the reaction of ozone with terpenes and other unsaturated organic compounds was demonstrated and confirmed in many test chamber experiments (e.g. Wainman et al., 2000; Fan et al., 2003; Sarwar et al., 2004; Liu et al., 2004b; Vartiainen et al., 2006; Destaillats et al., 2006; Aoki & Tanabe, 2007). In two office rooms, for instance, there was an increase in the particle mass and the PNC (Fig. 6) with realistic ozone and limonene contents (Weschler et al., 2003). Ozone was in these experiments the limiting factor in the formation of SOAs.

3.5 Outdoor air as a source

The contribution of outdoor air to the amount of PM concentration in indoor air depends, in addition to the particle fraction, in particular on the ventilation behaviour of the room user, the tightness of the building envelope, the dust deposition rates indoors, the resuspension effects in the room and the coagulation behaviour of the particles. The ventilation behaviour itself is naturally dependent to a large extent on the season and the meteorology (Nazaroff, 2004). Through the windows and doors but also through leakages of the building envelope there is an exchange of air between the indoor air and the outdoor air. This results in a highly variable share of outdoor air in the amount of particle concentration in the indoor air. Other factors such as the building geometry (e.g. floor height) and location (e.g. close to a heavily trafficked road) can have a significant influence on the exposure situation. Cyrys et al. (2004) report in respect of the examination of two model rooms without an indoor activity that 75% of the indoor air contents of $PM_{2.5}$ but only 43% of the PNCs can be explained by corresponding outdoor air contents. During the parallel measurements of particle distribution in rooms without indoor source and outdoor air there were in the event of closed windows and doors in the indoor environments significantly lower contents in the particle size classes than outdoors (Franck et al., 2003). Fig. 7 shows results which represent the ventilation-related influencing of PM from outside to residential indoor environments (Riley et al., 2002).

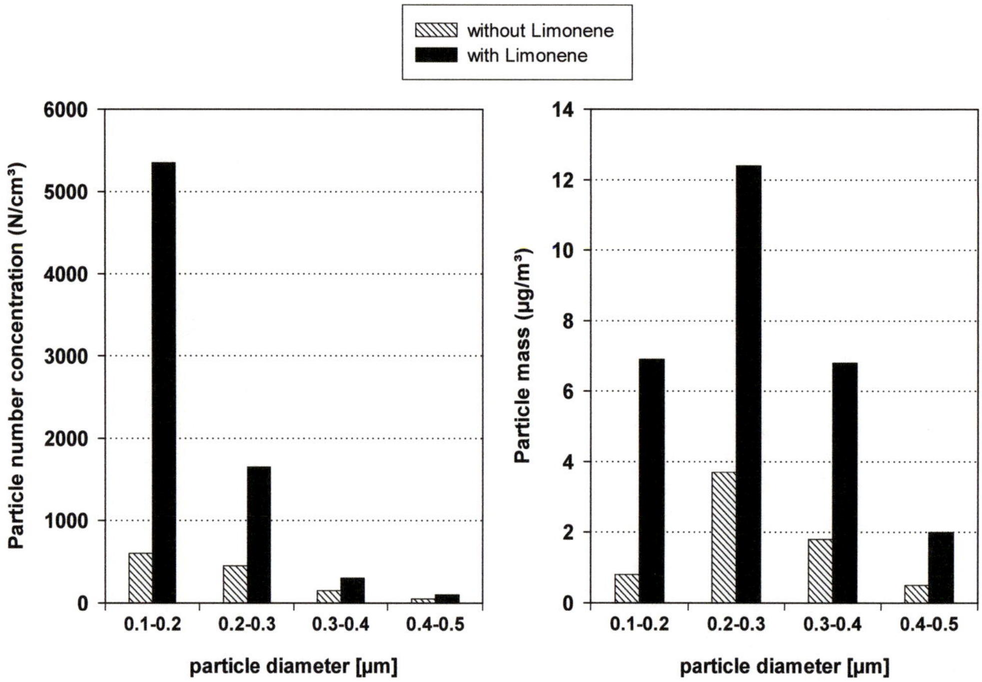

Fig. 6. Comparison between the concentrations of particles (left: number, PNC; right: mass) in an office with a limonene source and one without (modified from Weschler et al., 2003)

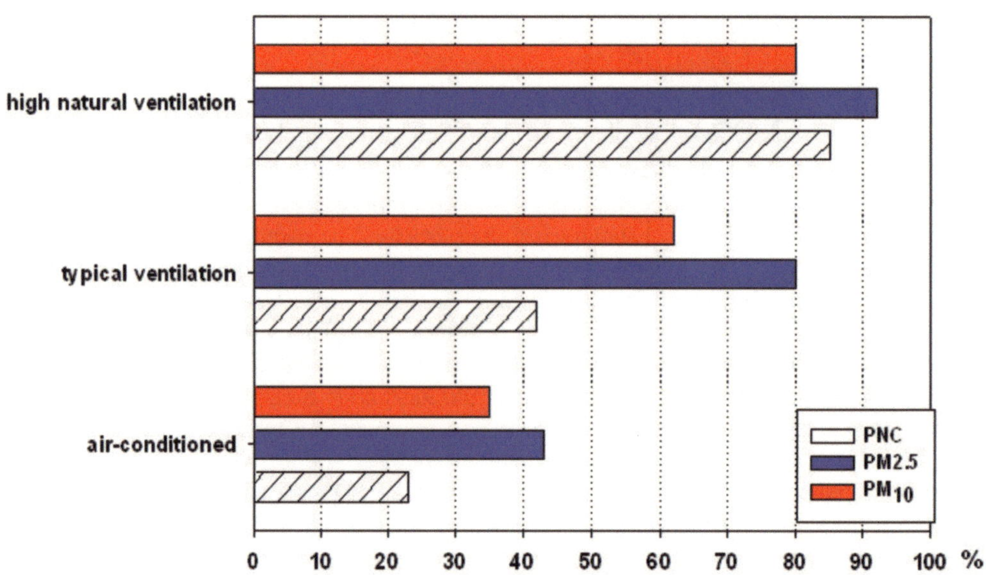

Fig. 7. Predicted proportion of outdoor particles in three urban residential scenarios (modified from Riley et al., 2002)

4. Occurrence of particles in indoor spaces

4.1 Particles in residences

In the scientific literature a large number of measurements of particle mass concentrations in indoor air are described. Table 1 shows the results for the mass-related measurements in residential indoor environments. It must be taken into account that due to different sampling and measurement methods the results can only be compared to a limited extent.

Reference	Concentration	Description
Europe		
Hänninen et al., 2004+	31 (A), 26 (B), 13 (H), 36 (P)	A: Athens, B: Basel, H: Helsinki, P: Prague; n: 186; 1996-2000
Lai et al., 2004	10	UK; n: 42; 1998-2000
Fromme et al., 2005*	30 (Wi), 27 (Su)	Berlin, Germany; n: 62; WI: 1997/98; SU: 2000
Raaschou-Nielsen et al., 2011	13	Denmark; n: 389; 1999-2002
Link et al., 2004	19	Germany; n: 126; 2001-2002
Franck et al., 2011	32	Germany; n: 129; 2001/2002
Stranger et al., 2007+	36	Belgium; n: 19; 2002-2003
Wichmann et al., 2010	10	Sweden; n: 29; 2003/2004
Osman et al., 2007	18	Scotland; n: 75; 2004/2005
Santen et al., 2009	3-15	Germany; n: 50; 2007
Cattaneo et al., 2011	23	Italy; n: 107; 2007/2008
America, Australia		
Wallace et al., 2003+	28	USA; n: 294; 7 cities
Meng et al., 2005	14	USA; n: 212; 1999-2001
Breysse et al., 2005+	26	USA; n: 90
Simons et al., 2007	35 (a), 10 (b)	USA; n: 100 city (a), 20 suburban (b)
Baxter et al., 2007	17	USA; n: 43; 2003-2005
Héroux et al., 2010	6	Canada; n: 96; 2007
Jung et al., 2010	14	USA; n: 286; 2005-2010
Asia		
Li & Lin, 2003+	39 (Wi), 37 (Su)	Taiwan; urban; n: 10; 1999-2000
Chao & Wong, 2002+	45	Hong Kong; n: 34; 1999-2000
Lim et al., 2011	48	Korea; n: 60; 2008

Wi: winter; Su: summer; S: smoker; NS: non smoker; *: PM_4; +: mean

Table 1. Median concentrations of $PM_{2.5}$ in the indoor air of residences in μg/m³

In different studies it could be shown that smoking is the most important influencing factor for the PM contents (e.g. Özkaynak et al., 1995; Wallace & Howard-Reed, 2002; Lai et al., 2004; Fromme et al., 2005; Breysse et al., 2005; Héroux et al., 2010; Franck et al., 2011). In Germany, the mean PM_4 concentrations in smoker households amounted in winter and summer, for instance, to 109 $\mu g/m^3$ and 59 $\mu g/m^3$ respectively, and in non smoker households they amounted during the two seasons only to 28 $\mu g/m^3$ (Fromme et al., 2005). Other important influencing factors for the indoor air contents are the season, the outdoor air, the ventilation behaviour, the age and the location of the buildings and indoor activities such as cooking, the use of ovens and the burning of incense sticks (Mönkkönen et al., 2005; Martuzevicius et al., 2008; Santen et al., 2009; Rodes et al., 2010; Héroux et al., 2010; Byun et al., 2010; Raaschou-Nielsen et al., 2011).

Studies on the ultrafine particles (as particle number concentration, PNC) in residences resulted in Germany in cities on average in 20,400 particles/cm^3 (Link et al., (2004) or to between 4,000 and 25,000 particles/cm^3 in a monthly median (n: 50) (Santen et al., 2009) and in an epidemiological study in 59 residences in the median of 9,000 particles/cm^3 (Franck et al., 2011). McLaughlin et al. (2005) report in seven Irish residences about mean PNCs between 4,900 and 105,200 particles/cm^3 with a maximum value of up to 485,300 particles/cm^3. In a Swedish study three residences were investigated with mean daily values between approximately 1,800 and 8,300 particles/cm^3 (Matson, 2005). The proportion of indoor to outdoor ranged between 0.7 and 2.5.

In the USA in an apartment in Boston mean PNCs of 16,000 particles/cm^3 (Levy et al., 2002) and in seven Californian homes values of 9,200 to 35,000 particles/cm^3 were measured (Bhangar et al., 2011). With indoor sources a mean value of 18,700 particles /cm^3 (maximum: 300,000 particles/cm^3) was found in a house; without indoor sources it only amounted to 2,400 particles/cm^3 (maximum: 58,000 particles/cm^3) (Wallace & Howard-Reed, 2002). In 36 houses in Canada mean contents of 21,600 particles/cm^3 were determined during the afternoon whereas during the night the average contents were only at 6,700 particles/cm^3 (Weichenthal et al., 2007). In another Canadian study median PNCs of 2,700 particles/cm^3 (summer) were determined in 94 flats, 3,700 particles/cm^3 (winter) and 2,600 particles/cm^3 (summer) (Kearney et al., 2011).

In Australia Morawska et al. (2003) measured mean PNCs of 18,200 particles/cm^3 (during indoor activities) and 12,400 particles/cm^3 (without corresponding activities) when examining kitchens in 15 flats in 1999.

4.2 Particles in schools

Figure 8 shows some examples of results from schools. In most of the studies the $PM_{2.5}$ contents ranged on average between 8 and 20 $\mu g/m^3$. Merely in a study in 27 Belgian schools 61 $\mu g/m^3$ were described, i.e. comparatively high concentrations (Stranger et al., 2007). By contrast, the PM_{10} contents at schools were highly variable with medians in the range of 50 - 100 $\mu g/m^3$. Significantly higher contents were determined in a Greek study in which there was, however, also a high outdoor air pollution (Diapouli et al., 2007). In a European survey of 45 schools contents between 14 and 260 $\mu g/m^3$ (PM_{10}) were measured (HESE, 2006).

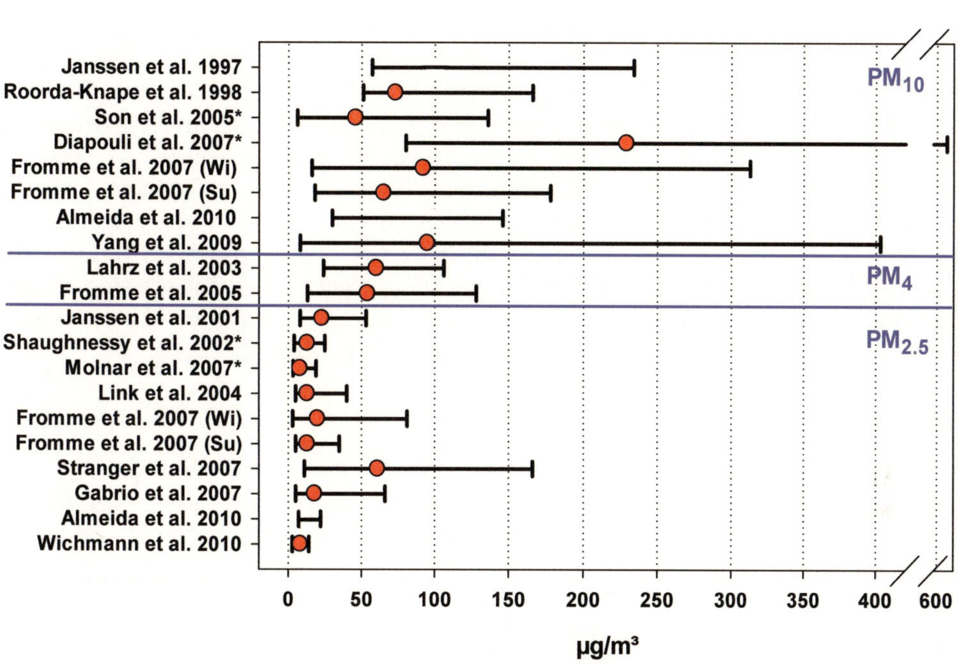

Fig. 8. Particulate matter in the indoor air of schools (minimum, median, maximum); *: mean; Su: summer; wi: winter

Different studies showed for $PM_{2.5}$ a ratio of indoor to outdoor air was in the range of 1 and a strong dependence on the outdoor air contents (Diapouli et al., 2007; Fromme et al., 2007; Wichmann et al., 2010; Guo et al., 2010). The situation is different if coarse particle fractions are considered. In a German study it was observed that 90% of the variability of the daily indoor medians of PM_1 were attributable to differences between the schools and / or the days but in this way only 45% of the PM_{10} variants could be explained (Fromme et al., 2007). Indoor sources themselves seem to be highly significant in this connection. In order to be able to assess the contribution of different sources and outdoor air, different studies also determined the elementary and/or ionic composition of PM (Diapouli et al., 2007; Molnar et al., 2007; Fromme et al., 2008). It turned out that in particular the coarse particles did not originate from the outdoor air but that the source was the classroom itself. Examinations of the filters by EDX (energy dispersive X-ray spectroscopy) suggested that the PM_{10} contents were mainly composed of floor particles and other mineral substances, attrition of building materials and chalk dust (Fromme et al., 2008). A further study revealed significantly more silicate particles (36% of all particles), organic particles (29%, probably from human skin) and Ca carbonate particles (12%, probably from paper) indoors, whereas in the corresponding outdoor filters in particular Ca sulphate containing particles (38%) were determined (Oeder et al., 2011).

The physical activity of the pupils and the associated whirling up of suspended particles from the floor seems to be the main reason for the high PM_{10} concentrations in classrooms (Fromme et al., 2007, 2008; Almeida et al., 2010; Guo et al., 2010; Oeder et al., 2011).

Measurements of the particle number concentration (PNC) have only been carried out so far in some cases at schools. In 36 German classrooms PNCs of 2,600 to 12,100 particles/cm^3 were measured (Fromme et al., 2007). In another German study the contents ranged between 2,400 and 75,500 particles/cm^3 (city) or 1,720 and 47,100 particles/cm^3 (rural area) (Link et al., 2004). In a study in Greece mean PNCs of 24,000 particles/cm^3 were determined during the class time at seven primary schools in Athens (Greece) which correlated well with the outdoor air contents (32,000 particles/cm^3) (Diapouli et al., 2007). In an Australian study 3,100 particles/ cm^3 were determined as mean value which increased within the framework of indoor activities such as cooking or cleaning of the floor surfaces to a maximum of 100,000 particles/cm^3 (Guo et al., 2010). Since in school classrooms these classical sources for ultra fine particles are as a rule missing, the exposure of pupils during class time is essentially determined by the pollution of the outdoor air.

4.3 Particles in offices

Table 2 represents the mass related contents in indoor air of office buildings. The study results are difficult to compare with one another, since it was partly not mentioned whether smoking was allowed in the rooms. The median $PM_{2.5}$ and PM_{10} values in non-smoker

Reference	Median (Min-Max)	Description
PM_{10}		
Phillips et al., 1998	53 (NS) *; 63 (S) *	France; n: 222 personal monitoring; 1995
Gemenetzis et al., 2006	103 (25- 370)	Greece; 40 rooms in 2 buildings; natural ventilated
Heavner et al., 1996	30 (<DL- 98)(NS) * 67 (18- 217) (S) *	USA, New Jersey, Pennsylvania; n: 52 (NS) and 28 (S); 1992
Burton et al., 2000	11 (3- 35) +	USA; n: 100; with AC; 1994-1998
Reynolds et al., 2001	14 to 36#	USA; n: 6; with AC; 1996/1997
Liu et al., 2004a	63 (14- 166)	China, Peking; n: 11; 2002/2003
$PM_{2.5}$		
Mosqueron et al., 2005	26 (5- 265)	France; n: 55; 1999/2000
Lahrz et al., 2002	29 (5- 120) (NS)	Germany; n: 25; natural ventilation; 2001
Gemenetzis et al., 2006	77 (11- 250)	Greece; 40 rooms in 2 buildings; natural ventilated
Vardavas et al., 2007	51 (39- 63)* (NS) 107 (39- 63) (S)	Greece; n: 6; natural ventilation; 2006
Horemans et al., 2007	11 (5- 28)	Belgium; n:9; natural ventilation; 2007
Burton et al., 2000	7 (1- 25) +	USA ; n: 100; with AC; 1994-1998
Liu et al., 2004a	28 (3- 103)	China, Peking; n: 11; 2002/2003

*: Mean; +: geometric mean; #: geometric mean per building; DL: detection limit; S: smoker; NS: non smoker

Table 2. Concentrations of particulate matter in the indoor air in office buildings in μg/m^3

offices ranged between 7 - 51 µg/m³ and 30 - 63 µg/m³, respectively. Noticeably low values resulted from the most extensive examination in 100 buildings with air conditioning systems in the USA (Burton et al., 2000). By contrast, particularly high concentrations were observed in Greek offices (Gemenetzis et al., 2006). These are attributed to the high outdoor air concentrations and the presence of smokers.

Concerning the ultra fine particles, higher PNCs were observed in offices exposed to tobacco smoke than in outdoor air whereby they ranged between approximately 1,000 and 13,000 particles/cm³ in offices with air conditioning (Matson, 2005). In an Australian study a mean concentration of 6,500 particles/cm³ was measured during and 1,200 particles/cm³ after working hours (He et al., 2007) in an open plan office with ventilation and air conditioning system and smoking ban. The highest measured concentration amounted to 38,000 particles/cm³ in this study.

4.4 Particles in hospitality venues

An overview of the exposure in pubs, restaurants and similar venues is provided by Table 3. The worldwide studies all reach the conclusion that in venues in which smoking is permitted very high concentrations have to be expected. A German study in discos (n = 10) resulted, for instance, for PM_{10} in a median of 1,014 µg/m³ and for $PM_{2.5}$ of 869 µg/m³ (Bolte et al., 2008). In pubs (n = 18) the same working group measured medians of 210 µg/m³ (PM_{10}) and 195 µg/m³ ($PM_{2.5}$). Figure 9 shows, for instance, the $PM_{2.5}$ concentration time course in three venues which were examined during the above mentioned study.

Results about the development of indoor air pollution in bars, restaurants and similar venues after the introduction of smoking bans are available so far to a larger extent from the USA, Italy, Ireland, Scotland and Norway (summary in Fromme et al., 2009). Overall, it turned out that a considerable reduction of the $PM_{2.5}$ contents between 70 and 97%, mostly above 90%, can be achieved through the implementation of a consistent smoking ban in these venues alone.

On the other hand, the published results proved that through spatially not completely separated smoking areas in pubs and with ventilation systems no or only a low decrease in particle pollution is achieved. This is confirmed in a position paper by the American Society of Heating, Refrigerating and Air Conditioning Engineers which does not see ventilation systems as a useful instrument to protect from passive smoking in these venues (ASHRAE, 2005).

So far there are hardly any study results on the number of ultrafine particles. Milz et al., (2007) investigated 2 restaurants in two American cities. Whereas in non-smoker restaurants the mean contents amounted to ca. 15,000 particles/cm³, 82,000 particles/cm³ and ca. 106,000 particles/cm³ were observed in smoker rooms. Concerning ultrafine particles, smoker rooms result in a pollution of areas nearby in which smoking is banned. In Germany very high median PNCs of 221,100 particles/cm³ were measured in 4 cafés/ restaurants, 119,100 particles/cm³ in 2 bars and 289,900 particles/cm³ in 7 discos (Bolte et al., 2008).

Reference	Median (Min-Max)	Description
Europe		
Bohanon et al., 2003	194 (56- 312)+	Restaurants; France
	75 (0- 277) +	Restaurants; Schwitzerland
	201 (62- 391) +	Restaurants; UK
Gee et al., 2006	94*+	59 pubs; England; 2001
Edwards et al., 2006a	167 (54- 1395)	33 pubs, with cooking; UK; 2004
	217 (15- 1227)	31 Pubs; no cooking; UK; 2004
Goodman et al., 2007	35,5 *	42 pubs; Ireland; 2004/2005
Valente et al., 2007	119	40 locations; Italy; 2005
Schneider et al., 2008	173 (22- 831) 131 (24- 1029) 378 (144- 2022)	38 restaurants; Germany; 2005 20 cafes; Germany; 2005 11 bars; Germany; 2005
Bolte et al., 2008	164 (55- 570)	11 restaurants, cafes; Germany; 2005/2006
	203 (103- 1250)	7 pubs and bars; Germany; 2005/2006
	869 (291- 4475)	10 discotheques; Germany; 2005/2006
Vardavas et al., 2007	268 (19- 612)*	31 bars, pubs, cafes, clubs; Greece; 2006
Semple et al., 2010	197 (8- 902)	42 bars; Scotland; 2006
	92 (5- 1005)	52 bars; England; 2007
	184 (16- 872)	12 bars; Wales; 2007
Rosen et al., 2007	465 (66- 862)+	6 bars, pubs; Israel; 2007
	52 (18- 557)+	8 cafes; Israel; 2007
Daly et al., 2011	83 (51- 108)*	70 bars, cafes, restaurants; Switzerland; 2008
America, Australia		
Maskarinec et al., 2000	66 (0- 233) +	Restaurants; USA; 1996/1997
	82 (0- 768) +	Bars; USA; 1996/1997
Brauer et al., 2000	(11- 163)	11 restaurants; Canada
	(47- 253)	4 bars; Canada
Repace et al., 2006	178 (43- 323)	6 pubs; USA; 2003
Connolly et al., 2005	206 (23- 727)	28 locations, USA; 2005
Brennan et al., 2010	61 (6- 338)	19 pubs; Australia; 2007
Jiang et al., 2011	63 (18- 183)	36 casinos, USA; 2008
Asia		
Baek et al., 1997	159 (33 - 475)+	6 restaurants; Korea; 1994/1995
Lee et al., 1999	400 - 1760	3 restaurants; China; 1996/1997
Bohanon et al., 2003	194 (0- 611) +	Restaurants; Japan
	107 (54- 172) +	Restaurants; Korea
Lee et al., 2010	92 (17- 565)*	55 restaurants; 7 countries; 2008/2009
	114 (14- 565)*	35 cafes; 7 countries; 2008/2009
	191 (33- 748)*	34 bars, clubs; 7 countries; 2008/2009
	169 (4- 881)*	44 entertainment venues; 7 countries; 2008/2009

*: mean; +: PM_4 or respirable particulate matter (RPM)

Table 3. Concentrations of particulate matter ($PM_{2.5}$) in indoor air of hospitality venues in μg/m^3

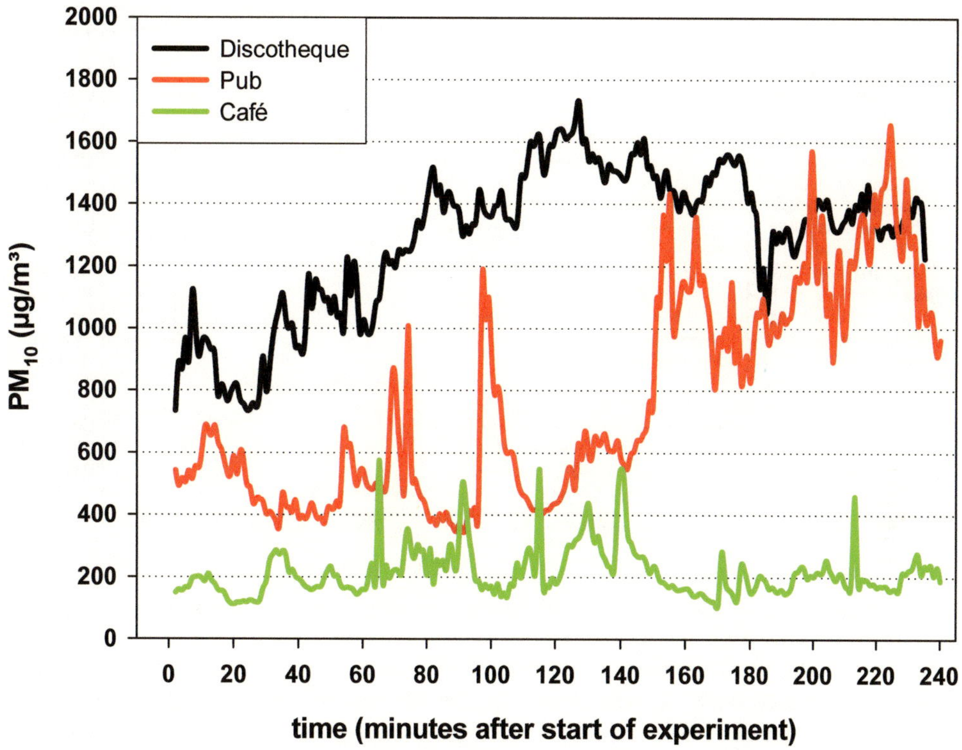

Fig. 9. Time course of $PM_{2.5}$ in three hospitality venues in Germany (modified from Bolte et al., 2008)

4.5 Particles in transportation systems

4.5.1 PM in aboveground transportation systems

The contents of $PM_{2.5}$ in above ground buses and cars are shown in Figure 10. The highest contents in cars and buses were observed in Asian cities; merely in one study in Mexico City, in Boston (Levy et al., 2002) and in Peru (Han et al., 2005) similar high concentrations were described. The other studies, in particular in Europe and Australia, refer, by contrast, to a mean exposure level for $PM_{2.5}$ of approximately 10 - 40 µg/m³; as a rule the concentrations are significantly higher indoors than in the ambient air. There was a dependency of the indoor air contents on the outdoor levels, the time of day and the day of week (e.g. Lee et al., 2010).

Table 4 shows the results of the measurements of ultrafine particles in cars and buses. The mean PNC ranges between 10,000 and 50,000 particles/cm³. By contrast, very high contents were described by Kaur et al. (2005) in the City of London which ranged on average between 90,000 and 100,000 particles/cm³. During a drive on the freeway with an open window high concentrations were likewise determined (Eiguren-Fernandez et al., 2005). Under special conditions, eg a diesel truck ahead, short term peak concentrations of up to 500,000 particles/cm³ were observed (Abraham et al., 2002; Eiguren-Fernandez et al., 2005).

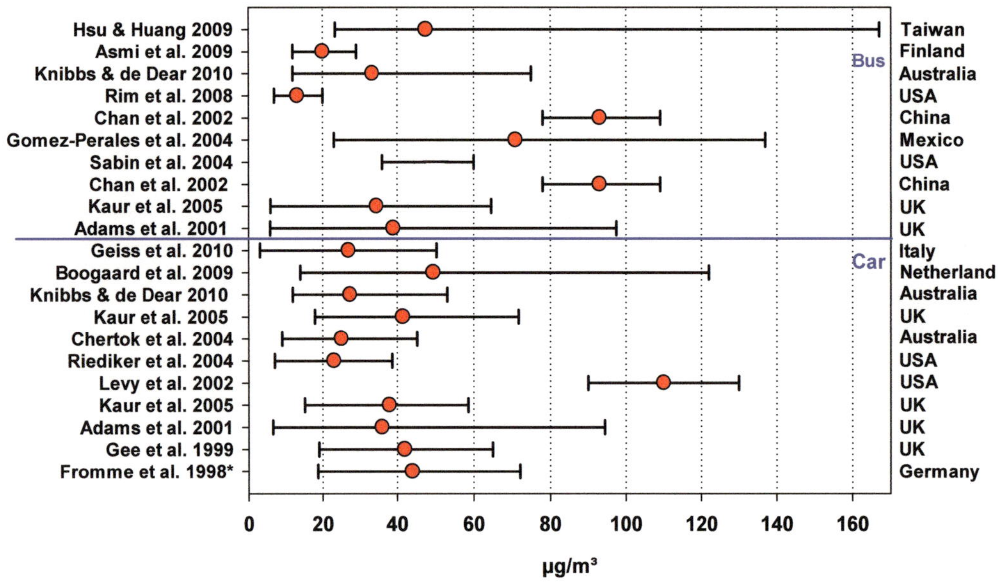

Fig. 10. Concentrations of $PM_{2.5}$ in cars and buses (*: PM_4) (minimum, median, maximum)

Overall, the exposure in the indoor environment of transportation systems is influenced by many different factors such as ventilation, quality of driving, traffic volume and traffic composition, built-up area and meteorology. Car passengers seem to be exposed to a slightly higher $PM_{2.5}$ and PNCs than cyclists (Adams et al., 2001; Kaur et al., 2005), whereas they are stated as the same or only slightly different in other studies (Gulliver & Briggs, 2004; Boogaard et al., 2009; Zuurbier et al., 2010; Int Panis et al., 2010). In this connection it must, however, be taken into account that due to the approximately 2 - 4.5 fold higher breathing volume cyclists have a significantly higher inhaled dose compared to car and bus users (Zuurbier et al., 2010; Int Panis et al., 2010).

In addition, air conditioning and filter systems, which can have a major influence on the contents in the indoor environment of transportation systems depending on the quality and the separation level, must be taken into account (Rim et al., 2008). Studies in an urban environment also showed a higher exposure for pedestrians than during car use or compared to the general outdoor air pollution (Kaur et al., 2005).

A special pollution situation results from passive smoking exposure. During a drive in Wellington with a fully opened side window mean $PM_{2.5}$ contents of 169 µg/m³ (maximum: 217 µg/m³) were measured while smoking a cigarette (Edwards et al., 2006b). With a closed window the mean values were 2,962 µg/m³ (maximum: 3,645 µg/m³) and in a Canadian study mean $PM_{2.5}$ contents between 790 and 4,626 µg/m³ (maximum: 7,635 µg/m³) were determined (Sendzik et al., 2006). Rees & Connolly (2006) determined in 45 measurements during smoking with closed windows mean $PM_{2.5}$ values of 271 µg/m³ (maximum: approximately 500 µg/m³) and with opened side windows approximately 50 µg/m³ (maximum: approximately 100 µg/m³). Liu & Zhu (2010) observed inside cars the tenfold PNC and 120-fold $PM_{2.5}$ contents compared to outdoor air.

Reference	Mean (Min-Max)		Description
Europe			
Dennekamp et al., 2002	53 (-) [a]	bus	Aberdeen, UK; n: 11
Mackay, 2004	44 (10-143) 58 (8-282)	bus car	Leeds, UK
Krausse & Mardaljevic, 2005	- (46-116)		Leicester, UK; n:133
Kaur et al., 2005	101 (65-159) 100 (37-152) 88 (52-114)	bus car taxi	London, UK; 2003
Diapouli et al., 2007	94 (25-217)	car	Athens, Greece; through city
Geiss et al., 2010	16 (8-30)	car	Italy; 18 cars; 2009
North America, Australia			
Abraham et al., 2002	30 [b] (4-190)		New York, USA; 3 city routes
Levy et al., 2002	~32 (12-80) ~39 (11-83)	bus car	Boston, USA; 2000
Eiguren-Fernandez et al., 2005	25 (X, AC) 55 (X, nAC) 69 (Y, AC) 246 (Y,nAC)	car	Los Angeles, USA, car; AC: air condition, nAC: windows open; X: small streets; Y: freeway
Rim et al., 2008	6-35	bus	Austin; USA; 6 busses; 2006
Wallace & Ott, 2011	29-34 (-)	car	USA; 17 trips; 2005-2009
Knibbs & de Dear, 2010	11 9	bus car	Sydney; Australia; 40 trips; 2004
Zhang & Zhu, 2010	7.3-34	bus	Texas; USA; school buses; 2008

a: median; b: mean of three cycles

Table 4. Particle number concentrations (PNC) in indoor air of transportation systems (10^3 of particles/cm^3)

4.5.2 PM in underground transportation systems

Studies in this micro environment show that the exposure is significantly above the values measured in above ground transportation systems. In the Berlin underground mean PM_4 contents of 141 (124-169 μg/m^3) were measured in winter and 153 μg/m^3 (121-176 μg/m^3) in summer (Fromme et al., 1998). Similar results were obtained in the London underground with mean $PM_{2.5}$ concentrations of 247 μg/m^3 (105 - 371 μg/m^3) (summer) and 157 μg/m^3 (12 - 263 μg/m^3) (winter) (Adams et al., 2001). In a more recent study conducted in London mean values of 180 - 200 μg/m^3 ($PM_{2.5}$) were measured (Hurley et al., 2004). By contrast, significantly lower mean pollutions were observed in Boston (70 μg/m^3 $PM_{2.5}$), Los Angeles (13.7 μg/m^3) and Helsinki (21 μg/m^3) (Levy et al., 2002; Aarnio et al., 2005; Kam et al., 2011). In the underground of Mexico City the mean values amounted to 61 μg/m^3 (31 - 99 μg/m^3)

(Gómez-Perales et al., 2004). Measurements in the Seoul underground and in a Chinese city, resulted in concentrations of 148 µg/m³ (Sohn et al., 2005) and 67 µg/m³ (26-123 µg/m³), respectively (Chan et al., 2002).

In the Berlin underground stations the PM_4 contents ranged between 128 and 311 µg/m³ during operation (Fromme et al., 1998), and in London the average $PM_{2.5}$ contents were 270 - 480 µg/m³ (Hurley et al., 2004). In Boston 130 µg/m³ were measured at the underground station (Levy et al., 2002). On the other hand the contents in Taipei were on average only 25-40 µg/m³ (Cheng & Yan, 2011) and in Paris (Raut et al., 2009) the contents were 61 µg/m³ (normal hours) and 93 µg/m³ (rush hours) correspondingly lower.

At present only a few measurements on ultrafine particles are available. In three underground lines in London 17,000 - 23,000 particles/cm³ (>50 nm) were measured on average whereas on the platforms of the three underground stations the average contents determined were 14,000 - 29,000 particles/cm³ (Hurley et al., 2004). Similar results were reached by measurements at the underground of Boston with mean values of approximately 21,000 particles/cm³ (Levy et al., 2002) and in Helsinki, with 27,000 particles/cm³ (14,000-50,000 Pt./cm³) (Aarnio et al., 2005).

5. References

Aarnio, P.; Yli-Tuomi, T.; Kousa, A.; Mäakelä, T.; Hirsikko, A.; Hämeri, K.; Räisänen, M.; Hillamo, R.; Koskentalo, T. & Jantunen, M. (2005). The concentrations and composition of and exposure to fine particles ($PM_{2.5}$) in the Helsinki subway system. *Atmos Environ*, 39, 5059–5066.

Abraham, J.L.; Siwinski, G. & Hunt, A. (2002). Ultrafine particulate exposures in indoor, outdoor, personal and mobile environments: effects of diesel, traffic, pottery kiln, cooking and HEPA filtration on micro-environmental particle number concentration. *Ann Occup Hyg*, 46 (suppl 1), 406-411.

Abt, E.; Suh, H.H.; Allen, G. & Koutrakis, P. (2000). Characterization of indoor particle sources: a study conducted in the metropolitan Boston area. *Environ Health Perspect*, 108, 35-44.

Adams, H.S.; Nieuwenhuijsen, M.J.; Colvile, R.N.; McMullen, M.A.S. & Khandelwal, P. (2001). Fine particle ($PM_{2.5}$) personal exposure levels in transport microenvironments, London, UK. *Sci Total Environ*, 279, 29-44.

Afshari, A.; Matson, U. & Ekberg, LE. (2005). Characterization of indoor sources of fine and ultrafine particles: a study conducted in a full-scale chamber. *Indoor Air*, 15, 141-150.

Almeida, S.M.; Canha, N.; Silva, A.; do Carmo Freitas, M.; Pegas, P.; Alves, C.; Evtyugina, M. & Pio, C.A. (2010). Children exposure to atmospheric particles in indoor of Lisbon primary schools. *Atmos Environ*, corrected proof, doi:10.1016/j.atmosenv.2010.11.052.

Aoki, T. & Tanabe, S. (2007). Generation of sub-micron particles and secondary pollutants from building materials by ozone reaction. *Atmos Environ*, 41, 3139-3150.

ASHRAE (American Society of Heating, Refrigerating and Air-Conditioning Engineers) (2005). Environmental Tobacco Smoke -Position Document Approved by ASHRAE Board of Directors. June 30, 2005, online: www.ashrae.org.

Asmi, E.; Antola, M.; Yli-Tuomi, T.; Jantunen, M.; Aarnio, P.; Mäkelä, P.; Hillamo, R. & Hämeri, K. (2009). Driver and passenger exposure to aerosol particles in buses and trams in Helsinki, Finland. *Sci Total Environ*, 2860–2867.

Baxter, L.K.; Clougherty, J.E.; Laden, F. & Levy, J.I. (2007). Predictors of concentrations of nitrogen dioxide, fine particulate matter, and particle constituents inside lower cocioeconimic status urban homes. *J Expos Sci Environ Epidemiol*, 17, 433-444.

Birmili, W.; Kinnersley, R.P. & Baker, J. (2003). Factors influencing human exoposure to fine and ultrafine particles in a city centre office. Abstract presented at the European Aerosol Conference.

Boogaard, H.; Borgman, F.; Kamminga, J. & Hoek, G. (2009). Exposure to ultrafine and fine particles and noise during cycling and driving in 11 Dutch cities. *Atmos Environ*, 43, 4234–4242.

Bohanon, H.R.; Piadé, J.J.; Schorp, M.K. & Saint-Jalm, Y. (2003). An international survey of indoor air quality, ventilation, and smoking activity in restaurants: a pilot study. *J Expo Anal Environ Epidemiol*, 13, 378-392.

Bolte, G.; Heitmann, D.; Kiranoglu, M.; Schierl, R.; Diemer, J.; Körner, W. & Fromme, H. (2008). Exposure to environmental tobacco smoke in German restaurants, pubs and discotheques. *J Expo Sci Environ Epidemiol*, 18, 262-271.

Brauer, M.; Hirtle, R.; Lang, B. & Ott, W. (2000). Assessment of indoor fine aerosol contributions from environmental tobacco smoke and cooking with a portable nephelometer. *J Expo Anal Environ Epidemiol*, 10, 136-144.

Brennan, E.; Cameron, M.; Warne, C.; Durkin, S.; Borland, R.; Travers, M.J.; Hyland, A. & Wakefield, M.A. (2010) Secondhand smoke drift: Examining the influence of indoor smoking bans on indoor and outdoor air quality at pubs and bars. *Nicotine & Tobacco Res*, 12, 271–277.

Breysse, P.N.; Buckley, T.J.; Williams, D.A.; Beck, C.M.; Jo, S.J.; Merriman, B.; Kanchanaraksa, S.; Swartz, L.J.; Callahan, K.A.; Butz, A.M.; Rand, C.S.; Diette, G.B.; Krishnan, J.A.; Moseley, A.M.; Curtin-Brosnan, J.; Durkin, N.B. & Eggleston, P.A. (2005). Indoor exposures to air pollutants and allergens in the homes of asthmatic children in inner-city Baltimore. *Environ Res*, 98, 167-176.

Burton, L.E.; Girman, J.G. & Womble, S.E. (2000). Airborne particulate matter within 100 randomily selected office buildings in the united states (BASE), *Proceedings of Healthy Buildings*, 1, 157-162.

Byun, H.; Bae, H.; Kim, D.; Shin, H. & Yoon, C. (2010). Effects of socioeconomic factors and human activities on children's PM_{10} exposure in inner-city households in Korea. *Int Arch Occup Environ Health*, 83, 867–878.

Cattaneo, A.; Peruzzo, C.; Garramone, G.; Urso, P.; Ruggeri, R.; Carrer, P. & Cavallo, D.M. (2011) Airborne particulate matter and gaseous air pollutants in residential structures in Lodi Province, Italy. *Indoor Air*, accepted.

Chan, L.Y.; Lau, W.L.; Lee, S.C. & Chan, C.Y. (2002). Commuter exposure to particulate matter in public transportation modes in Hong Kong. *Atmos Environ*, 36, 3363-3373.

Chao, Y.H.; Tung, T.C.W. & Burnett, J. (1998). Influence of different indoor activities on the indoor particulate levels in residential buildings. *Indoor Built Environ*, 7, 110-121.

Chao, C.Y. & Wong, K.K. (2002). Residential indoor PM_{10} and $PM_{2.5}$ in Hong Kong and the elemental composition. *Atmos Environ*, 36, 265-277.

Chen, C. & Zhao, B. (2011). Review of relationship between indoor and outdoor particles: I/O ratio, infiltration factor and penetration factor. *Atmos Environ*, 45, 275-288.

Cheng, YH. & Yan, J.W. (2011). Comparisons of particulate matter, CO, and CO_2 levels in underground and ground-level stations in the Taipei mass rapid transit system. *Atmos Environ*, 45, 4882-4891.

Chertok, M.; Voukelatos, A.; Sheppeard, V. & Rissel, C. (2004). Comparison of personal exposures to air pollutants by commuting mode in Sydney. BTEX & NO_2. Report prepared for NSW Department of Health, Australia.

Connolly, G.N.; Carpenter, C.; Alpert, H.R.; Skeer, M. & Travers, M. (2005). Evaluation of the Massachusetts smoke-free workplace law. A preliminary report of the Division of Public Health Practice, Harvard School of Public Health, Tobacco Research Program. Online: http://www.hsph.harvard.edu/php/pri/tcrtp/Smoke-free_Workplace.pdf.

Cyrys, J.; Pitz, M.; Bischof, W.; Wichmann, H.E. & Heinrich, J. (2004). Relationship between indoor and outdoor levels of fine particle mass, particle number concentrations and black smoke under different ventilation conditions. *J Expo Anal Environ Epidemiol*, 14, 275-284.

Daly, B.J.; Schmid, K. & Riediker, M. (2011). Contribution of fine particulate matter sources to indoor exposure in bars, restaurants, and cafes. *Indoor Air*, 20, 204–212.

Dennekamp, M.; Howarth, S.; Dick, C.A.J.; Cherrie, W.; Donaldson, K. & Seaton, A. (2001). Ultrafine particles and nitrogen oxides generated by gas and electric cooking. *Occup Environ Med*, 58, 511-516.

Dennekamp, M.; Mehenni, O.; Cherrie, J. & Seaton, A. (2002). Exposure to ultrafine particles and $PM_{2.5}$ in different microenvironments. *Ann Occup Hyg*, 46 (suppl. 1), 412-414.

Dermentzoglou, M.; Manoli, E.; Voutas, D. & Samara, C. (2003). Sources and patterns of polycyclic aromatic hydrocarbons and heavy metals in fine indoor particulate matter of Greek houses. *Fresenius Environ Bull*, 12, 1511-1519.

Destaillats, H.; Lunden, M.M.; Singer, B.C.; Coleman, B.K.; Hodgson, A.T.; Weschler, C.J. & Nazaroff, W.W. (2006). Indoor secondary pollutants from household product emissions in the presence of ozone: a bench-scale chamber study. *Environ Sci Technol*, 40, 4421-4428.

Diapouli, E.; Chaloulakou, A.; Mihalopoulos, N. & Spyrellis, N. (2008). Indoor and outdoor PM mass and number concentrations at schools in the Athens area. *Environ Monit Assess*, 136, 13-20.

Edwards, R.; Hasselholdt, C.P.; Hargreaves, K.; Probert, C.; Holford, R.; Hart, J.; Van Tongeren, M. & Watson, A.F.R. (2006a). Levels of second hand smoke in pubs and bars by deprivation and food-serving status: a cross-sectional study from North West England. *BMC Public Health*, 6, 42

Edwards, R.; Wilson, N. & Pierse, N. (2006b). Highly hazardous air quality associated with smoking in cars: New Zealand pilot study. NZMJ, 119, 27 October 2006.

Eiguren-Fernandez, A.; Miguel, A.H.; Zhu, Y.F. & Hering, S.V. (2005). In-cabin passenger exposure to ultrafine and nano-particles during daily commute in Los Angeles roads and freeways: evaluation of a HEPA filtration system. *Proceedings: Indoor Air*, 1763-1767. Peking.

Fan, Z.H.; Lioy, P.; Weschler, C.J.; Fiedler, N.; Kipen, H. & Zhang, J.F. (2003). Ozone-initiated reactions with mixtures of volatile orghanic compounds under simulated indoor conditions. *Environ Sci Technol*, 37, 1811-1821.

Ferro, A.R.; Kopperud, R.J. & Hildemann, L.M. (2004). Elevated personal exposure to particulate matter from human activities in a residence. *J Expo Anal Environ Epidemiol*, 14, S34-S40.

Fine, P.M.; Cass, G.R. & Simoneit, B.R.T. (1999). Characterization of fine particle emissions from burning church candles. *Environ Sci Technol*, 33, 2352-2362.

Franck, U.; Herbarth, O.; Wehner, B.; Wiedensohler, A. & Manjarrez, M. (2003). How do the indoor size distributions of airborne submicron and ultrafine particles in the absence of significant indoor sources depend on outdoor distributions? *Indoor Air*, 13, 174-181.

Franck, U.; Herbarth, O.; Röder, S.; Schlink, U.; Borte, M.; Diez, U.; Krämer, U. & Lehmann, I. (2011). Respiratory effects of indoor particles in young children are size dependent. *Sci Total Environ*, 409, 1621–1631.

Fromme, H.; Oddoy, A.; Lahrz, T.; Krause, M. & Piloty, M. (1998). Polycyclic aromatic hydrocarbons (PAH) and diesel engine emission (elemental carbon) inside a car and a subway train. *Sci Total Environ*, 217, 165-173.

Fromme, H.; Lahrz, T.; Hainsch, A.; Oddoy, A.; Piloty, M. & Rüden, H. (2005). Elemental carbon and respirable particulate matter in the indoor air of apartments and nursery schools and outdoor air in Berlin (Germany). *Indoor Air*, 15, 335-341.

Fromme, H.; Twardella, D.; Dietrich, S.; Heitmann, D.; Schierl, R.; Liebl, B. & Rüden, H. (2007). Particulate matter in the indoor air of classrooms - exploratory results from Munich and surrounding. *Atmos Environ*, 41, 854-866.

Fromme, H.; Diemer, J.; Dietrich, S.; Cyrys, J.; Heinrich, J.; Lang, W.; Kiranoglu, M. & Twardella, D. (2008). Chemical and morphological properties of particulate matter (PM_{10}, $PM_{2.5}$) from indoor of schools and outdoor air. *Atmos Environ*, 42, 6597–6605.

Fromme, H.; Kuhn, J. & Bolte, G. (2009). Secondhand smoke in hospitality venues. Exposure, body burden, economic and health aspects in conjugation with smoking bans. *Das Gesundheitswesen*, 71, 242-257 [in German].

Gabrio, T.; Volland, G.; Baumeister, I.; Bendak, J.; Flicker-Klein, A.; Gickeleiter, M.; Kersting, G.; Maisner, V. & Zöllner, I. (2007). Messung von Feinstäuben in Innenräumen. *Gefahrstoffe - Reinhaltung der Luft*, 67, 96-102.

Gee, I.L. & Raper, D.W. (1999). Commuter exposure to respirable particles inside buses and by bicycle. *Sci Total Environ*, 235, 403-405.

Gee, I.L; Watson, A.F.; Carrington, J.; Edwards, P.R.; van Tongeren, M.; McElduff, P. & Edwards, R.E. (2006). Second-hand smoke levels in UK pubs and bars: do the English Public Health White Paper proposals go far enough? *J Public Health*, 28, 17-23.

Geiss, O.; Barrero-Moreno, J.; Tirendi, S. & Kotzias, D. (2010). Exposure to particulate matter in vehicle abins of private cars. *Aerosol Air Qual Res*, 10, 581–588.

Gemenetzis, P.; Moussas, P. & Arditsoglou, (2006). A. Mass concentration and elemental composition of indoor $PM_{2,5}$ and PM_{10} in University rooms in Thessaloniki, northern Greece. *Atmos Environ*, 40, 3195-3206.

Gomez-Perales, J.E.; Colvile, R.N.; Nieuwenhuijsen, M.J.; Fernandez-Bremauntz, A.; Gutierrez-Avedoy, V.J.; Paramo-Figueroa, V.H.; Blanco-Jimenez, S.; Bueno-Lopez, E.; Mandujano, F.; Bernabe-Cabanillas, R. & Ortiz-Segovia, E. (2004). Commuters` exposure to $PM_{2.5}$, CO, and benzene in public transport in the metropolitan area of Mexico City. *Atmos Environ*, 38, 1219-1229.

Goodman, P.; Agnew, M.; McCaffrey, M.; Paul, G. & Clancy, L. (2007). Effects of the Irish smoking ban on respiratory health of bar workers and air quality in Dublin pubs. *Am J Respir Crit Care Med*, 175, 840-845.

Gulliver, J. & Briggs, D.J. (2004). Personal exposure to particulate air pollution in transport microenvironments. *Atmos Environ*, 38, 1-8.

Guo, H.; Morawska, L.; He, C.; Zhang, Y.L.; Ayoko, G. & Cao, M. (2010). Characterization of particle number concentrations and $PM_{2.5}$ in a school: influence of outdoor air pollution on indoor air. *Environ Sci Pollut Res*, 17, 1268–1278.

Han, X.; Aguilar-Villalobos, M.; Allen, J.; Charlton, C.S.; Ronbinson, R.; Bayer, C. & Naeher, L. (2005). Traffic-related occupational exposures to $PM_{2.5}$, CO and VOCs in Trujillo, Peru. *Int J Occup Environ Health*, 11, 276-288.

Hänninen, O.O.; Lebret, E.; Ilacqua, V.; Katsouyanni, K.; Künzli, N.; Sráme, R.J. & Jantunen, M. (2004). Infiltration of ambient $PM_{2.5}$ and levels of indoor generated non-ETS $PM_{2.5}$ in residences of four European cities. *Atmos Environ*, 38, 6411-6423.

He, C., Morawska, L.; Hitchins, J. & Gilbert, D. (2004). Contribution from indoor sources to particle number and mass concentrations in residential houses. *Atmos Environ*, 38, 3405-3415.

He, C.; Morawska, L. & Tablin, L. (2007). Particle emission characteristics of office printers. *Environ Sci Technol*, 41, 6039-6045.

Heavner, D.L.; Morgan, W.T. & Ogden, M.W. (1996). Determination of volatile organic compounds and respirable suspended particulate matter in New Jersey an Pennsylvania homes and workplaces, *Environ Int*, 22, 159-183.

Héroux, M.E.; Clark, N.; Van Ryswyk, K.; Mallick, R.; Gilbert, N.L.; Harrison, I.; Rispler, K.; Wang, D.; Anastassopoulos, A.; Guay, M.; MacNeill, M. & Wheeler, A.J. (2010). Predictors of indoor air concentrations in smoking and non-smoking residences. *Int J Environ Res Public Health*, 7, 3080-3099.

HESE (Health Effects of School Environment) (2006). Final Scientific Report prepared for the Health & Consumer Protection Directorate General. Siena, Italy. Online: http://ec.europa.eu/health/ph_projects/2002/pollution/fp_pollution_2002_frep_04.pdf

Horemans, B.; Worobiec, A.; Buczynska, A.; Van Meel, K. & Van Grieken, R. (2008). Airborne particulate matter and BTEX in office environments. *J Environ Monit*, 10, 867–876.

Hsu, D.J. & Huang, H.L. (2009). Concentrations of volatile organic compounds, carbon monoxide, carbon dioxide and particulate matter in buses on highways in Taiwan. *Atmos Environ*, 43, 5723–5730.

Hu, B.; Freihaut, J.D.; Bahnfleth, W.; Gomes, C.A.S. & Brandolyn. T. (2005). Literatur review and parametric study: indoor particle resuspension by human activity. *Proceedings: Indoor Air*, 1541-1545.

Hurley, F.; Cherrie, J.; Donaldson, K.; Seaton, A. & Tran, L. (2004). Assessment of health effects of long-term occupational exposure to tunnel dust in the London underground. Universitiy of Aberdeen. Research report TM/02/04.

Hussein, T.; Glytsos, T.; Ondrácek, J.; Dohányosová, P.; Zdimal, V.; Hämeri, K.; Lazaridis, M.; Smolik, J. & Kulmala, M. (2006). Particle size characterization and emission rates during indoor activities in a home. *Atmos Environ*, 40, 4285-4307.

Hussein, T.; Hruska, A.; Dohanyosová, P.; Dzumbová, L.; Hemerka, J.; Kulmala, M.; Smolík, J. (2009). Deposition rates on smooth surfaces and coagulation of aerosol particles inside a test chamber. *Atmos Environ*, 43, 905-914.

Int Panis, L.; de Geus, B.; Vandenbulcke, G.; Willems, H.; Degraeuwe, B.; Bleux, N.; Mishra, V.; Thomas, I. & Meeusen, R. (2010). Exposure to particulate matter in traffic: A comparison of cyclists and car passengers. *Atmos Environ*, 44, 2263-2270.

Janssen, N.A.H.; van Vliet, P.H.N.; Aaarts, F.; Harssema, H. & Brunekreef, B. (2001). Assessment of exposure to traffic related air pollution of children attending schools near motorways. *Atmos Environ*, 35, 3875-3884.

Jetter, J.J.; Guo, Z.; McBrian, J.A. & Flynn, M.R. (2002). Characterization of emissions from burning incense. *Sci Total Environ*, 295, 51-67.

Jiang, R.T.; Cheng, K.C.; Acevedo-Bolton; V.; Klepeis, N.E.; Repace, J.L.; Ott, W.R. & Hildemann, L.M. (2011). Measurement of fine particles and smoking activity in a statewide survey of 36 California Indian casinos. *J Expo Sci Environ Epidemiol*, 21, 31-41.

Jung, K.H.; Patel, M.M.; Moors, K.; Kinney, P.L.; Chillrud, S.N.; Whyatt, R.; Hoepner, L.; Garfinkel, R.; Yan, B.; Ross, J.; Camann, D.; Perera, F.P. & Miller, R.L. (2010). Effects of heating season on residential indoor and outdoor polycyclic aromatic hydrocarbons, black carbon, and particulate matter in an urban birth cohort. *Atmos Environ*, 44, 4545-4552,

Kam, W.; Cheung, K.; Daher, N. & Sioutas, C. (2011). Particulate matter (PM) concentrations in underground and ground-level rail systems of the Los Angeles Metro. *Atmos Environ*, 45, 1506-1516.

Kaur, S.; Nieuwenhuijsen, M. & Colvile, R. (2005). Personal exposure of street canyon intersection users to $PM_{2.5}$, ultrafine particle counts and carbon monoxide in central London, UK. *Atmos Environ*, 39, 3629-3641.

Kearney, J.; Wallace, L.; MacNeill, M.;Xuc, X.; Van Ryswyk, K.; Youa, H.; Kulka, R. & Wheeler, A.J. (2010). Residential indoor and outdoor ultrafine particles in Windsor, Ontario. *Atmos Environ*, 45, 7583-7593.

Knibbs, L.D. & de Dear, R.J. (2011). Exposure to ultrafine particles and $PM_{2.5}$ in four Sydney transport modes. *Atmos Environ*, 44, 3224-3227.

Krausse, B. & Mardaljevic, J. (2005). Patterns of driver´s exposure to particulate matter. In: Williams K (Ed.). Spatial planning, urban form and sustainable transport. Ashgate, Aldershot, UK.

Lahrz, T.; Piloty, M.; Pfeiler, P. & Honigmann, I. (2002). Messungen von Schadstoffen an Berliner Büroarbeitsplätzen, Bericht des Institutes für Lebensmittel, Arzneimittel und Tierseuchen, Fachbereich Umwelt- und Gesundheitsschutz, Berlin.

Lahrz, T.; Piloty, M.; Oddoy, A. & Fromme, H. (2003). Gesundheitlich bedenkliche Substanzen in öffentlichen Einrichtungen in Berlin. Untersuchungen zur Innenraumluftqualität in Berliner Schulen. Bericht des Instituts für Lebensmittel, Arzneimittel und Tierseuchen, Fachbereich Umwelt- und Gesundheitsschutz. Berlin.

Lai, H.K.; Kendall, M.; Ferrier, H.; Lindup, I.; Alm. S.; Hänninen, O.; Jantunen, M.; Mathys, P.; Colvile, R.; Ashmore, M.R.; Cullinan, P. & Nieuwenhuijsen, M.J. (2004). Personal exposures and microenvironment concentrations of $PM_{2.5}$, VOC, NO_2 and CO in Oxford, UK. *Atmos Environ*, 38, 6399-6410.

Lee, J.; Lim, S.; Lee, K.; Guo, X.; Kamath, R.; Yamato H.; Abas, L.E.; Nandasena, S.; Nafees, A.A. & Sathiakumar, N. (2010). Secondhand smoke exposures in indoor public places in seven Asian countries. International *J Hyg Environ Health,* 213, 348–351.

Levy, J.I.; Dumyahn, T. & Sprengler, J.D. (2002). Particulate matter and polycyclic hydrocarbon concentrations in indoor and outdoor microenvironments in Boston, Massachusetts. *J Expo Anal Environ Epidemiol,* 12, 104-14.

Li, S.C. & Lin, C.H. (2003). Carbon profile of residential indoor PM_1 and $PM_{2.5}$ in the subtropical region. *Atmos Environ,* 37, 881-888.

Link, B.; Gabrio, T.; Zöllner, I.; Schwenk, M.; Siegel, D.; Schultz, E.; Scharring, S. & Borm, P. (2004). Feinstaubbelastung und deren gesundheitliche Wirkungen bei Kindern. Bericht des Landesgesundheitsamtes Baden-Württemberg.

Lim, J.M.; Jeong, J.H.; Lee, J.H.; Moon, J.H.; Chung, Y.S. & Kim, K.H. (2011). The analysis of $PM_{2.5}$ and associated elements and their indoor/outdoor pollution status in an urban area. *Indoor Air,* 21, 145–155.

Liu, D.L. & Nazaroff, W.W. (2003). Particle penetration through building cracks. *Aerosol Sci Technol,* 37, 565-573.

Liu, Y.; Chen, R.; Shen, X. & Mao, X. (2004a). Wintertime indoor air levels of PM_{10}, $PM_{2.5}$ and PM_1 at public places and their contributions to TSP. *Environ Int,* 30, 189-197.

Liu, X.Y.; Mason, M.; Krebs, K. & Sparks, L. (2004b). Full-scale chamber investigation and simulation of air freshener emissions in the presence of ozone. *Environ Sci Technol,* 38, 2802-2812.

Liu, S. & Zhu, Y. (2010). A case study of exposure to ultrafine particles from secondhand tobacco smoke in an automobile. *Indoor Air,* 20, 412-423.

Long, C.M.; Suh, H.H. & Koutrakis, P. (2000). Characterization of indoor particle sources using continuous mass and size monitors. *J Air & Waste Manage Assoc,* 50, 1236-1250.

Long, C.M.; Suh, H.H.; Catalano, P.J. & Koutrakis, P. (2001). Using time- and size-resolved particulate data to quantify indoor penetration and deposition behavior. *Environ Sci Technol,* 35, 2089-2099.

Mackay, E. (2004). An investigation of the variation in personal exposure to carbon monoxide and particulates on the A660 in Leeds. M.Sc. thesis, University of Leeds.

Martuzevicius, D.; Grinshpun, S.A.; Lee, T.; Hu, S.; Biswas, P.; Reponen, T. & LeMasters, G. (2008). Traffic-related $PM_{2.5}$ aerosol in residential houses located near major highways: indoor versus outdoor concentrations. *Atmos Environ,* 42, 6575-6585.

Maskarinec, M.P.; Jenkins, R.A.; Counts, R.W. & Dindal, A.B. (2000). Determination of exposure to environmental tobacco smoke in restaurant and tavern workers in one US city. *J Expo Anal Environ Epidemiol,* 10, 36-49.

Matson, U. (2005). Indoor and outdoor concentrations of ultrafine particles in some Scandinavian rural and urban areas. *Sci Total Environ,* 343, 169-176.

Matson, U. & Ekberg, L.E. (2005). Prediction of ultrafine particle concentrations in various indoor enviroments. *Proceedings of Indoor Air,* 1581-1585.

McLaughlin, J.; Hogg, C. & Guo, L.Y. (2005). Ultrafine and coarse mode aerosol measurements in selected dwellings in Ireland. *Proceedings Indoor Air,* 698-701.

Meng, Q.J.; Turpin, B.J.; Korn. L.; Weisel, C.P.; Morandi, M.; Colome, S.; Zhang, J.; Stock, T.; Spektor, D.; Winer, A.; Zhang, L.; Lee, J.H.; Giovanetti, R.; Cui, W.; Kwon, J.; Alimokhtari, S.; Shendell, D.; Jones, J.; Farrar, C. & Maberti, S. (2005). Influence of ambient (outdoor) sources on residential indoor and personal $PM_{2.5}$ concentrations: Analyses of RIOPA data. *J Expo Anal Environ Epidemiol*, 15, 17-28.

Miguel, A.F.; Aydin, M. & Reis, A.H. (2005). Indoor deposition and forced re-suspension of respirable particles. *Indoor Built Environ*, 14, 391-396.

Milz, S.; Akbar-Khanzadeh, F.; Ames, A.; Spino, S.; Tex, C. & Lanza, K. (2007). Indoor air quality in restaurants with and without designeted smoking rooms. *J Occup Environ Hyg*, 4, 246-252.

Mönkkönen, P.; Pai, P.; Maynard, A.; Lehtinen, K.E.J.; Hämeri, K.; Rechkemmer, P.; Ramachandran, G.; Prasad, B. & Kulmala. M. (2005). Fine particle number and mass concentration measurements in urban Indian households. *Sci Total Environ*, 347, 131-147.

Molnár, P.; Bellander, T.; Sällsten, G. & Boman, J. (2007). Indoor and outdoor concentrations of $PM_{2.5}$ trace elements at homes, preschools and schools in Stockholm, Sweden. *J Environ Monit*, 9, 348-57.

Morawska, L.; He, C.; Hitchins, J.; Mengersen, K. & Gilbert, D. (2003). Characteristics of particle number and mass concentrations in residential houses in Brisbane, Australia. *Atmos Environ*, 37, 4195-4203.

Morawska, L. & Salthammer, T. (2003). Fundamentals of indoor particles and settled dust. In: Morawska L, Salthammer T. (Eds.) Indoor environment. Airborne particles and settled dust. Wiley-VCH Verlag Weinheim, Germany.

Mosqueron, L.; Momas, I. & Moullec, Y. (2002). Personal exposure of Paris office workers to nitogen, dioxide and fine particeles. *Occup Environ Med*, 59, 550-556.

Naeher, L.P.; Smith, K.R.; Leaderer, B.P.; Mage, D. & Grajeda, R. (2000). Indoor and outdoor $PM_{2.5}$ and CO in high- and low-density Guatemalan villages. *J Expo Anal Environ Epidemiol*, 10, 544-551.

Nazaroff, W.W. (2004). Indoor particle dynamics. *Indoor Air*, 4 (Suppl. 7), 175-183.

Neas, L.M.; Dockery, D.W., Ware, J.H.; Spengler, J.D.; Ferris, B.G. & Speizer, F.E. (1994). Concentration of indoor particulate matter as a determinant of respiratory health in children. *Am J Epidemiol*, 139, 1088-1099.

Oeder, S.; Weichenmeier, I.; Dietrich, S.; Schober, W.; Pusch, G.; Jörres, R.A.; Schierl, R.; Nowak, D.; Fromme, H.; Behrendt, H. & Buters, J. (2011). Toxicity and elemental composition of particulate matter from outdoor and indoor air of elementary schools in Munich, Germany. *Indoor Air* Accepted manuscript online: DOI: 10.1111/j.1600-0668.2011.00743.x.

Ogulei, D.; Hopke, P.K. & Wallace, L.A. (2006). Analysis of indoor particle size distributions in an occupied townhouse using positive matrix factorization. *Indoor Air*, 16, 204-215.

Osman, L.M.; Douglas, J.G.; Garden, C.; Reglitz, K.; Lyon, J.; Gordon, S. & Ayres, J.G. (2007). Indoor air quality in homes of patients with chronic obstructive pulmonary disease. *Am J Respir Crit Care Med*, 176, 465-472.

Özkaynak, H.; Xue, J., Weker, R.; Butler, D.; Koutrakis, P. & Spengler, J.D. (1995). The Particle TEAM (PTEAM) Study: analysis of the data. Final report. Vol. III, US - Environmental Protection Agency.

Phillips, K.; Bentley, M.C.; Howard, D.A. & Alvan, G. (1998). Assesment of air quality in Paris by personal monitoring of non-smokers for respirable suspended particles and environmental tobacco smoke. *Environ Int*, 24, 405-425.

Raaschou- Nielsen, O.; Sørensen, M.; Hertel, O.; Chawes, B.L.K.; Vissing, N.; Bønnelykke, K. & Bisgaard, H. (2011). Predictors of indoor fine particulate matter in infants' bedrooms in Denmark. *Environ Res*, 111, 87–93.

Raut, J.C.; Chazette, P. & Fortain, A. (2009). Link between aerosol optical, microphysical and chemical measurements in an underground railway station in Paris. *Atmos Environ*, 43, 860–868.

Rees, V.W. & Connolly, G.N. (2006). Measuring air quality to protect children from second hand smoke in cars. *Am J Prev Med*, 31, 363-368.

Repace, J.; Hyde, J.N. & Brugge, D. (2006). Air pollution in Boston bars before and after a smoking ban. *BMC Public Health*, 6, 266.

Reynolds, S.J.; Bleck, D.W.; Borin, S.S.; Breuer, G.; Burmeister, L.F.; Fuortes, L.J.; Smith, T.F.; Stein, M.A., Subramanian, P.; Thorne, P.S. & Whitten, P. (2001). Indoor Enviromental Quality in Six Commercial Office Buildings in the Midwest United States. *App Occup Environ Hyg*, 16, 1065-1077.

Riediker, M.; Cascio, W.E.; Griggs, T.R.; Herbst, M.C.; Bromberg, P.A.; Neas, L.; Williams, R.W. & Devlin, R.B. (2004). Particulate matter exposure in cars is associated with cardiovascular effects in healthy young men. *Am J Respir Crit Care Med*, 169, 934-940.

Riley, W.J.; McKone, T.E.; Lai, A.C. & Nazaroff, W.W. (2002). Indoor particulate matter of outdoor origin: importance of size-dependent removal mechanisms. *Environ Sci Technol*, 36, 200-207. Erratum in: *Environ Sci Technol*, 36, 1868.

Rim, D.; Siegel, J.; Spinhirne, J.; Webb, A. & McDonald-Buller, E. (2008). Characteristics of cabin air quality in school buses in Central Texas. *Atmos Environ*, 42, 6453–6464.

Rodes, C.E.; Lawless, P.A.; Thornburg, J.W.; Williams, R.W. & Croghan, C.W. (2010). DEARS particulate matter relationships for personal, indoor, outdoor, and central site settings for a general population, *Atmos. Environ.*, 44, 1386-1399.

Rojas-Bracho, L.; Suh, H.H. & Koutrakis, P. (2000). Relationships among personal, indoor, and outdoor fine and coarse particle concentrations for individuals with COPD. *J Expo Anal Environ Epidemiol*, 10, 294-306.

Rosen, L.J.; Zucker, D.M.; Rosen, B.J. & Connolly, G.N. (2010). Second-hand smoke levels in Israeli bars, pubs and cafes before and after implementation of smoke-free legislation. *European J Public Health*, 21, 15–20.

Sabin, L.D.; Behrentz, E.; Winer, A.M.; Jeong, S.; Fitz, D.R.; Pankratz, D.V.; Colome, S.D. & Fruin, S.A. (2005). Characterizing the range of children's air pollutant exposure during school bus commutes. *J Expo Anal Environ Epidemiol*, 15, 377–387.

Santen, M.; Wesselmann, M.; Fittschen, U.; Cremer, R.; Braun, P.; Lüdecke, A. & Moriske, H.J. (2009). Measurements of fine and ultrafine particles in indoor environment of living rooms. Gefahrstoffe-Reinhaltung der Luft, 69, 63-70 [in German].

Sarwar, G.; Olson, D.A.; Corsi, R.L. & Weschler, C.J. (2004). Indoor fine particles: the role of terpene emissions from consumer products. *J Air & Waste Manage Assoc*, 54, 367-377.

Schneider, S.; Seibold, B.; Schunk, S.; Jentzsch, E.; Dresler, C.; Travers, M.J.; Hyland, A. & Pötschke-Langer, M. (2008). Exposure to secondhand smoke in Germany: Air contamination due to smoking in German restaurants, bars, and other venues. *Nicotine & Tobacco Res*, 10, 547–555.

Semple, S.; Van Tongeren, M.; Galea, K.S.; Maccalman, L.; Gee, I.; Parry, O.; Naji, A. & Ayres, J.G. (2010). UK smoke-free legislation: changes in $PM_{2.5}$ concentrations in bars in Scotland, England, and Wales. *Ann Occup Hyg*, 54, 272–280.

Sendzik, T.; Fong, G.; Travers, M. & Hyland, A. (2006). The hazard of tobacco smoke pollution in cars: evidence from an air quality monitoring study. 13th World Conference on Tobacco and Health, Washington DC.

Shaughnessy, R.J.; Turk, B.; Evans, S.; Fowler, F.; Casteel, S. & Louie, S. (2002). Preliminary study of flooring in school in the U.S.: airborne particulate exposure in carpeted vs. uncarpeted classrooms. *Proceedings of Indoor Air*, 974-979.

Simons, E.; Curtin-Brosnan, J.; Buckley, T.; Breysse, P. & Eggleston, PA. (2007). Indoor environmental differences between inner city and suburban homes of children with asthma. *J Urban Health*, 84, 577-590.

Singer, B.C.; Coleman, B.K.; Destaillats, H.; Hodgson, A.T.; Lunden, M.M.; Weschler, C.J. & Nazaroff, W.W. (2006). Indoor secondary pollutants from cleaning products and air freshner use in the presence of ozone. *Atmos Environ*, 40, 6696-6710.

Son, B.S.; Song, M.R. & Yang, W.H. (2005). A study on PM_{10} and VOCs concentrations of indoor environment in school and recognition of indoor air quality. *Proceedings: Indoor Air*, 827-832.

Stranger, M.; Potgieter-Vermaak, S.S. & Van Grieken, R. (2007). Comparative overview of indoor air quality in Antwerp, Belgium. *Environ Int*, 33, 789–797.

Thatcher, T.L. & Layton, D. (1995). Deposition, Resuspension, and Penetration of Particles within a Residence, *Atmos Environ*, 29, 1487-1497.

Thatcher, T.L.; McKone, T.E.; Fisk, W.J.; Sohn, M.D.; Delp, W.W.; Riley, W.J. & Sextro, R.G. (2001). Factors affecting the concentration of outdoor particles indoors (COPI): identification of data needs and existing data. Lawrence Berkeley National Laboratory (LBNL). Report under contract No. DW-89938748.

Valente, P.; Forastiere, F., Bacosi, A.; Cattani, G., Di Carlo, S.; Ferri, M.; Figà-Talamanca, I.; Marconi, A., Paoletti, L., Perucci, C. & Zuccaio, P. (2007). Exposure to fine and ultrafine particles from secondhand smoke in public placet bifore and after the smoking ban, Italy 2005. *Tobacco Control*, 16, 312-317.

Vardavas, C.I.; Kondilis, B.; Travers, M.J.; Petsetaki, E.; Tountas, Y. & Kafatos, A.G. (2007). Environmental tobacco smoke in hospitality venues in Greece. *BMC Public Health*, 7, 302.

Vartiainen, E.; Kulmala, M.; Ruuskanen, T.M.; Taipale, R.; Rinne, J. & Vehkamäki, H. (2006). Formation and growth of indoor air aerosol particles as a result of d-limonene oxidation. *Atmos. Environ*, 40, 7882-7892.

Vette, A.F.; Rea, A.W.; Law, P.A.; Rodes, C.E.; Evans, G.; Highsmith, V.R. & Sheldon, S. (2001). Characterization of indoor-outdoor aerosol concentration relationships during the Fresno PM Exposure Studies. *Aerosol Sci Technol*, 34, 118 - 126.

Wainman, T.; Zhang, J.; Weschler, C.J. & Lioy, P.J. (2000). Ozone and limonene in indoor air: a source of submicron particle exposure. *Environ Health Perspect*, 108, 1139-1145.

Wallace, L. & Howard-Reed, C. (2002). Continuous monitoring of ultrafine, fine and coarse particles in a residence for 18 months in 1999-2000. *J Air Waste Manage Assoc*, 52, 828-844.

Wallace, L.A.; Mitchell, H.; O'Connor, G.T.; Neas, L.; Lippmann, M.; Kattan, M.; Koeng, J., Stout, J.W., Vaughn, B.J.; Wallace, D., Walter, M.; Adams, K. & Liu, L.-J.S. (2003). Particle concentrations in inner-city homes of children with asthma: the effect of smoking, cooking, and outdoor pollution. *Environ Health Perspect*, 111, 1265-1272.

Wallace, L. & Ott, W. (2011). Personal exposure to ultrafine particles. *J Expos Sci Environ Epidemiol*, 21, 20–30.

Weichenthal, S.; Dufresne, A.; Infante-Rivard, C. & Joseph, L. (2007). Indoor ultrafine particle exposure and home heating systems: A cross-sectional survey of Canadian homes during the winter months. *J Expo Sci Environ Epidemiol*, 17, 288-297.

Weschler, C.J. (2003). Indoor chemistry as a source of particles. In: Morawska, L. & Salthammer, T. (Eds.) Indoor environment. Airborne particles and settled dust. Wiley-VCH Verlag Weinheim, 2003.

Weschler, C.J.; Wells, J.R.; Poppendieck, D.; Hubbard, H. & Pearce, T.A. (2006). Workgroup report: indoor air chemistry and health. *Environ Health Perspect*, 114, 442-446..

Wichmann, J.; Lind, T.; Nilsson, M.A.M. & Belland, T. (2010). $PM_{2.5}$, soot and NO_2 indoor outdoor relationships at homes, pre-schools and schools in Stockholm, Sweden. *Atmos Environ*, 44, 4536-4544.

Yang, W.; Sohn, J.; Kim, J.; Son, B. & Park, J. (2009). Indoor air quality investigation according to age of the school buildings in Korea. *J Environ Manage*, 90, 348e-354e.

Zhang, Q. & Zhu, Y. (2010). Measurements of ultrafine particles and other vehicular pollutants inside school buses in South Texas. *Atmos Environ*, 44, 253-261.

Zuurbier, M.; Hoek, G.; Oldenwening, M.; Lenters, V.; Meliefste, K.; van den Hazel, P. & Brunekreef, B. (2010). Commuters' Exposure to Particulate Matter Air Pollution Is Affected by Mode of Transport, Fuel Type, and Route. *Environ Health Perspect*, 118, 783–789.

7

A Methodology of Estimation on Air Pollution and Its Health Effects in Large Japanese Cities

Keiko Hirota[1], Satoshi Shibuya[2],
Shogo Sakamoto[2] and Shigeru Kashima[2]
[1]Japan Automobile Research Institute
[2]Faculty of Science and Engineering, Chuo University
Japan

1. Introduction

The correlation between air pollution and health effects in large Japanese cities presents a great challenge owing to the limited availability of data on the exposure to pollution, health effects and the uncertainty of mixed causes. A methodology for quantitative relationships (between the emission volume and air quality, and the air quality and health effects) is analysed with a statistical method in this chapter; the correlation of air pollution reduction policy in Japan from 1974 to 2007.

This chapter discusses a step-by-step methodology of determining the direct correlation between emission volumes, air quality, and health effects. Figure 1 shows the Japanese compensation system with two hypotheses in order to clarify the correlation. Hypothesis 1 states that the total emission volume affects air quality. Hypothesis 2 states that the air quality influences the number of certified patients.

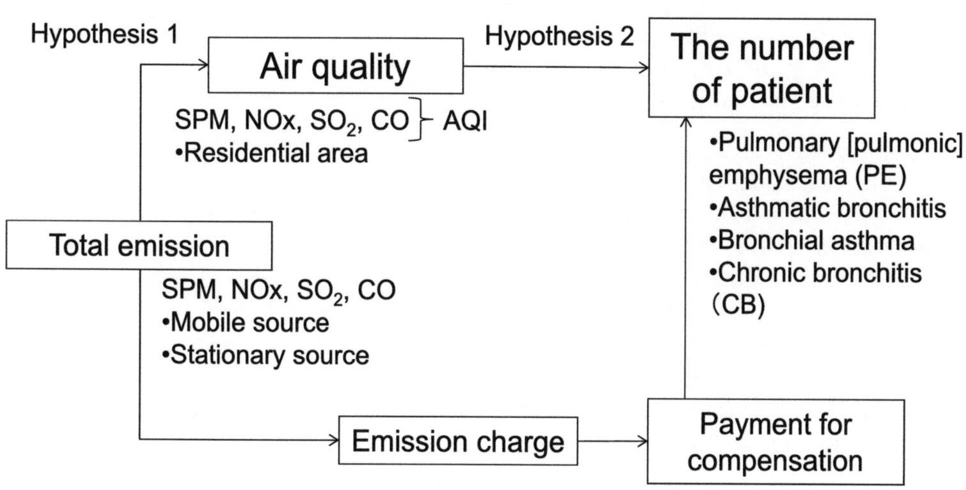

Fig. 1. The relationship between emissions, air quality and the number of certified patients (above) and flow of compensation (below)

The relationship between total emissions (NOx, PM) (from both stationary and mobile sources) and air quality (NO2, SPM) was found to be significant, which supports hypothesis 1. The correlation analysis of emission volume, and air quality suggests that NOx and PM levels worsen according to increases in NO2.

When the correlation between the air pollutant and the type of health effects (certified, mortality, recovery, and newly registered) was examined according to the certified area, an inverse relationship was observed. The relationship between air quality (NO2) and health effect was found to be significant, which support hypothesis 2. When NO2 worsens, certified patients, mortality rates and newly certified patients increase, according to the data from 1989 to 2007 with dummy variable analysis.

2. Methodology

2.1 Pollution-related health damage compensation system and the law concerning special measures to reduce the total amount of nitrogen oxides and particulate matter emitted from motor vehicles in specified areas

Mie prefecture rapidly deteriorated as a result of rapid economic growth in the early 1960s, the air quality in the industrial sector of Yokkaichi. Residents had higher rates of morbidity and higher prevalence rates of respiratory and circulatory diseases in Yokkaichi than in other areas.

Environmental laws and standards were established to address those issues by the government in order to mitigate this deterioration of air quality in the late 1960s to early 1970s. The Basic Law of Pollution Prevention was established in 1967. This law determined the permissible limits of each contaminant in the air (SO2 in 1969, CO in 1970, suspended particulate matter (SPM) in 1972, and NO2 in 1973). The policies initially focused on stationary sources such as industries for reducing air pollution.

The health effects caused by air pollution became a social issue. A major issue was the apportioning of the compensation among the potential polluters. There was strong opposition from the industrial sector when discussions began among the polluters [Matsuura, 1994 a, 1994 b, 1994 c]. The affected patients sought compensation from the Japanese government and the industrial sector.

To address the social issues associated with the health effects observed since the early 1960s, the Pollution-related Health Damage Compensation Law (Abbreviation: the compensation law) was enacted in 1973 to support the affected patients by providing prompt and fair compensation to cover damages and by implementing the necessary programs for the welfare of these patients. The Pollution-related Health Damage Compensation and Prevention System (Abbreviation: the compensation system) was established in June 1974 in accordance with the compensation law. The certified regions and diseases for the compensation system were determined by the government in 1974 [Amagasaki City Government 1994, Kawasaki City Government 1994, Kita Kyushu City Government 1994 & Osaka City Government 1974] (See (6) of Figure 2).

The compensation system implements welfare services to facilitate the recovery of patients with health impairments based on the compensation law. The compensation law is responsible for determining the groups that should bear the expenses in the designated areas "Class 1", where health damage has occurred. The compensation law also classifies "certifying patients" who are affected by one of the four pollution-related respiratory and circulatory diseases (pulmonary emphysema, asthmatic bronchitis, bronchial asthma, and chronic bronchitis). The patients with one of these four diseases are considered as certified patients if they satisfy certain other conditions (such as extended residence in the certified area, non-smoker status, or illness diagnosed by a medical doctor) in the certified areas [ERCA Homepage http://www.erca.go.jp]. The compensation system collects emission charge from 9,000 factories and companies on the basis of their SO2 emission volumes. SO2 emission charge from polluting industries (80%) and a vehicle tax (20%) covered the welfare services for the certified patients.

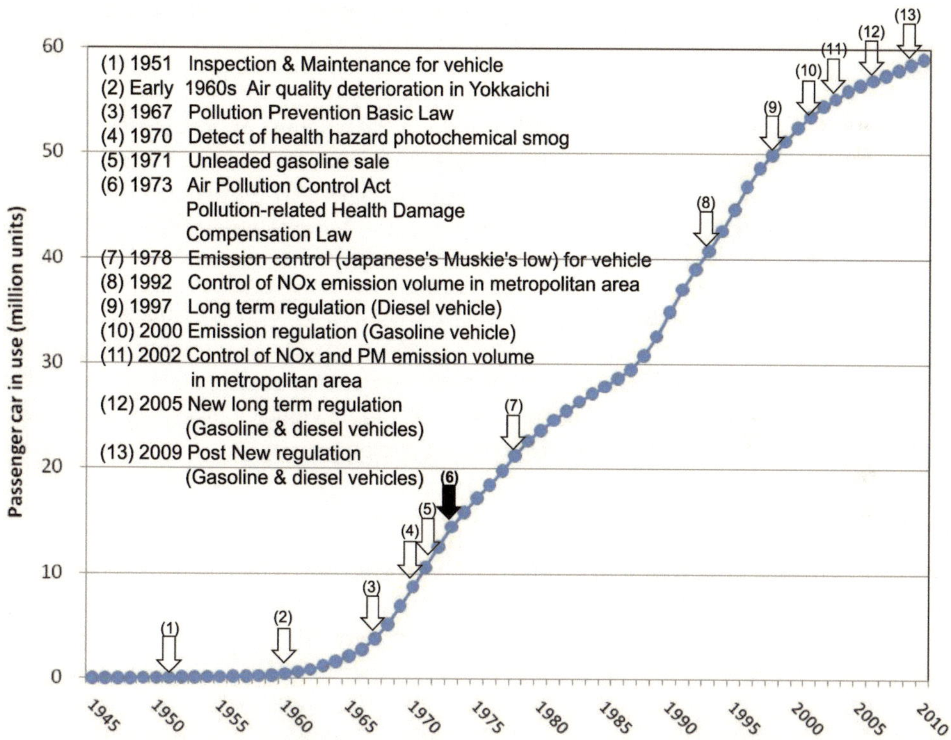

Fig. 2. Motorization and emission reduction measures from the automobile sector in Japan

The industrial sector promoted the government to amend the compensation law in the beginning of 1980s. The committee of air pollution reduction measures reported that certain areas could have had higher rates of morbidity and higher prevalence rates of chronic obstructive pulmonary disease (Abbreviation: COPD) caused by air pollution in 1950s and 1960s. However, it is not the same situation as those of 1950s and 1960s. In September 1987, amendments were made to the compensation system based on the report of the committee.

In March 1988, the Association of Environmental Restoration and Conservation Agency of Japan started an additional project to prevent health damage caused by air pollution on the basis of the revision of the compensation law. This new project intended to assist local public organizations with health consultations, health examinations, and functional training to carry out research and to distribute knowledge leading to health maintenance in local communities. The Law was revised in 1988 to remove the Class 1 specification (compensation disbursement was continued) and to start health damage prevention projects [ERCA Homepage http://www.erca.go.jp].

The focus of emission reduction policies shifted from the industrial sector to non-industrial sources, namely, mobile sources in 1990s due to motorization. The Ministry of the Environment adopted a total emission volume control measure by the "Law Concerning Special Measures to Reduce the Total Amount of Nitrogen Oxides Emitted from Motor Vehicles in Specified Areas", called in short "The Motor Vehicle NOx Law" in 1992. The law copes with NOx pollution problems from existing vehicle fleets in highly populated metropolitan areas. Under the Motor Vehicle NOx Law, several measures had to be taken to control NOx from in-use vehicles, including enforcing emission standards for specified vehicle categories.

The regulation was amended in June 2001 to tighten the existing NOx requirements and to add PM control provisions. The amended rule is called the "Law Concerning Special Measures to Reduce the Total Amount of Nitrogen Oxides and Particulate Matter Emitted from Motor Vehicles in Specified Areas", or in short "the Motor Vehicle NOx and PM Law". The amended regulation became effective in October 2002.

The law concerning Special Measures for Total Emission Reduction of Nitrogen Oxides and Particulate Matter (Abbreviation: The NOx-PM Law) introduces emission standards for specified categories of in-use highway vehicles including commercial goods (cargo) vehicles such as trucks and vans, buses, and special purpose motor vehicles, irrespective of the fuel type. The regulation also applies to diesel powered passenger cars.

2.2 Literature survey

This section explains why this research focuses on correlation analysis by comparing some previous studies. Iwai K, Mizuno S, Miyasaka Y, and Mori T [2005] analysed the odds ratio from the 2001 data for particulate matters PM2.5 converted from suspended particulate matter (SPM) for the whole of Japan. The authors tried to find out the direct correlation between respiratory diseases and air quality. Their contribution showed a strong correlation between the classification of diseases and air quality. Following this methodology of the direct correlation analysis, this chapter focuses on the diseases caused by air pollution and air quality.

Makino [1996] focused on 23 Tokyo wards from 1958 to 1989, analysing SOx, NOx, and dust and comparing those to the total registered vehicles. In this research, a time lag of 15 to 20 years from the time of exposure to the appearance of the health effect was found. This demonstrates the importance of long-term analysis.

Sunyer J et al. [2006] focused their research on 21 European cities from 1991 to 1993. The author analysed the correlation (cross section) between the prevalence ratio of chronic bronchitis and air quality. The prevalence ratio was adjusted for attributes such as sex, place of residence, smoking habits, income level, and social status based on a questionnaire (n = 6824). Higher NO2 levels resulted in a higher prevalence ratio of female patients.

These results suggest that more variables, such as lifestyle, age, and sex, need to be included for further research. The data set of the Japanese compensation system is employed because this system certified patients suffering from diseases caused by air pollution.

2.3 Precondition

2.3.1 Certified areas of the compensation system and the NOx-PM law

The compensation system covers 25 certified areas (Figure 3). The NOx-PM law designated a total of 276 communities in the Tokyo, Saitama, Kanagawa, Chiba, Aichi, Mie, Osaka, and Hyogo Prefectures as certified areas with significant air pollution due to nitrogen oxides emitted from motor vehicles. Kita Kyushu area is not the NOx-PM certified area.

Five certified areas including 19 wards in Tokyo , Kawasaki (Saiwai, Kawasaki), the whole of Osaka city, Amagasaki (East region and South region), and Kita Kyushu (Wakamatsu, Tobata, Yawata, Kokura, Kokusetsu-Kita Kyushu, Nijima, Kurosaki, and Higashi-Kokura) were selected for the analysis of Hypothesis 1 & 2.

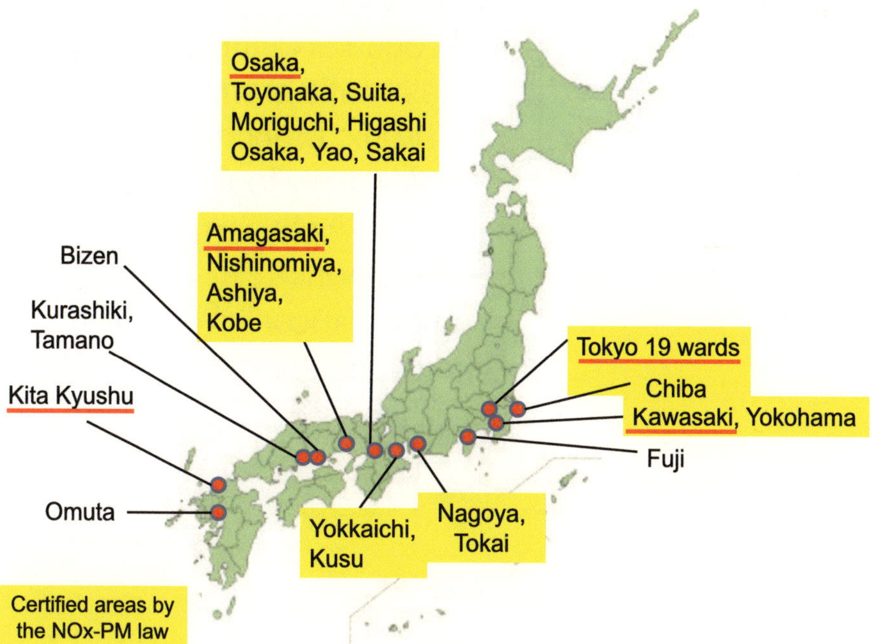

Fig. 3. Areas of certified (Class 1) under the Pollution-Related Health Damage Prevention system (the research areas assessed in this paper are underlined) and the NOx-PM area (Highlighted in yellow)

2.3.2 Air pollutants and health effects

Figure 4 shows the air pollutants and their affected diseases. USEPA and WHO considered respiratory and cardiovascular diseases caused by air pollution. PM is a representative air pollutant for analysis according to USEPA and WHO. [UPEPA 1999, WHO 2002, UK DH 2006].

The Japanese definition of disease caused by air pollution is narrower compared to those of USEPA and WHO. NO2 and SO2 affect the four certified respiratory diseases in the compensation system in Japan. Emission fees are determined by the volume of SO2 emissions in the compensation system because SO2 measurement method was already established. NO2 measurement did not seem to be well-established in the early 1970s. The relationship between NO2 and health effects has been a representative indicator for the evaluation of health effects in epidemiological research in Japan.

First, we used both SPM and NO2, but we found that the analysis with NO2 has more significant results than SPM. Second, the Air Quality Index (abbreviation: AQI) data were also revised in order to include all the pollutants. The results of the AQI data confirm that PM and NOx are the main indicators for the assessment of the health effects of air pollution. Finally, NOx was selected for the analysis based on previous literature and on its strong relationship to health effects.

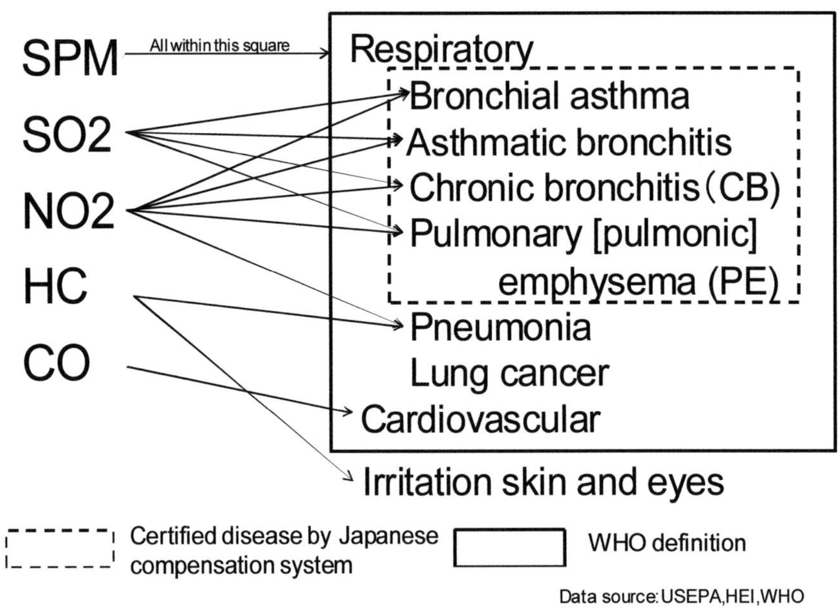

Fig. 4. Air pollutants and their affected diseases

2.3.3 Health effects caused by air pollution

The data of the compensation system has two benefits. First, local governments issue white paper. The white paper shows the number of certified, mortality, newly registered and recovered patients in public. These consistency data is useful to check if it is reliable by calculation. Second, a certified patient is required to be non-smoker as a rule. The data is classified patients of certified 4 diseases and non-smokers in the compensation system.

There are two different trends according to statistical data of certified patients in the compensation system; one before 1988 and one after 1989 (Figure 5). The air quality dramatically diminished and the number of patients increased before 1988. The number of certified patients reduced because no new patients could be registered after 1989. Two time series will be considered in order to present the correlation. One is for the whole time period (1974–2007), and the other is for two separate time series (1974–1988 and 1989–2007).

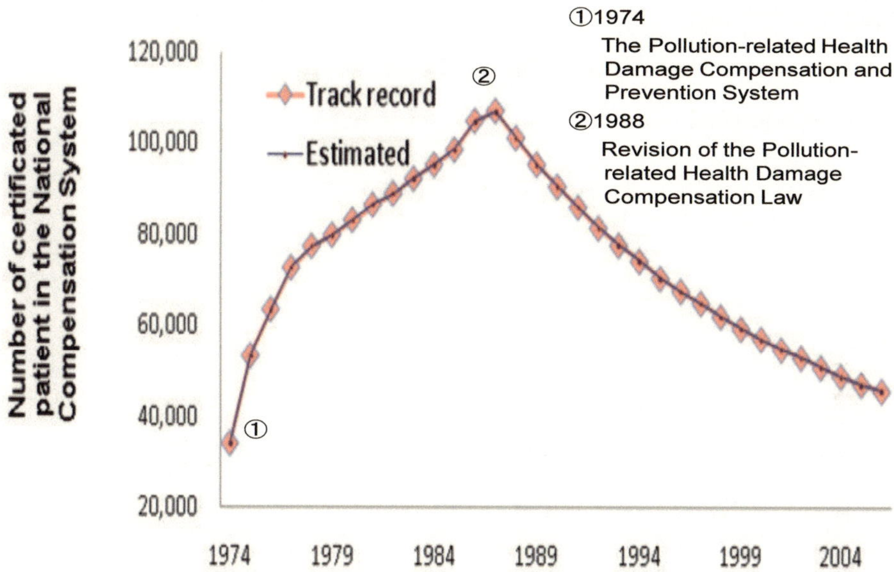

Fig. 5. Number of certificated patient in the pollution-related health damage compensation and prevention system (compensation system)

2.3.4 Estimation models

These two hypotheses are expressed by the following 5 equations. The annual average value NO2 of residential areas depends on emission volumes including those from stationary and mobile sources (hypothesis 1) in Eq. 1. The health effects (certified patients, mortality rates, recovery, and newly registered patients) depend on NO2 level (hypothesis 2) from Eq. 2 to Eq. 5.

The emission volume is divided by the surface area in order to compare the health effects. The number of certified patients is divided by the total population because this group becomes patient group from normal population group. Certified patient includes newly registered and re-registered patients. The numbers of deceased and recovered are divided by the number of certified patients because these groups change status from certified patients. The number of newly registered patients is divided by the number of population because this group becomes patient group from normal population group.

Methodology of hypothesis 1 (Emission-air quality)

$$M_{t,c} = \alpha \ln X_{t,c} + c \tag{1}$$

M: Annual average value of NO_2 in residential area (ppm)
X: Emission volume of NOx (stationary + mobile sources) (t/year/km^2)

t: year (t = 33)
Tokyo: 1974–1976,1980,1985,1990,1995,1997
Kawasaki:1985,1993,1995,1997,2000,2005
Osaka:1972,1978,1980,1988,1990,1992,1994,1996,1998,2000,2002–2006
c: certified area (n = 1–3)

Methodology of hypothesis 2 (Air quality and health effects)

$Q, D, R, N,$ are estimated from the coefficients and NO_2

$$Q_{t,r} = a_1 M_{t,r} + b_1 \tag{2}$$

$$D_{t,r} = a_2 M_{t,r} + b_2 \tag{3}$$

$$R_{t,r} = a_3 M_{t,r} + b_3 \tag{4}$$

$$N_{t,r} = a_4 M_{t,r} + b_4 \tag{5}$$

Q: certified patients per 1000 population of certified areas
D: mortality per 1000 certified patients of certified areas
R: recovery per 1000 certified patients of certified areas
N: newly certified patients per 1000 population of certified areas
M: annual average value of NO_2 in residential areas
t: year (t = 1–33)
a: coefficient (n = 1–4)
b: constant (n = 1–4)
c: certified area (n = 1–5)

3. Data description

3.1 Emission volume

The NOx, PM, SO2, and CO emission volumes from stationary sources have been measured since the early 1970s in Tokyo, Kawasaki, Osaka, Amagasaki, and Kita Kyushu. The data was in public by environmental white paper of each local government. The NOx and PM emission volumes from mobile sources have been measured since the 1970s primarily to evaluate policy implementation of the NOx-PM law.

The total emission data (stationary and mobile sources) were estimated for certain years in Tokyo, Kawasaki, and Osaka (Table 1). They were not continuous data of every year. The Amagasaki data were included in the total emissions from Hyogo prefecture according to the Hyogo prefecture environment white paper [Hyogo Prefecture Government 2003]. There was no separate data set available for only Amagasaki. There was no data set of emissions from mobile sources in Kita Kyushu because Kita Kyushu is not certified of the NOx-PM law [Kita Kyushu City Government, 1982]. Only the data set for emission volumes from stationary sources was available in Kita Kyushu.

There appear to be some outliers in the correlation analysis. There are four estimated values of emission volumes for 1985 in four different reports of the Tokyo metropolitan government. The outlier for Osaka was 1972. It is also the highest level of emission volume among the Osaka data.

These data were checked by the Tokyo metropolitan government and the Osaka city local government for verification. There are multiple data sources from different research projects for the same year. The criterion for selection is that the estimation must represent the actual situation without any policy implementation assumption.

Data on total NOx emission volumes (stationary and mobile sources) in Tokyo (23 wards) [Environmental Bureau of Tokyo Metropolitan Government, 1971, 1973, 1984, 1994, 1996, 2000a, 2000b, 2004, 2005, 2009, Tokyo Metropolitan Government, 1983, 1995, 1998], Kawasaki [Kawasaki City Government 2008, 2009], and Osaka [Osaka City Government, 1975, 1976, 1991, 2001, 2004, 2007] from 1972 to 2007 are available from various annual environmental reports written by local governments.

	Data source	Emission Volume (NOx, PM)	Air quality (NO2,SPM, CO,HC, SO2)	Certified Patient	Mortality	Recovery	Newly registered patient
Tokyo (19 wards)	Environmental Bureau of Tokyo Metropolitan Government	● (Tokyo 23 wards)	●	●			
Kawasaki	Kawasaki City Government, Kanagawa Prefecture Government	● (Whole city)	●	●	●	●	●
Osaka	Osaka City Government	● (Whole city)	●	●	●	●	●
Amagasaki	Amagasaki City Government. Environment in Amagasaki, Hyougo Prefecture Government	● (Hyogo prefecture)	●	●	●	●	●
Kita Kyushu	Kita Kyushu Government, Fukuoka Prefecture Government		●	●	●	●	●

Emission volume = stationary + mobile sources

Table 1. Data sources of each item (1974-2007)

3.2 Air quality

The annual average values of the pollutants (NO2) in residential areas in Tokyo (All 23 wards), [National Institute for Environmental Studies. http://www-gis.nies.go.jp/], Kawasaki [Kawasaki City Government 2008 a, 2008 b, 2009, Kanagawa Prefecture Government 1983], and Osaka [Osaka City Government, 1975, 1976, 1991, 2001, 2004, 2007] from 1960 to 2007 were used.

3.3 Certified, mortality, recovered and newly registered patients

We used the records of certified, deceased, recovered patients, and newly registered for correlation analysis. As certified areas, the data from Tokyo (19 wards), Kawasaki (Saiwai, Kawasaki wards), Osaka city, Amagasaki city (East region and South regions), and Kita Kyushu city (Wakamatsu, Tobata, Yawata, Nijima, Kurosaki, and Higashi-Kokura) were used.

Some data regarding mortality rates, newly registered patients, and recovered patients in the 19 wards of Tokyo are missing. The term 'recovered patients' includes actual recovery, non-renewal of registration, and cancelation of registration.

4. Correlation analysis

4.1 Hypothesis 1

To support hypothesis 1, the NOx and PM emission volume and their air quality data are analyzed to determine a direct relationship.

Figure 6 shows the correlation between the total NOx emission (stationary and mobile sources) and NOx air quality (residential area, annual average) in Tokyo, Kawasaki, and Osaka. The data for emission volumes (stationary and mobile sources) were obtained from the respective local governments. A log-linear regression model is assumed. The t-value of NOx emission volume is 5.660, which is sufficiently high.

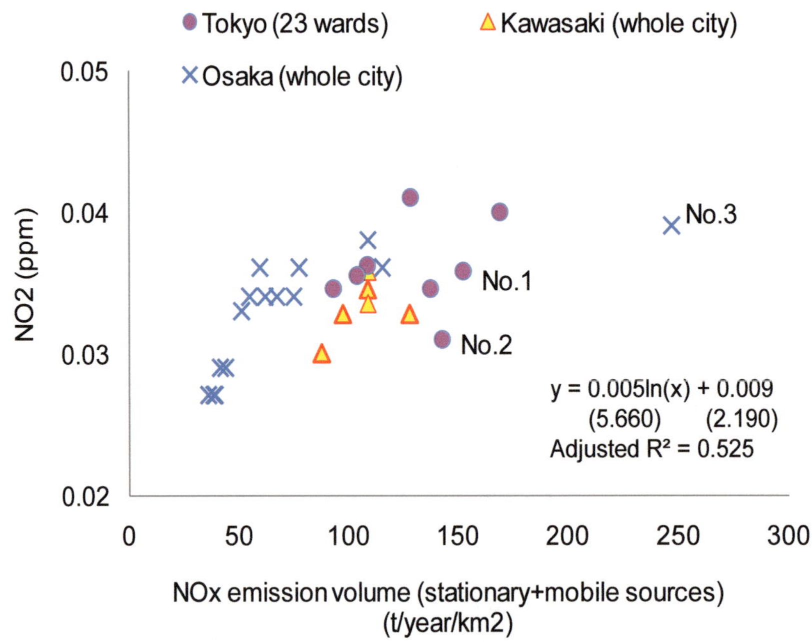

Fig. 6. Correlation between total NOx emissions and NO2 in Tokyo (23 wards), Kawasaki-city, and Osaka-city

Figure 7 shows the correlation between total PM emission (stationary and mobile sources) and SPM in Tokyo, Kawasaki and Osaka. The data for Tokyo (B) were estimated in 1975. The data in (A) were estimated in 1990. Both PM emission and SPM have improved from (B) to (A). The t-value of PM emission volume is 2.225, which is sufficiently high. The Osaka data of 1998 (C) seem to be lower than those of Tokyo.

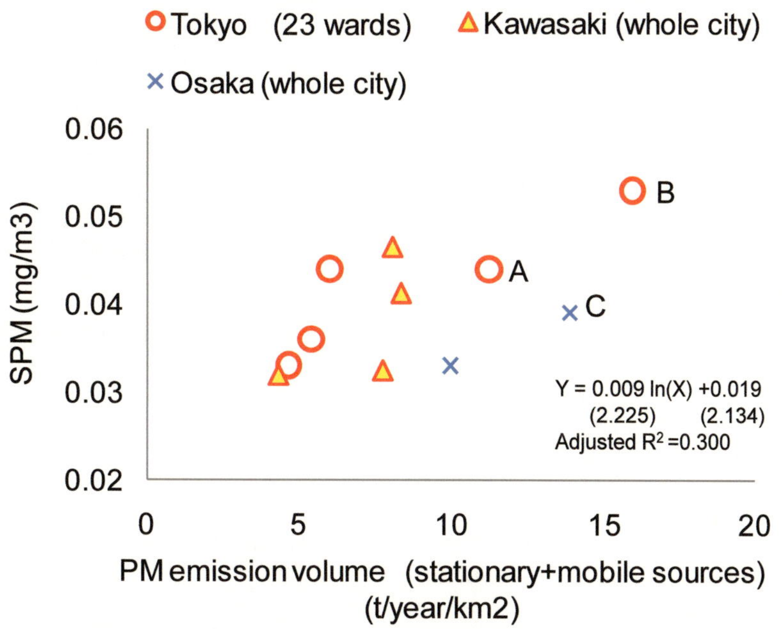

Fig. 7. Correlation between total PM emissions and SPM in Tokyo (23 wards), Kawasaki-city, and Osaka-city

4.2 Hypothesis 2

The second hypothesis focuses on the correlation between air quality and health effects (number of certified patients, recovered patients, new patients, and total mortality figures) in the areas that were certified as Class 1 under the compensation law. Figure 8 shows NO2 and certified patients in the 5 Japanese cities. The range of NO2 is from 0.02 ppm to 0.06 ppm for the Tokyo data. The Ohta ward (1974), the Itabashi ward (1987, 1988, 1991-1995, 2001), and the Kita ward (1975, 1991, 1992, 1995) have higher annual averages compared to the rest of the Tokyo wards, Kawasaki, Osaka, Amagasaki, and Kita Kyushu. The highest number of certified patients per 1000 people was in the Ohta ward (26 persons, 1974). The data from 1974 and 1975 are outliers; in the early 1970s.

Some values are outside the general trend because of the data instabilities caused by the introduction of the compensation system and air quality measurements. In fact, many experts have pointed out that the methodology for air quality measurements was not well-established around 1974, when the permissible limits were introduced. Owing to the variability of the measurements, different results were recorded in different governmental documents. Further, the data for certified patients, mortality, recovery, and new patients for that year were not stable as a result of the introduction of the compensation system in 1974.

An NO2 dummy variable is applied when NO2 exceeds 0.05 ppm and when the number of certified patients exceeds 5 persons per 1000 population. When the dummy variable is applied, the inverse relationship and R2 improve.

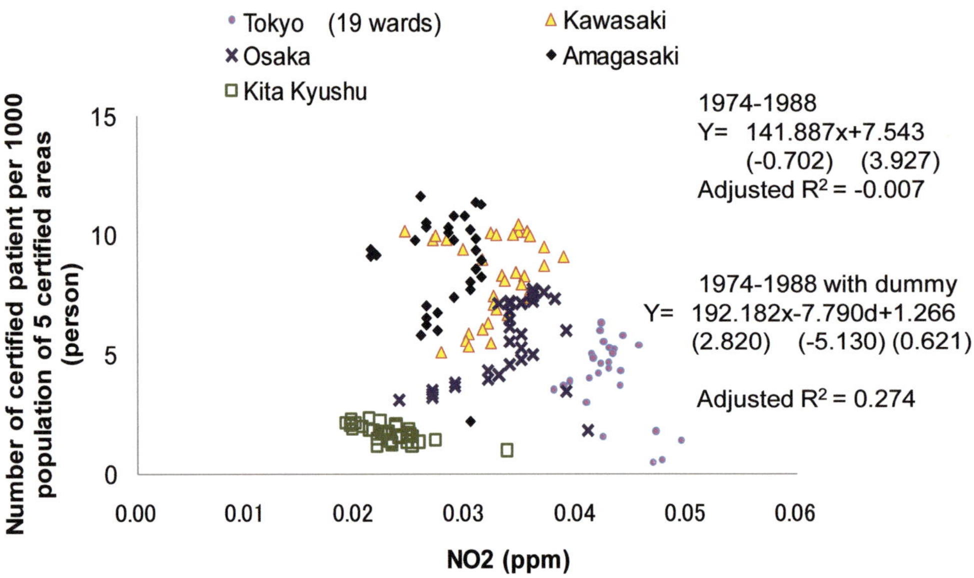

Fig. 8. Correlation: NO2 and certified patients in Tokyo (19 wards), Kawasaki-city, Osaka-city, Amagasaki-city and Kita Kyushu-city (1974-2007)

Figure 9 shows NOx and mortality values of certified patients in Kawasaki, Osaka, Amagasaki, and Kita Kyushu. The air quality is between 0.02 ppm and 0.04 ppm. The range of annual mortality numbers per 1000 certified patients is 4-32 in Kawasaki. There were 4 per 1000 certified patients in 1974 in Osaka. The number increased to 11 per 1000 certified patients in 1975. The mortality rate became stable at approximately 30 per 1000 certified patients after 1976. Osaka has three outlier years: 1974, 1975, and 1976. The Amagasaki data from 1983 to 1999 are missing. The mortality figure was 33 per 1000 certified patients in 1974. The number stabilised to around 20 from 1975 to 1982. The range of mortality increased from 27 to 33 between 2001 and 2007. The mortality figures were between 13 and 24 in Kita Kyushu. When the dummy variables were introduced for the three Osaka outliers from 1974 to 1988, the R2 and t-values improved and the correlation estimation reached a significant level.

Figure 10 shows NOx and recovery values of certified patients in Kawasaki, Osaka, Amagasaki, and Kita Kyushu. The recovery rates increased from 2 people per 1000 certified patients in 1975 to 47 people per 1000 certified patients in 1976 in Kawasaki. The number of recovery cases decreased, reaching 6 recoveries in 1997. The number of recoveries went up (to 22 recoveries) and down (to 2 recoveries) after 1998. There were 29 recovery cases in 1974 in Osaka. The number decreased to 6 in 1975. The number increased to 80 recovery cases per 1000 patients in 1980. There is no 'recovery' data in Amagasaki. It includes only non-renewal of registration and retirement cases. Recovery decreased from 30 people in 1970 to 0 in 2002. The data exhibit a significant reduction—from 74 people in 1974 to 10 people in 2004 in Kita

Kyushu. A dummy variable was introduced when the recovery figure was more than 40 and the level of NO2 was more than 0.035 ppm (Osaka, 1979, 1980, 1981, and 1982; Kita Kyushu, 1977, 1980, and 1983). The R2 and t-values improved when the dummy variable was applied.

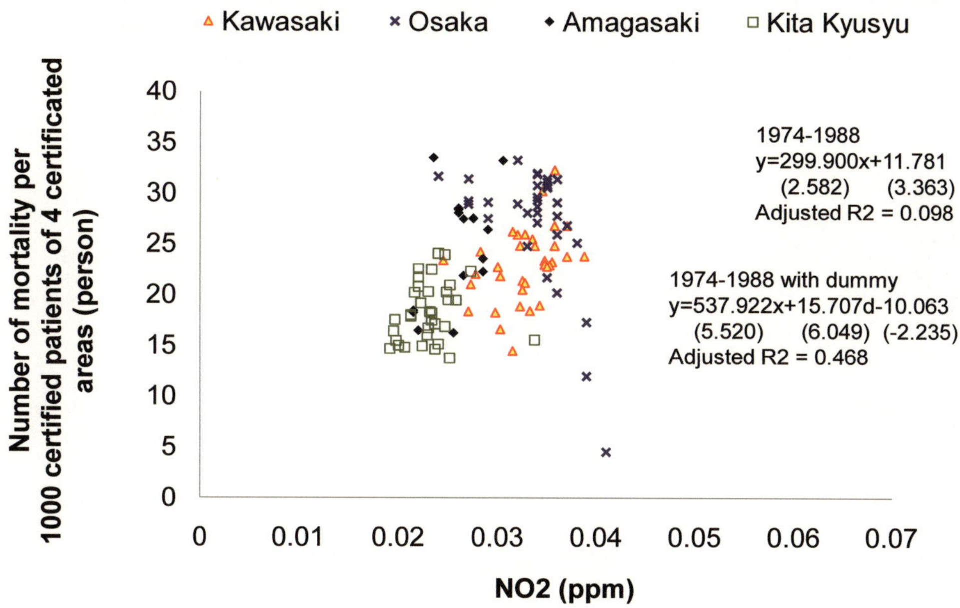

Fig. 9. Correlation: NOx and mortality in Kawasaki-city, Osaka-city, Amagasaki-city and Kita Kyushu-city

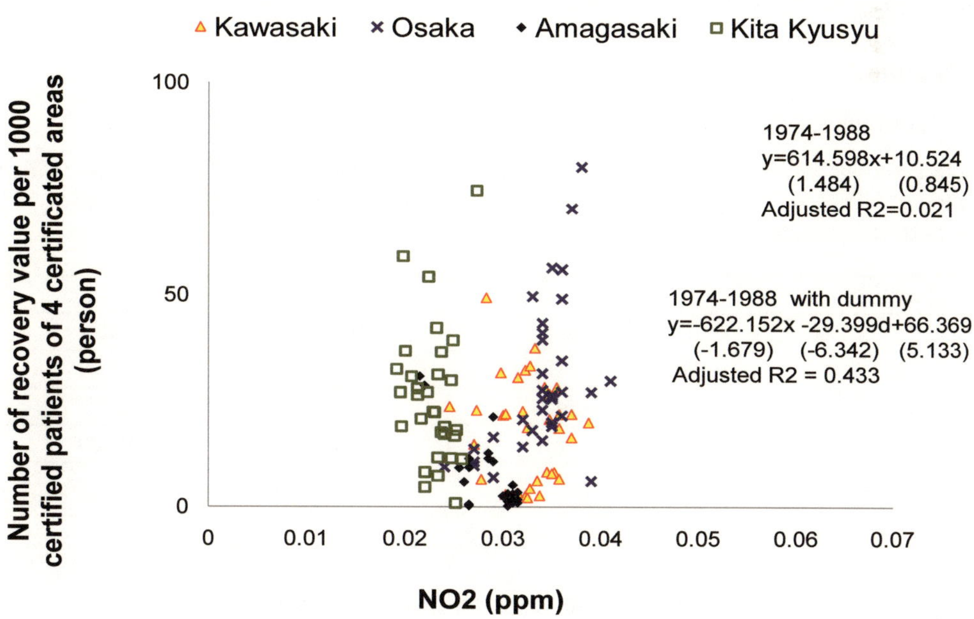

Fig. 10. Correlation: NO2 and recovery in Kawasaki-city, Osaka-city, Amagasaki-city and Kita Kyushu-city (1974–2007)

Figure 11 shows NO2 levels and the number of new patients. The number of newly certified patients tended to increase before the compensation law was amended in 1989. In Kawasaki, Osaka, and Amagasaki, new patients decreased from 1974 to 1986. There were sharp increases in new patients in 1987 and 1988. A dummy value is applied for Kawasaki 1975, Osaka 1976, Osaka 1989 and Amagasaki 1975. R2 and t-values improve when the dummy value is applied.

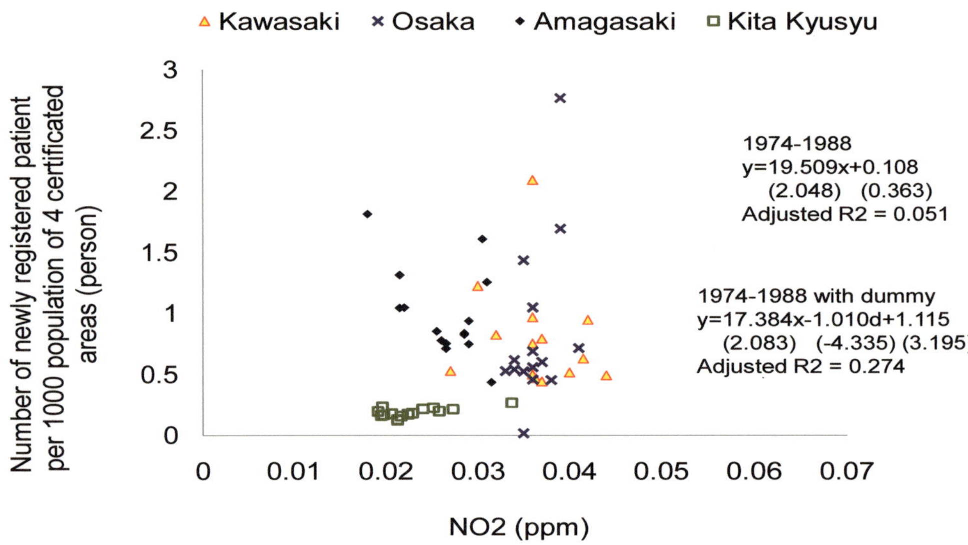

Fig. 11. Correlation: NO2 and newly certified patients in Kawasaki-city, Osaka-city, Amagasaki-city and Kita Kyushu-city (1974–1988)

5. Results

5.1. Hypothesis 1 (Correlation between emission volume and air quality)

Table 2 shows the results from Eq 1. Without adjustment by dummy, NOx concentrations related to NOx emission are estimated to be significant. Without adjustment by dummy, PM concentrations related to PM emission are estimated to be significant.

		NOx	PM
1974–2007	Constant	0.009	0.001
	NO_2/PM pollutant	0.005	0.009
	t value	5.660***	2.225**
	Adjusted R^2	0.525	0.300

Significance at * 20%, **5%, ***1%

Table 2. Result of correlation (Eq 1)

5.2 Hypothesis 2 (Correlation between air quality and health effects)

Table 3 shows the results from Eq. 2 to Eq. 5 for the whole period 1974-2007. R2 tends to be low without adjustment with the dummy value. The correlation between the air quality and the number of certified patients is not estimated to be significant. There is an inverse relationship between recovery rates and air quality owing to the large variance. The number of newly registered patients increased before the amendment of the compensation law in 1988.

Table 4 represents the results from Eq. 2 to Eq. 5 for the period before and the period after the system amendment (1974–1988 and 1989–2007, respectively). When the data are separated, the relationship between NO2 and the number of certified patients is indicated clearly, even though the t-value is still low. The number of recovered patients may affect the number of certified patients. Mortality rates and the numbers of newly registered patients are estimated to be significant.

Table 5 represents the result from Eq 4 to Eq 7 with a dummy variable in separate periods (1974-1988). All explanatory variables are estimated significantly.

		Certified patients	Newly registered	Mortality	Recovery
1974–2007	Constant	5.426		11.763	13.819
	NO_2	-10.115		386.387	242.825
	t value	-0.652		4.439***	0.889
	Adjusted R^2	-0.001		0.138	-0.002

Significance at * 20%, **5%, ***1%

Table 3. Results of correlation 1974–2007 (Eq 2-Eq5)

		Certified patients	Newly registered	Mortality	Recovery
1974–1988	Constant	6.600	-42.715	11.781	10.524
	NO_2	-14.396	5396.263	299.900	614.598
	t value	-0.371	2.768***	2.582***	1.484*
	Adjusted R^2	-0.007	0.105	0.098	0.021
1988–2007	Constant	4.417		10.011	19.177
	NO_2	5.926		517.227	-150.001
	t value	0.484		4.862***	-0.553
	Adjusted R^2	-0.002		0.261	-0.010

Significance at * 20%, **5%, ***1%

Table 4. Results of correlation 1974–1988 and 1989–2007 (Eq 2-Eq5)

		Certified patients	Newly registered	Mortality	Recovery
1974–1988	Constant	1.317	1.115	-10.063	66.369
	NO_2	68.730	17.384	537.922	-622.152
	t value	1.466*	2.083**	5.520***	-1.679*
	Dummy	3.071	-1.010	15.707	-29.399
	t value	2.981***	-4.335***	6.049***	-6.342***
	Adjusted R^2	0.052	0.274	0.468	0.433

Significance at * 20%, **5%, ***1%

Table 5. Results of correlation 1974–1988 with dummy variable (Eq 2-Eq5)

6. Resume of the chapter

The correlation between air pollution and health effects in the large Japanese cities presents a great challenge owing to the limited availability of data on the exposure to air pollution, health effects and the uncertainty of mixed causes. In this paper, methodology for quantitative relationships (between emission volume and air quality, and air quality and health effects) is analyzed with a statistical method: the correlation of air pollution reduction policy in Japan from 1974 to 1988.

6.1 Hypothesis 1 (Correlation between emission volume and air quality)

Hypothesis 1 states that the total volume of emissions affects the air quality. Emissions volume data from stationary and mobile sources were collected only from the 23 Tokyo wards, Kawasaki (2 wards), and Osaka whole city. Amagasaki were not available. Kita Kyushu was not covered by the NOx-PM certified area. The emission volumes data from mobile sources were not available in Kita Kyushu.

For air quality data, the annual averages were collected from environmental white papers of the areas of hypothesis 1. The health effect data—Tokyo (23 wards), Kawasaki (2 wards), Osaka, Amagasaki, and Kita Kyushu—was collected.

The relationship between emission volumes and air quality was clarified using data on emission volumes and air quality in Tokyo, Kawasaki, and Osaka. The air quality (NOx, SPM) tended to worsen with increases in emission volumes (from both stationary and mobile sources). The relationship between total emissions and air quality was found to be significant, which supports hypothesis 1. The correlation analysis of emission volume, and air quality suggests that NOx levels worsen according to increases in NO2.

6.2 Hypothesis 2 (Correlation between air quality and health effects)

Hypothesis 2 states that the air quality influences the number of certified patients. The health effect data are adopted from the compensation system for two reasons. First, local governments issue environmental white papers which indicate the number of certified, deceased, newly registered, and recovered patients in the compensation system. These consistent data are useful for checking the reliability of the calculations. Second, to be a certified patient, a patient must be a non-smoker. The data include classified patients with one of the four certified diseases. In addition, the data include classified patients affected by diseases caused by air pollution.

In the compensation system, patients who suffer from one of the four certified respiratory diseases in a certified area can be registered with the local government. For air quality data, the annual averages were collected from environmental white papers of each local government. The health effect data—Tokyo (19 wards), Kawasaki (2 wards), Osaka whole city, Amagasaki, and Kita Kyushu—was collected. The annual averages were collected from environmental white papers of the areas of hypothesis 2 for air quality data.

Based on previous literature and on its strong relationship to health effects, NOx was selected for the analysis. In order to include all the pollutants, the Air Quality Index (AQI) data were also revised. The results of the AQI data confirm that PM and NOx are the main indicators for the assessment of the health effects of air pollution.

An amendment in the compensation system (1988) clearly affects the data trends, so this period is divided into two stages: 1974–1988 (before the amendment) and 1989–2007 (after the amendment). To increase accuracy, the coefficients were re-examined by separating the period into two stages, namely, 1974–1988 and 1989–2007.

The correlation is estimated by using a simple linear model to support hypothesis 2. Most of the coefficients would be expected to be positive; however, there were some inverse relationships. Estimates based only on air quality data may lead to overestimation or underestimation of the number of certified, of mortality, recovered patients, or newly registered patient rates.

The relationship between certified patients and NO2 was estimated to be significant. The relationship between NO2 and the number of death was also estimated to be significant. When the air quality worsens, mortality increases. When the three outliers of Osaka were excluded, R2 and the t-value improved. The recovery data exhibited a large variance among the four areas; therefore, the relationship between NO2 and recovery rate was not significant. When the outliers of Osaka and Kita Kyushu from the end of the 1970s and the beginning of the 1980s were excluded, the estimate was significant. The recovery data may have affected the number of certified patients. The relationship between NO2 and the number of new certified patients was estimated to be significant.

The relationship between air quality (NO2) and health effects was found to be significant, which support hypothesis 2. When NO2 worsens, certified patients, mortality rates and newly certified patients increase, according to the data from 1989 to 2007 with dummy variable analysis.

6.3 Conclusion and further issues

The discussion focuses on how to choose the methodology for correlation analysis of the area, time period, pollutants, air quality data, emission volume, and health effects in the large Japanese cities. In order to capture the data related to health problems caused by air pollution, we focused on the data from the Japanese compensation system because these data represent respiratory diseases caused by air pollution.

This chapter establishes a methodology by paying attention to historical events of the compensation system. The previous studies suggest that the data set of respiratory disease caused by air pollution were not separated from the other risks. This research collected existent data of health effects caused by air pollution. The methodology is designed for application to large Asian cities.

This research also indicates some improvement points. The further issues are the followings.

- Only outliner was detected for revision of data. It also needs to check all the data.
- The area used for analysis should cover all the compensation areas in Japan. Increasing the analysis area can also raise the accuracy of the coefficients.
- Air Quality Index (AQI) is an index for reporting air quality with regard to public health [USEPA http://www.airnow.gov/]. A lower AQI indicates better air quality. PM and NO2 values were converted to AQI values in all five regions.
- For the correlation analysis, it must be required adjustment on health effect data by age, gender and population.

- Inverse relationships are observed in the data from Japan, but the replacement of outliers by dummy variables yields significant estimation results. These results suggest the possibility of introducing other variables—such as weather, lifestyle, and age—for further analysis. The area used for analysis should cover all the compensation areas in Japan. Increasing the analysis area can also raise the accuracy of the coefficients for large Asian cities.

7. Acknowledgment

This research was conducted as a part of the research for "Economic analysis for emission reduction measures from automobile sector in large Asian cities". I am grateful to Mr. Hirokazu Sugizaki (ERCA), Mr. Toshiro Kokaji (Safety and Environmental Bureau, Osaka Prefecture Government), Mr. Yamada (Environmental Protection Division, Kita Kyushu City Government), Mr. Satoshi Kakurai (NEXCO), Mr. Michiya Kousaka (Library of Health and Safety Center, Tokyo Metropolitan Government) for data, librarians Tokyo Metropolitan Institute for Environmental Protection for advice of data collection, many colleagues of Kashima Seminar (Chuo University), Dr. Tazuko Marikawa (JARI), Dr. Naomi Tsukue (JARI), Dr. Koichiro Ishii (Tokyo Metropolitan Institute for Environmental Protection) for discussion. I am solely responsible for all the errors in this paper.

8. References

Air Pollution Division Pollution Regulation Bureau Tokyo Metropolitan Government. Air Pollution Emission Survey (in Japanese). Tokyo Metropolitan Government, Tokyo, 1977, 41-44.

Amagasaki City Government. Amagasaki Statistical Yearbook (in Japanese). Amagasaki City Government, Amagasaki, 2008, 129-130.

Amagasaki City Government. Environment in Amagasaki (in Japanese). Amagasaki City Government, Amagasaki, 2008, 23. Available at < http://www.city.amagasaki.hyogo.jp/sogo_annai/toukei/033amakan.html> (Accessed 20 June 2009).

Amagasaki City Government. Environment in Amagasaki. (in Japanese). Amagasaki City Government, Amagasaki, 1998, 125-128.

Amagasaki City Government. Environment in Amagasaki. (in Japanese). Amagasaki City Government, Amagasaki, 2001, 197.

Amagasaki City Government. Environment in Amagasaki. (in Japanese). Amagasaki City Government, Amagasaki, 2006, 129-130.

Amagasaki City Government. Pollution in Amagasaki (Amagasaki Environmental White Paper) (in Japanese), Amagasaki City Government, Amagasaki, 1974, 192-195.

Amagasaki City Government. Pollution in Amagasaki. (in Japanese). Amagasaki City Government, Amagasaki, 1970, 221-228.

Amagasaki City Government. Pollution in Amagasaki. (in Japanese). Amagasaki City Government, Amagasaki, 1976, 9.

Amagasaki City Government. Pollution in Amagasaki. (in Japanese). Amagasaki City Government, Amagasaki, 1985, 164-165.

Environmental Bureau of Tokyo Metropolitan Government. Total Automotive Emission Reduction Projection in the Future (in Japanese). Tokyo Metropolitan Government, Tokyo, 2000.

Environmental Bureau of Tokyo Metropolitan Government. Total Emission Reduction Plan Promotion Management Survey (in Japanese). Tokyo Metropolitan Government, Tokyo, 1994, 129.

Environmental Bureau of Tokyo Metropolitan Government. Total Emission Reduction Plan Promotion Management Survey (in Japanese). Tokyo Metropolitan Government, Tokyo, 1996, 128.

Environmental Bureau of Tokyo Metropolitan Government. Total Emission Reduction Plan Promotion Management Survey (in Japanese). Tokyo Metropolitan Government, Tokyo, 2000, 10.

Environmental Bureau of Tokyo Metropolitan Government. Total Emission Reduction Plan Promotion Management Survey (in Japanese). Tokyo Metropolitan Government, Tokyo, 2004, 6.

Environmental Bureau of Tokyo Metropolitan Government. Total Emission Reduction Plan Promotion Management Survey (in Japanese). Tokyo Metropolitan Government, Tokyo, 2005, 8.

Environmental Bureau of Tokyo Metropolitan Government. Total Emission Reduction Plan Promotion Management Survey (in Japanese). Tokyo Metropolitan Government, Tokyo, 2009, 10.

Environmental Bureau of Tokyo Metropolitan Government. Total PM Emission Survey 2003 (in Japanese). Tokyo Metropolitan Government, Tokyo, 2000, 6-12.

Environmental Protection Bureau of Tokyo Metropolitan Government. Trend of Pollution from Automobile (in Japanese). Tokyo Metropolitan Government, Tokyo, 1984. 6.

Environmental Restoration and Conservation Agency of Japan (ERCA). The Pollution-related Health Damage Prevention Program. Available at <http://www.erca.go.jp/english/activities/ac_02.html> (Accessed 23 July 2008).

ERCA. Pollution-related Health Damage Compensation. ERCA annual report. Chapter 3.

Fukuoka Prefecture Government. Fukuoka Environmental White Paper (in Japanese). Fukuoka Prefecture, Fukuoka, 1970, 15.

Fukuoka Prefecture Government. Fukuoka Environmental White Paper (in Japanese). Fukuoka Prefecture, Fukuoka, 1974, 26.

Fukuoka Prefecture Government. Fukuoka Environmental White Paper (in Japanese). Fukuoka Prefecture, Fukuoka, 1978, 17.

Hyougo Prefecture Government (2003). Automobile NOx and PM Emission Reduction Plan. Hyougo Prefecture Government. Hyougo, 1.

Iwai K, Mizuno S, Miyasaka Y and Mori T, Correlation between suspended particles in the environmental air and causes of disease among inhabitants: Cross-sectional studies using the vital statistics and air pollution data in Japan, Environmental Research, 99, 2005, 106-117.

Kanagawa Prefecture Government. Kanagawa Environmental White Paper (in Japanese). Kanagawa Prefecture Government, Kanagawa, 1973, 24.

Kanagawa Prefecture Government. Kanagawa Environmental White Paper (in Japanese). Kanagawa Prefecture Government, Kanagawa, 1982, 209.

Kanagawa Prefecture Government. Kanagawa Environmental White Paper (in Japanese). Kanagawa Prefecture Government, Kanagawa, 1983, 398.

Kanagawa Prefecture Government. Kanagawa Environmental White Paper (in Japanese). Kanagawa Prefecture Government, Kanagawa, 1985, 187.

Kanagawa Prefecture Government. Kanagawa Environmental White Paper (in Japanese). Kanagawa Prefecture Government, Kanagawa, 1988, 105.

Kanagawa Prefecture Government. Kanagawa Environmental White Paper (in Japanese). Kanagawa Prefecture Government, Kanagawa, 1994,129.

Kawasaki City Government. Environment in Kawasaki (in Japanese). 2008, 12.

Kawasaki City Government. Kawasaki City Environmental White Paper (in Japanese). Kawasaki City Government, Kawasaki, 2009, 12.

Kawasaki City Government. Pollution in Kawasaki (Kawasaki Environmental White Paper) (in Japanese), Kawasaki City Government, Kawasaki, 1994, 103.

Kawasaki City Government. Pollution in Kawasaki (Kawasaki Environmental White Paper) (in Japanese). Kawasaki City Government, Kawasaki, 1993, 26.

Kawasaki City Government. Statistics in Kawasaki (in Japanese). Kawasaki City Government, Kawasaki, 2008 Available at <http://www.city.kawasaki.jp/20/20tokei/home/tokeisyo/tokeisyo20/html/mokuji.htm> (Accessed 2 May 2009).

Kita Kyushu City Government. Environment in Kita Kyushu (in Japanese). Appendix, Kita Kyushu City Government, Kita Kyushu, 2008, 11.

Kita Kyushu City Government. Environment in Kita Kyushu (in Japanese). Kita Kyushu City Government, Kita Kyushu, 2008, Appendix 11-73.

Kita Kyushu City Government. Kita Kyushu Air Pollution Prevention Plan (in Japanese). 2007. Available at <http://www.city.kitakyushu.jp/pcp_portal/PortalServlet?DISPLAY_ID=DIRECT&NEXT_DISPLAY_ID=U000004&CONTENTS_ID=1964> (Accessed 20 June 2009).

Kita Kyushu City Government. Pollution in Kita Kyushu (in Japanese). Appendix, Kita Kyushu City Government, Kita Kyushu, 1987, 215.

Kita Kyushu City Government. Pollution in Kita Kyushu (Kita Kyushu Environmental White Paper) (in Japanese), Kita Kyushu City Government, Kita Kyushu, 1974, 246-247.

Kita Kyushu City Government. Pollution in Kita Kyushu. Kita Kyushu City Government, Kita Kyushu, 1982, Appendix 19.

Kita Kyushu City Government. Pollution in Kita-Kyushu (in Japanese). Kita Kyushu City Government, Kita Kyushu, 1981, Appendix 243.

Kita Kyushu City Government. Pollutiont in Kita-Kyushu (in Japanese). Kita Kyushu City Government, Kita Kyushu, Appendix 1974, 87.

Makino,K. Mortality from Lung Cancer and Air pollution in Tokyo. Environmental Science, 4,1, 1996, 025-036.

Matsuura, I. Process of establishment of Pollution-related Health Damage Compensation Law, Jurist, 1984, No.821, 29-35.

Matsuura, I. Process of establishment of Pollution-related Health Damage Compensation Law, Jurist, No.822, 80-85.

Matsuura, I. Process of establishment of Pollution-related Health Damage Compensation Law, Jurist, No.824, 91-97.

National Institute for Environmental Studies. Available at < http://www-gis.nies.go.jp/ > (Accessed 27 August 2008).

Osaka City Government. Osaka City Statistical Yearbook (in Japanese). Osaka, Osaka City Government, Chapter 20, Table 25.

Osaka City Government. Osaka City White Paper (in Japanese). Osaka, Osaka City Government, 1975, 48-49.

Osaka City Government. Osaka City White Paper (in Japanese). Osaka, Osaka City Government, 1976, 76.

Osaka City Government. Osaka City White Paper (in Japanese). Osaka, Osaka City Government, 1991, 25.

Osaka City Government. Osaka City White Paper (in Japanese). Osaka, Osaka City Government, 2001, 36.

Osaka City Government. Osaka City White Paper (in Japanese). Osaka, Osaka City Government, 2004, 31.

Osaka City Government. Osaka City White Paper (in Japanese). Osaka, Osaka City Government, 2007, 36.

Osaka City Government. Pollution in Osaka (Osaka Environmental White Paper) (in Japanese), Osaka City Government, Osaka, 1974, 169.

Osaka City Government. Statistical data of certificated Patients and cancelation. (Personal contact: 04 July 2009).

Pollution Bureau of Tokyo Metropolitan Government. Air Pollutants Emission Estimation Survey (in Japanese). Tokyo Metropolitan Government, March, 1971, 54-55.

Pollution Regulation Bureau of Tokyo Metropolitan Government. Air Pollutant Emission Factor Survey (in Japanese). Tokyo Metropolitan Government, Tokyo,July 1973, 149.

Sunya J, D Jarvis and T Gotschi, et al. Chronic bronchitis and urban air pollution in an international study. Occupational Environmental Medicine, 2006, 63, 836-843.

Tokyo Metropolitan Government. Number of Certificated Patients and Mortality, Tokyo Health Annual Report. Tokyo Metropolitan Government, Tokyo, 1985, No.37, 247-262.

Tokyo Metropolitan Government. Number of Certificated Patients and Mortality. Tokyo Health Annual Report. Tokyo Metropolitan Government, Tokyo, 1995, No.47, 276-277.

Tokyo Metropolitan Government. Number of Certificated Patients and Mortality. Tokyo Health Annual Report. Tokyo Metropolitan Government, Tokyo, 1999, No.51, 272-273.

Tokyo Metropolitan government. Pollution Prevention Plan in Tokyo Region (in Japanese), Tokyo Metropolitan Government, Tokyo, 1978.

Tokyo Metropolitan Government. Tokyo Environmental White Paper (in Japanese). Tokyo Metropolitan Government, Tokyo, 1983, 79.

Tokyo Metropolitan Government. Tokyo Environmental White Paper (in Japanese). Tokyo Metropolitan Government, Tokyo, 1995, 75.

Tokyo Metropolitan Government. Tokyo Environmental White Paper (in Japanese). Tokyo Metropolitan Government, Tokyo, 1998, 32.

Tokyo Metropolitan Government. Tokyo Regional Pollution Prevention Plan (in Japanese). Tokyo Metropolitan Government, Tokyo, 1978, 6.

Tokyo Metropolitan Government. Tokyo Statistical Yearbook. Tokyo Metropolitan Government, 1998, 32.

UK Department of Health (UK DH). Cardiovascular Disease and Air Pollution. A report by the Committee on the Medical Effects of Air Pollutants. UK DH, London, 2006, 21-137.

US EPA, Air Quality Index, USEPA, Washington D.C. Available at < http://www.airnow.gov/index.cfm?action=aqibasics.aqi > (Accessed 30 June 2009).

USEPA. The Benefits and Costs of the Clean Air Act 1990 to 2010. USEPA, Washington D.C.1999, D-58.

World Health Organization. The World Health Report. WHO, Geneva

8

Evaluation of Dry Atmospheric Deposition in Two Sites in the Vicinity of Fuel Oil-Fired Power Plants in Mexico

Cerón-Bretón Rosa María[1],
Cerón-Bretón Julia Griselda[1], Cárdenas-González Beatriz[2],
Ortínez-Álvarez José Abraham[2], Carballo-Pat Carmen Guadalupe[1],
Díaz-Morales Berenice[1] and Muriel-García Manuel[3]
[1]Research Center on Environmental Sciences, Autonomous University of Carmen, Botanical Garden, Carmen City, Campeche,
[2]National Center for Training and Environmental Research, National Institute of Ecology, México City,
[3]Mexican Institute of Petroleum, Carmen City, Campeche, México

1. Introduction

Airborne pollutants are involved in a complex cycle in the atmosphere that begins with their release and finishes with their removal and deposition on terrestrial and aquatic ecosystems. Some pollutants as dust or crustal particles remain without changes during this cycle; however, gaseous components undergo physical and chemical changes. This is the case of combustion gases such as SO_2 and NO_x. It is well known that atmospheric particles and gases including those pollutants emitted from natural (biogenic, volcanic activity, marine aerosol, and so on) and anthropogenic sources (vehicular and industrial emissions) are removed by wet and dry deposition processes. The phenomenon of acidic atmospheric deposition has been largely studied in different sites and locations around the world during the last years (Tsitouridou & Anatolaki, 2007), and it has been found that episodes of acid precipitation are directly related to industrial emissions, this is the case of coal and fuel burning power plants (Flues et al., 2002).

In Mexico, 72 % of generated energy comes from burning of fossil fuels ("combustoleo"[1], natural gas and coal). Particularly, in the last decade, "combustoleo" has been the fuel more used in the main productive sectors. Currently, 66.8% of energy is generated from power plants burning "combustoleo", resulting in fly ash emission, and large amounts of gases considered as acid precursors (such as SO_2 and NO_x) that contribute to the acid precipitation in the vicinity of the plants (Zuk et al., 2006). Considering that the residence times in the atmosphere for both acid precursors are different, their deposition rates are different too. Thus,

[1] It is a blend of residua and gas oil cutter stock and meets the viscosity requirements of ASTM D396. It is commonly used in power plants in large steam boilers, drying kilns and ovens.

NO_x are removed in an efficient way by wet deposition, being deposited around the emission source. On the other hand, since SO_2 has a residence time of 13 days, undergoes long-range transport, being deposited in distant places from the emission point. For this reason NO_x are referred as local pollutants whereas SO_2 is considered as a regional pollutant.

Qualitative and quantitative assessment of atmospheric deposition in regions probably impacted by anthropogenic sources is essential for understanding regional variations, to determine if critical loads are exceeded, to assess time-trends in a long-term, and to relate sources contributing to the acidity with local and regional meteorology. Long-term changes in air quality and atmospheric deposition are not obvious, in some cases due to daily and seasonal variations in parameters such as wind, temperature, precipitation and atmospheric circulation patterns which influence on dispersion, transport and deposition of pollutants. For this reason, it is important to measure wet and dry deposition fluxes at various sites all over the country during long-time periods to provide an outline of the main chemical characteristics of the deposition and to assess their trends along time. Concerning monitoring networks, air quality data are useful to determine if the current regulations are effective in the improvement of the air quality, or if it is necessary to promulgate new ones.

In 2009, National Institute of Ecology [INE] sponsored the design and operation (during two years) in a preliminary phase of the Mexican Atmospheric Deposition Network, integrated by four stations (considering both polluted and natural sites). In addition, in 2010, the Mexican Environmental Agency [SEMARNAT] supported a research project focused on the study of atmospheric deposition in five sites located in the surroundings of power plants burning "combustoleo". The aim of this work was to assess the chemical composition of dry deposition in two of these sites: Los Petenes and Tula de Allende, both sites are located in the vicinity of power plants. The first one is a Ramsar site and a Biosphere Reserve located in Yucatan Peninsula at the southeast of the country, and the second one is a polluted site located in an important industrial region of Central Mexico, in whose vicinity there are important historic monuments.

2. Sampling sites description

Dry deposition samples were collected during 2009 and 2010 at two sampling sites: Los Petenes Biosphere Reserve, Campeche and Tula de Allende, Hidalgo.

Sites	Classification area	Altitude (m asl)	Latitude	Longitude
Los Petenes, Campeche	Ramsar Site Biosphere Reserve	4	20° 51´ 30″ N	90° 45´15″ W
Tula de Allende, Hidalgo	Urban and Industrial, with the presence of historic monuments in the vicinity	2020	20° 05´08″ N	99° 33´05″ W

Table 1. Main features of the sampling sites.

Sampling sites were selected considering proper conditions of accessibility and safety. The specific location and the main characteristics of these sampling sites are shown in Figure 1 and Table 1.

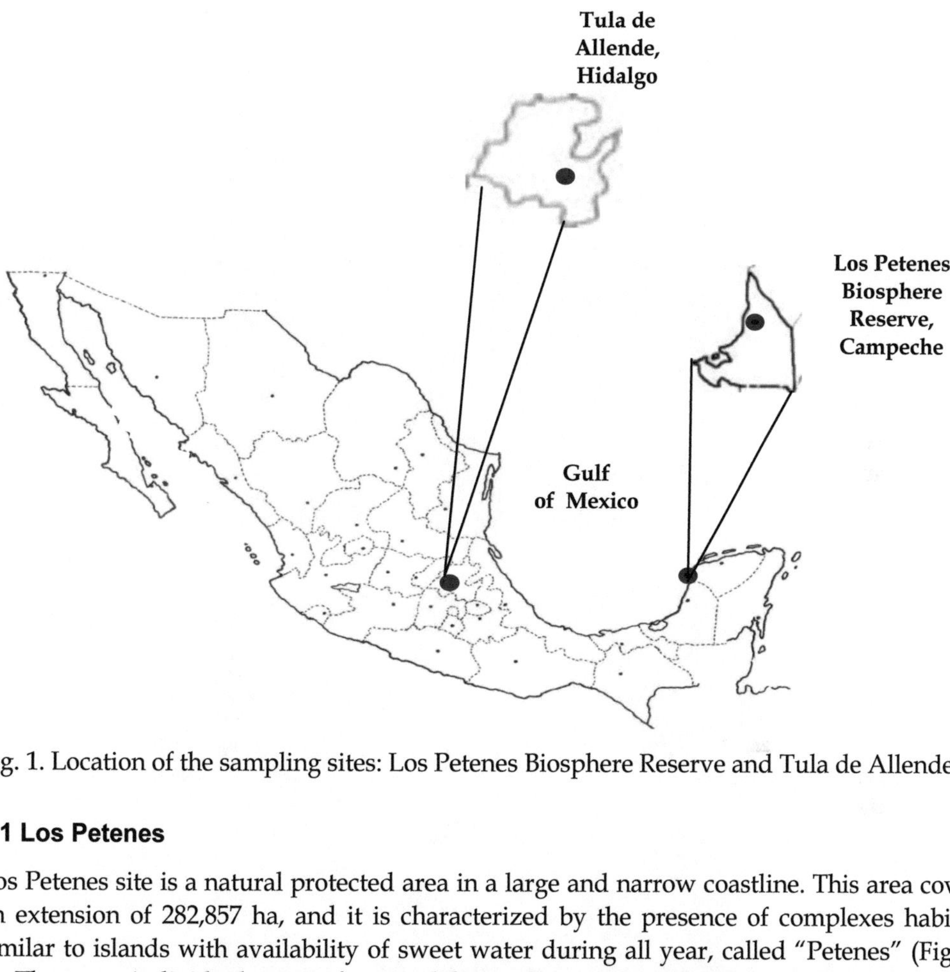

Fig. 1. Location of the sampling sites: Los Petenes Biosphere Reserve and Tula de Allende.

2.1 Los Petenes

Los Petenes site is a natural protected area in a large and narrow coastline. This area covers an extension of 282,857 ha, and it is characterized by the presence of complexes habitats similar to islands with availability of sweet water during all year, called "Petenes" (Figure 2). These are individual areas of tropical forest that reach a diameter ranging from a few meters to kilometers surrounded by wetlands, with a great diversity of flora and fauna. These habitats are only found in Cuba, Florida and Mexico, generally having a "Cenote" in its central part, originating vegetation that flourishes in rings of similar centers (Figure 2). Los Petenes is located within an eco-region that includes Ria Celestún Biosphere Reserve and the Natural Protected Area El Palmar in Yucatan State. Los Petenes is a Ramsar Site and constitutes a representative bio-geographic area of an ecosystem not significantly perturbed, in which representative species of national biodiversity live, including endemic, threatened and in risk of extinction species.

The climate is warm sub-humid (A_w) with summer rains with the presence of the mid-summer drought, and forest fires occurring frequently. Annual rainfall ranges from 729 to 1049 mm and temperature ranges from 26.1 and 28.8 °C. Los Petenes are characterized by four types of soils: a) Sandy soils at the coast (calcareous regosols), b) holomorphic soils in marshes (histosols), c) shallow and rocky soils (rendzina type) and clay soils (gleysols). Specific sampling site was located within the Center for Conservation of Wildfire Life of Autonomous

University of Campeche at 15 km from Campeche City and at 10 km from the fuel oil-fired power plant "Lerma". Characteristics and capacity of this power plant are shown in Table 2.

Fig. 2. Panoramic view of Los Petenes Biosphere Reserve.

Power Plant, Location	Units	Operation starting date	Installed effective capacity (MW)
Lerma; Campeche, Mexico	4 (combined cycle)	September 9, 1976	150
Francisco Pérez Ríos; Tula, Hidalgo	9 (steam and combined cycle technologies)	June 30, 1991	2000

Source: Zuk et al., 2006.

Table 2. Main characteristics of the power plants located in the vicinity of the sampling sites considered in this study.

2.2 Tula de Allende

The second study site is Tula de Allende, Hidalgo located at the Central Region of Mexico, at 20°03′ N and 99° 20′ W with an altitude of 2020 m asl, at 70 km from Mexico City and 3 km from the fuel oil-fired power plant "Francisco Pérez Ríos" located along the Tula-Vito-Apaxco industrial complex. Due to large volumes of atmospheric pollutants generated by the Tula-Vito-Apaxco corridor, this region has been classified as a critical zone by the Mexican environmental authorities. Industries located in this area such as an oil refinery, a plant energy (which is one of biggest power plants in Latin America), and two important cement plants, contribute to the development of the region (Figure 3).

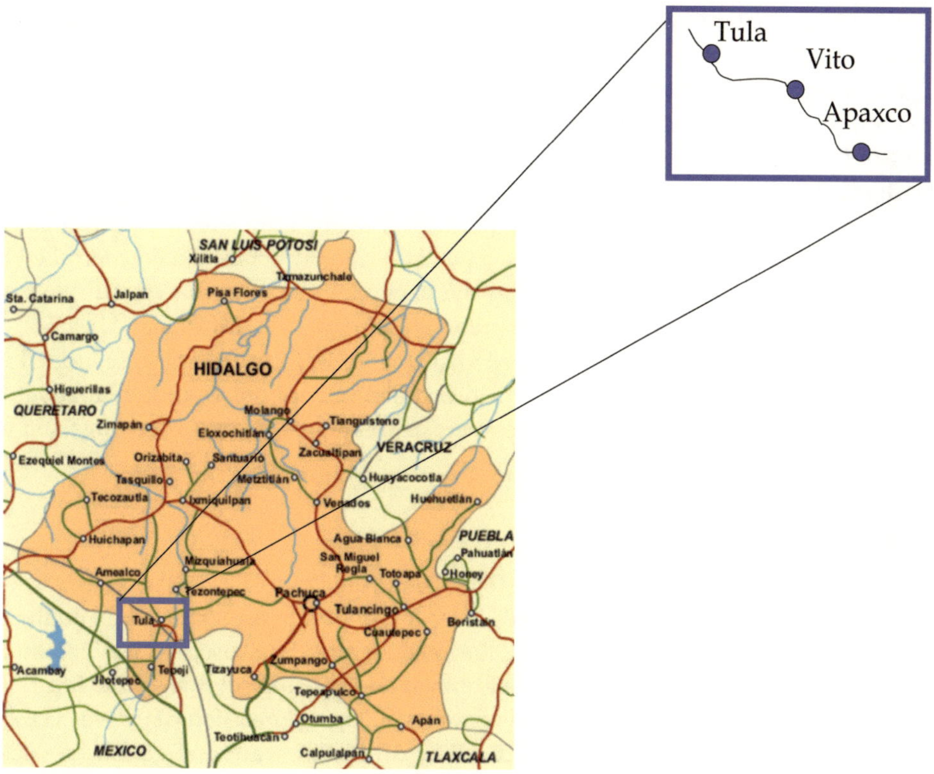

Fig. 3. Location of the Tula-Vito-Apaxco Industrial Complex in the vicinity of Tula de Allende.

According to the latest information from the environmental authorities, 323, 000 tons per year of SO_2 and 44, 000 tons per year of NO_x are released in this region. The main sources are the "Miguel Hidalgo" Oil Refinery and the "Francisco Pérez Ríos" Power Plant. "Miguel Hidalgo" Refinery processes 296,000 barrels per day of crude oil, representing 20% of the total refining capacity in the country. Other industries such as cement plants, open-sky mines and agricultural activities are also responsible of emission of important quantities of particulate matter. Therefore, the coexistence in the area of these industries could produce secondary compounds that can be potentially dangerous and risky for public health, ecosystems, and historic monuments. Nearby to this site, the ruins of the ancient capital city of Toltecs, also known as "Tula" or as "Tollan", are located. This site is characterized by a semi-arid climate (BSh) with an annual average rainfall of 600 mm, and an average temperature ranging between 16 and 17°C.

3. Experimental

3.1 Sampling

Dry deposition samples were collected on a weekly basis for a period of two years (from August 2009 to December 2010). Procedures to collect wet deposition are available anywhere since several years (Galloway & Likens, 1976), nevertheless, standard procedures to collect dry deposition are not available. Quantifying dry atmospheric input to natural surfaces is a difficult task due to the complex chemical processes involved (Balestrini et al., 2000). However, investigations using plastic materials, Teflon® and glass surfaces to collect dry deposition have demonstrated the importance of the physical, chemical and geometrical features of the surrogate surface in selecting the depositing material (Dasch, 1985). In this study, dry deposition was assessed using a surrogate surface according to the methodology used by Alonso et al (2005). Sampling was performed using automated wet-dry deposition collectors. Samplers were equipped with two polyethylene buckets and a lid controlled by a humid sensor, which moves depending on the beginning and the end of the rain event. Three nylon filters (Nylasorb, Pall Corporation, Gelman Lab, Michigan, USA; 47 mm diameter, 1 µm pore-size) per site were used as surrogate surfaces. Filters were located on a support within the dry deposition bucket and exposed in horizontal position at about 1 m height above ground. Dry deposition fluxes were calculated based on surface area and time exposure.

3.2 Analysis

After exposition, filters were stored at -18 °C until analysis. Three nylon filters were extracted with 80 ml of deionized-distilled water for 15 min in an ultrasonic bath, the first one was used to determine cations, and the second filter was used to analyze anions. Ammonium, pH and conductivity were determined on extracted solution of the third filter. pH and conductivity measurements were obtained by using precision pH meter (TERMO ORION 290) and a conductivity meter (CL 135). All plastic ware and glassware used to prepare standard solutions, for digestion of samples and for chemical analysis were rigorously washed, brushed and rinsed with distilled water. Plastic ware used for digestion of samples for analysis of cations by atomic absorption spectrophotometer was completely immersed during 24 h in a 20% ultrapure nitric acid bath (J.T. Baker, AA Grade), then rinsed several times with deionized water type I (Hycel) and sealed into double plastic bags. Before using, all material was again rinsed with deionized water. Extract from the first filter was digested in Teflon® closed flasks (Cole-Parmer) of 100 ml, using an autoclave equipment as energy source; subsequently, Na^+, K^+, Ca^{2+} and Mg^{2+} were analyzed using an atomic absorption spectrophotometer (Thermoscientific ICE 3300). Extract from the second filter was analyzed for SO_4^{2-}, Cl^- and NO_3^- by ion chromatography (Agilent 1100) with 200 µl sample loop. Finally, extracted solution of the third filter was used to determine NH_4^+ colorimetrically by means of a spectrophotometer UV (HACH DR 2800). Standard solutions were prepared by dilution of certified standards (J.T. Baker). The detection limits were calculated as three times the standard deviation of six blank samples.

3.3 Quality assurance

Repeatability was guaranteed by 3 replicate measurements for each sample. Results showed a coefficient of variation < 5% for all elements measured. To verify whether major components had been measured in the analysis, two kinds of quality control were applied. If linear regression analysis between measured anions and cations gives a regression coefficient minor than 0.9, it indicates that anions and organic acids (formate and acetate) were probably present but were not measured.

On the other hand, if linear regression analysis between the measured and the calculated conductivity gives a regression coefficient major than 0.9, it indicates that practically the majority of the ions were analyzed. Balance cation-anion obtained for Tula de Allende showed differences less than 10%, therefore, we considered that all the major ions were analyzed. In addition, measured conductivities in Tula de Allende were in agreement with calculated conductivities. Figure 4 shows the quality of the analytical data in Tula de Allende.

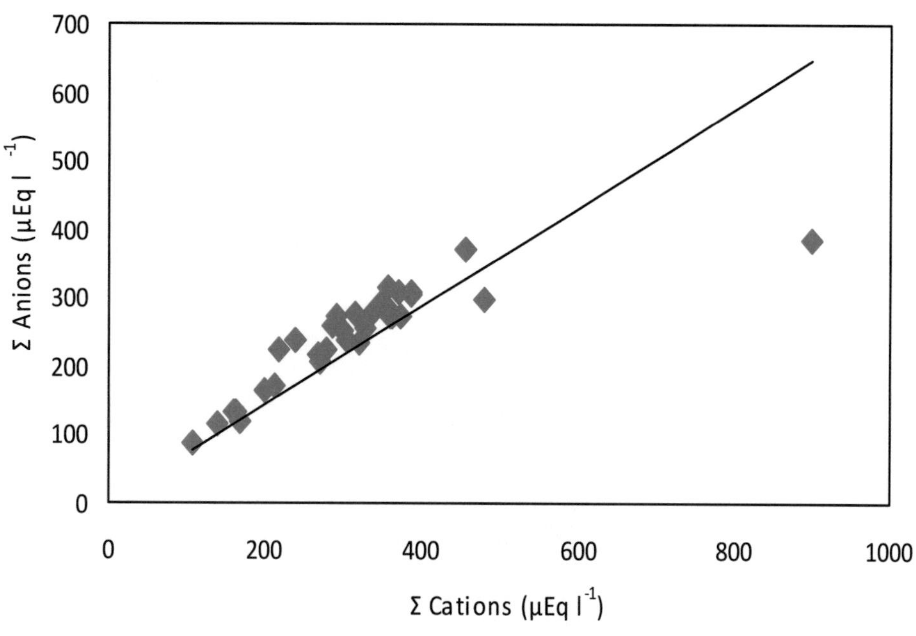

Fig. 4. Quality control (ionic balance) of chemical analysis for dry deposition samples collected at Tula de Allende.

In the case of Los Petenes Biosphere Reserve, conductivity differences were greater than 10%, and ionic balance suggests an ion deficit, probably due to the missing HCO_3^- and the presence of organic acids that are commonly found in natural sites. These short chain compounds were not specifically analyzed in this work.

3.4 Meteorological data

Two portable meteorological stations (Davies Inc, n.d.) that were operating during the whole study period in both sampling sites provided needed meteorological information. From these meteorological data, wind roses were constructed for each site by means of WRPLOT VIEW 6.5.2 (Lakes Environmental Inc, 2011).

To trace the origin of air masses for the whole study period, air-masses backward trajectories were calculated (48 h before) for both sampling sites by means of HYSPLIT-Hybrid Single Particle Lagrangian Integrated Trajectory Model (NOAA, n.d.). From calculated backward trajectories, it was observed that prevailing winds came from N-NE for Tula de Allende (Figures 5, 7) and from E for Los Petenes Biosphere Reserve (Figures 6,8).

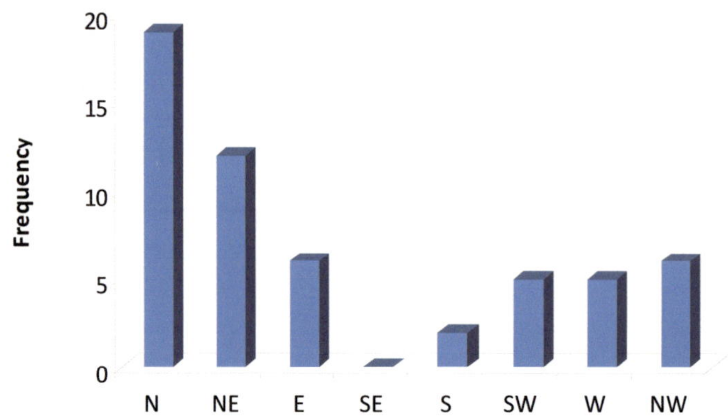

Fig. 5. Wind direction frequency distribution for Tula de Allende during the study period.

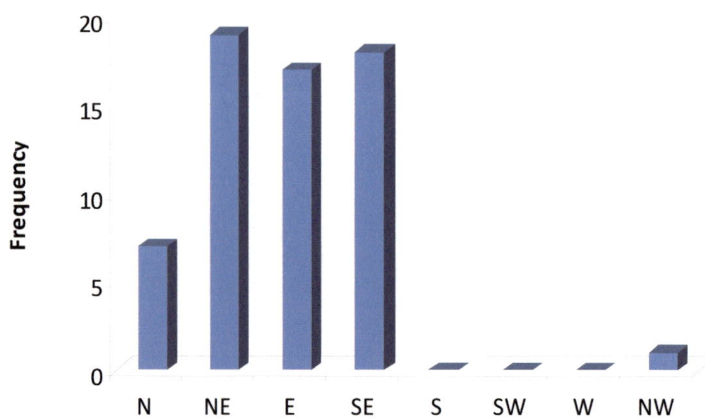

Fig. 6. Wind direction frequency distribution for Los Petenes Biosphere Reserve during the study period.

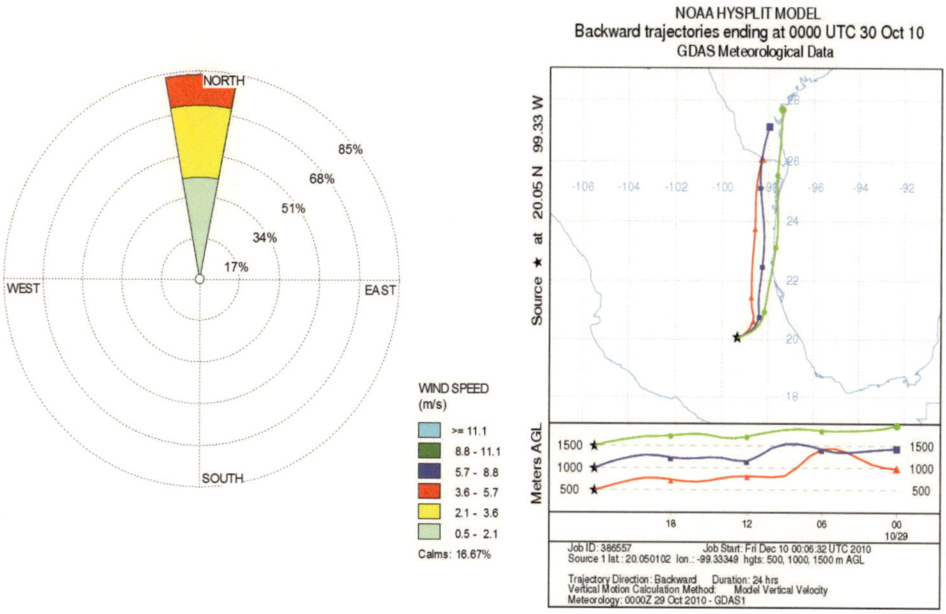

Fig. 7. Typical wind rose and backward air mass trajectory for Tula de Allende during the study period.

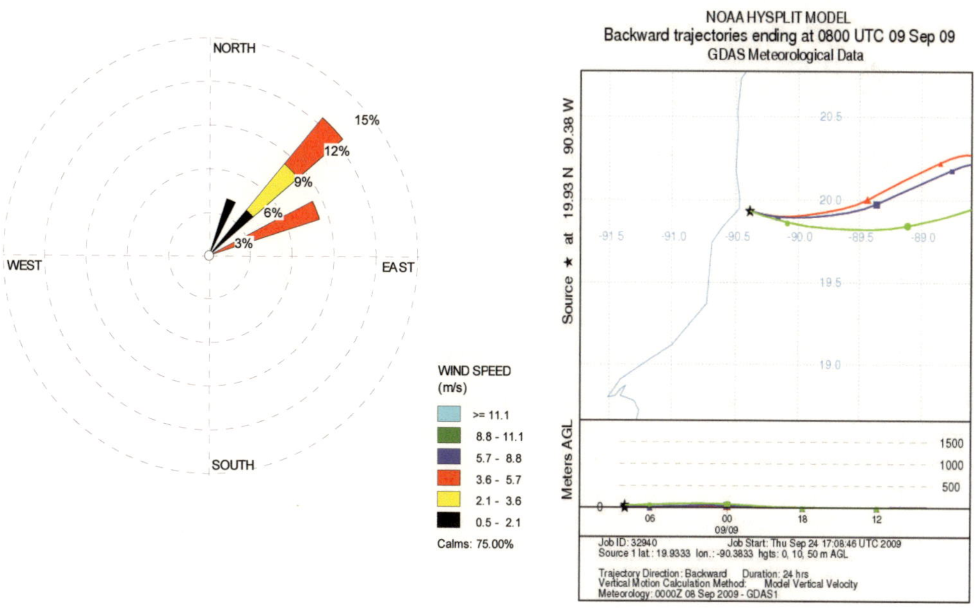

Fig. 8. Typical wind rose and backward air mass trajectory for Los Petenes Biosphere Reserve during the study period.

3.5 Statistical analysis

Pearson´s correlation analysis was applied to test the relationship among the total trace element concentrations for each sampling site. ANOVA (GLM) was performed to test the differences between each element.

Means comparison was also performed using the average concentration values for trace elements as well as pH values. The calculations were carried out using SAS 8.0 software (SAS Inc, 1998).

4. Results and discussion

4.1 Chemical composition

Mean concentration values and ionic abundance of principal ions in dry deposition at Los Petenes Biosphere Reserve and Tula de Allende are shown in Table 3.

Site	pH	Na^+	K^+	Ca^{2+}	Mg^{2+}	NH_4^+	Cl^-	SO_4^{2-}	NO_3^-
Los Petenes	5.9	2.00	5.79	70.51	16.15	132.75	60.49	14.61	90.63
	Ionic Abundance: $NH_4^+ > NO_3^- > Ca^{2+} > Cl^- > SO_4^{2-} > Mg^{2+} > K^+ > Na^+$								
Tula de Allende	5.77	153.6	4.30	109.50	26.20	20.50	103.50	83.30	56.40
	Ionic Abundance: $Na^+ > Ca^{2+} > Cl^- > SO_4^{2-} > NO_3^- > Mg^{2+} > NH_4^+ > K^+$								

Table 3. Mean values of pH and concentration (μEq l^{-1}) of dry deposition samples.

4.1.1 Los Petenes Biosphere Reserve

Ammonium was the most abundant ion in Los Petenes, probably due to agriculture practices and swine husbandry in the zone and biological decay. On the other hand, nitrate was second more abundant ion, probably attributed to biomass burning activities and emissions from a local power plant. The influence of forest fires on nitrate and ammonium levels has been reported in other works (Hegg et al, 1988; Galbally & Gillel, 1988). It is well known that forest fires produce great amounts of nitrogen oxides, which are rapidly converted to nitric acid in the atmosphere. Since marine aerosol does not contribute to the nitrate levels, it can be assumed that total NO_3^- content present in samples was in excess, and that it had an anthropogenic origin. In spite of Los Petenes is an important natural protected area, it cannot be considered a remote site, for this reason, nitrate levels (63.38 μEq l^{-1}) exceeded the background hemispheric levels reported for remote sites (2.8 μEq l^{-1}).

On the other hand, since Los Petenes can be considered as a coastal site (8 km from the coastline), it was required to estimate marine aerosol contribution to the sulfate levels, in order to determine the marine and anthropogenic fractions (commonly known as sulfate excess, $[SO_4^{2-}]_{xs}$). Sulfate excess levels found (12.66 μEq l^{-1}) were in agreement with the background hemispheric levels reported by Galloway et al (1982) for remote sites (10 μEq l^{-1}), suggesting that there was a minimal influence of local sources that contributed to this ion; this is in agreement to the residence time for SO_2 in the atmosphere. However, $(SO_4^{2-})_{xs}/NO_3^-$ was lower than other coastal sites (0.031), indicating a significant enrichment of nitrates. Clearly, there was a local source influencing on nitrate levels, probably forest fires occurring as a result of the mid-summer drought, and emissions coming from the power plant located at 10 km from the sampling site. Considering the residence times for SO_2 (13 days) and NO_2 (1 day), it can be assumed that Los Petenes site was not subjected to the input of long-range transported pollutants and probably only had the influence of local and temporal sources.

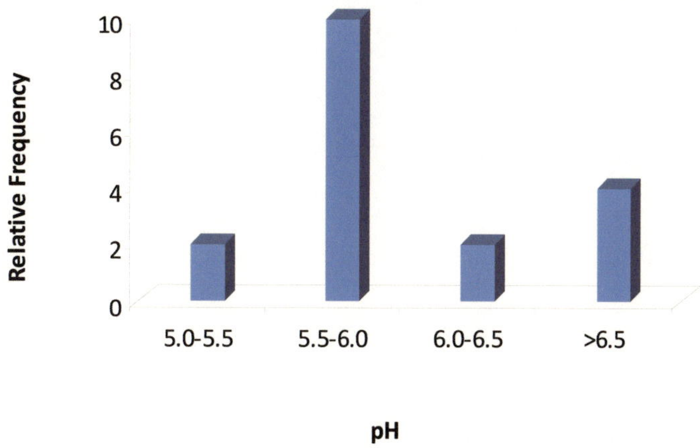

Fig. 9. pH frequency distribution for dry deposition samples collected at Los Petenes Biosphere Reserve.

pH values found in dry deposition samples showed a great variability, 22% of the samples were below of 5.6 (with a mean value of 5.9), and a negligible percentage (5.5%) was in the acid range (pH<5.0) (Table 3, Figure 9). In spite of that nitrate levels exceeded background hemispheric levels, pH values suggest that this site does not have acidity problems, owing to some cations probably played an important role in the neutralization process. This behavior is in agreement with the calcareous soils commonly found in Yucatan Peninsula which is constituted by alkaline particles participating in active way in the neutralization process.

4.1.2 Tula de Allende

Sulfate was the most abundant ion in Tula de Allende, probably attributed to anthropogenic sources. Tula de Allende is not a coastal site; therefore, it can be assumed that the total content of sulfate and nitrate is in excess (anthropogenic origin). Sulfate excess, $[SO_4^{2-}]_{xs}$) levels found (68.78 μEq l^{-1}) exceeded almost seven times the background hemispheric levels reported by Galloway et al (1982) for remote sites (10 μEq l^{-1}). In addition, nitrate levels (56.37 μEq l^{-1}) also exceeded the background hemispheric levels reported for remote sites (2.8 μEq l^{-1}). Both, nitrate and sulfate contributed to the acidity. Considering the residence times for SO_2 (13 days) and NO_2 (1 day), it can be assumed that this site had the influence of local sources (anthropogenic sources such as a local power plant and oil refinery) as well as the input of pollutants subjected to long-range transport process. From Table 3 and Figure 10, it can be observed that pH values found in dry deposition samples ranged from 5.3 to 6.7, with a mean value of 5.77. A great percentage of the samples (59%) was below of 5.6, but any sample was in the acid range (pH<5.0).

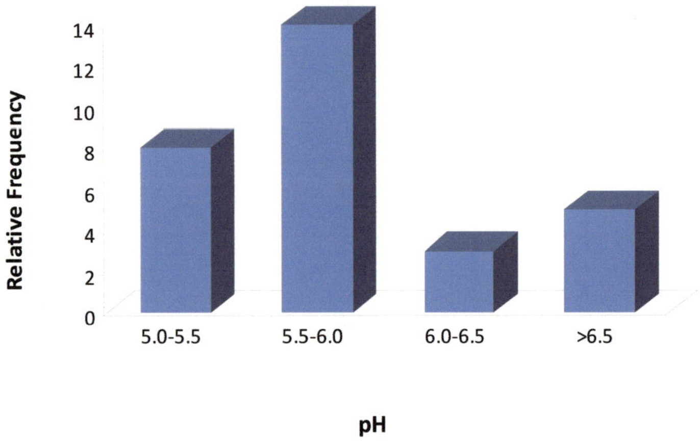

Fig. 10. pH frequency distribution for dry deposition samples collected at Tula de Allende.

In despite of nitrate levels exceeded the background hemispheric levels, it can be concluded that this site has enough buffer capacity to neutralize the acidity, probably due to Ca^{2+} (attributed to Portland cement plants located near this sampling site) playing an important role in the neutralization process.

4.2 Correlation analysis and time trends

The origin of elements in atmospheric deposition samples can be inferred by correlations between elements and let us to identify common sources, similar removal processes or a strong acid-base relationship.

4.2.1 Los Petenes Biosphere Reserve

From Pearson matrix (Table 4), it can be observed that nitrate ion showed a good correlation with Ca^{2+} ($r=0.89$ at $p<0.001$) and Mg^{2+} ($r=0.54$ at $p<0.001$), being evident a strong acid-base relation. In addition, Ca^{2+} was the third more abundant ion, indicating that this alkaline element, derived from crustal, was the main responsible of the neutralization process, resulting in pH values above of 5.6. NO_3^- - SO_4^{2-} correlated significantly ($r=0.63$ at $p<0.001$), suggesting that both ions had a common source.

Figure 11 shows the variations of ionic concentrations in dry deposition with time at Los Petenes Biosphere Reserve. Backward air masses trajectories analysis carried out using the HYSPLIT Model from NOAA demonstrates that winds during the sampling period had an East component. Therefore, it can be inferred that a specific source located in this direction (the power plant near to this site) possibly contributed to the significant levels of nitrate and sulfate.

Some temporal emissions as forest fires occurring as a result of the mid-summer drought could also contribute to these high levels of nitrate and sulfate. In Yucatan Peninsula, the mid-summer drought is associated with warm temperatures and the lack of humidity in soils favoring the conditions for forest fires. It can be observed that NO_3^- and NH_4^+ showed the same behavior, it suggests that both ions had a common source (forest fires). In addition, it can be observed that SO_4^{2-} and NO_3^- showed the same pattern, suggesting that both ions had a common origin, and being evident an enrichment of nitrates derived from local sources (power plant and biomass burning).

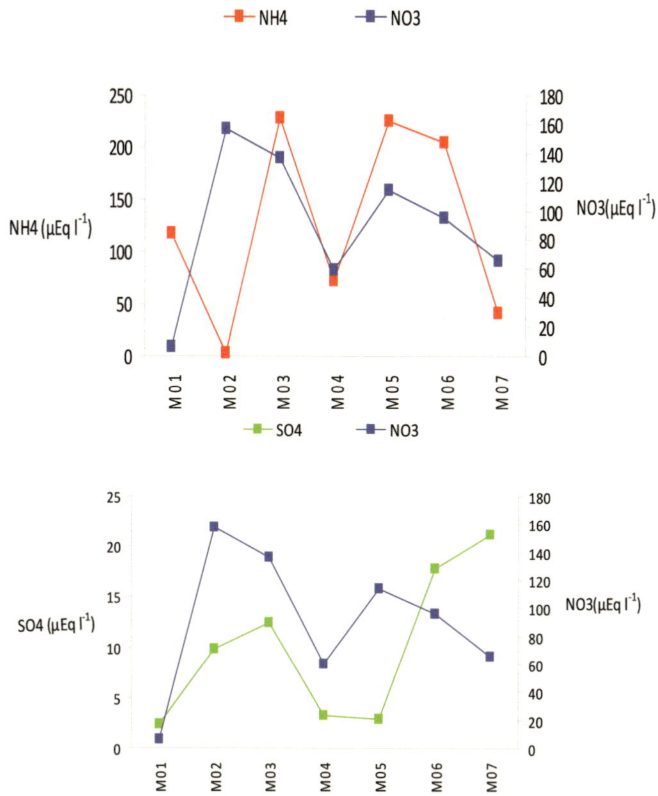

Fig. 11. Time-trends of ionic concentrations in dry deposition in Los Petenes.

	Na^+	K^+	Ca^{2+}	Mg^{2+}	NH_4^+	NO_3^-	Cl^-	SO_4^{2-}
Na^+	1							
K^+	0.41	1						
Ca^{2+}	0.52	0.41	1					
Mg^{2+}	0.02	0.09	0.65	1				
NH_4^+	0.26	0.21	0.26	0.14	1			
NO_3^-	0.51	0.27	0.89	0.54	0.18	1		
Cl^-	0.78	0.67	0.77	0.77	0.33	0.71	1	
SO_4^{2-}	0.51	0.37	0.55	0.04	0.26	0.63	0.21	1

Table 4. Pearson´s correlation matrix among trace elements for Los Petenes Biosphere Reserve.

4.2.2 Tula de Allende

Pearson matrix (Table 5) indicates that nitrate ion correlated significantly with SO_4^{2-} ($r=0.95$ at $p<0.001$) and Ca^{2+} ($r=0.74$ at $p<0.001$); whereas SO_4^{2-} correlated with Ca^{2+} ($r=0.79$ at p<0.001). It was completely evident a strong relationship acid-base between these ion pairs, suggesting that these ions were deposited as calcium nitrate and sulfate. These results support the fact that the co-existence of cement plant and power plant in the same location resulted in pH values near to neutrality value reported for natural rainwater.

	Na^+	K^+	Ca^{2+}	Mg^{2+}	NH_4^+	NO_3^-	Cl^-	SO_4^{2-}
Na^+	1							
K^+	0.29	1						
Ca^{2+}	0.10	0.74	1					
Mg^{2+}	0.20	0.77	0.91	1				
NH_4^+	0.38	0.06	0.18	0.11	1			
NO_3^-	0.14	0.44	0.74	0.51	0.10	1		
Cl^-	0.59	0.27	0.10	0.20	0.41	0.13	1	
SO_4^{2-}	0.13	0.47	0.79	0.60	0.13	0.95	0.13	1

Table 5. Pearson´s correlation matrix among trace elements for Tula de Allende.

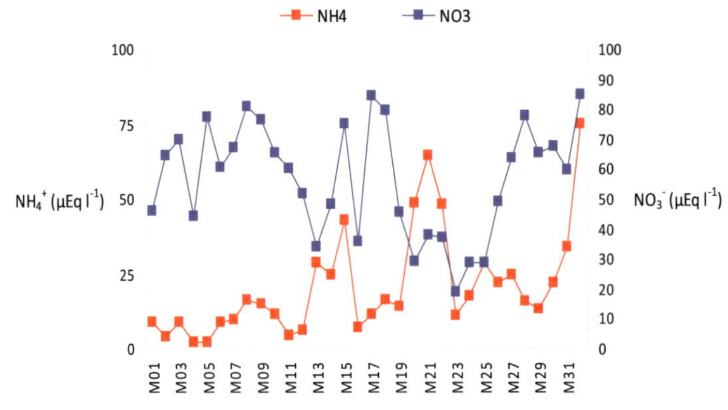

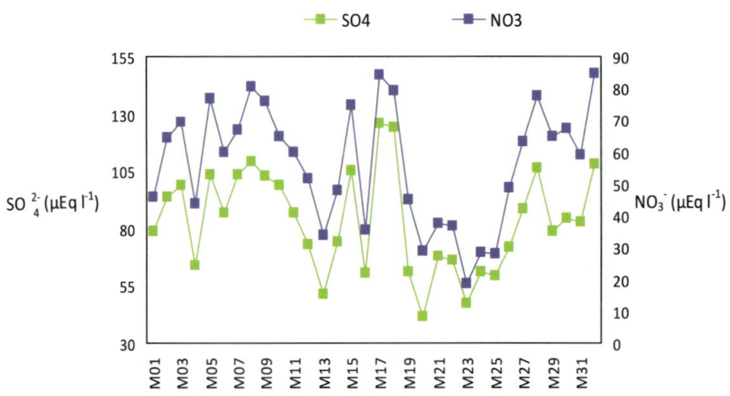

Fig. 12. Time-trends of ionic concentrations in dry deposition in Tula de Allende.

Backward air masses trajectories analysis demonstrated that winds during the sampling period had a N-NE component. These results imply that a specific source located in this direction (the power plant near to this site) contributed to the significant levels of nitrate and sulfate. Figure 12 shows the variations of ionic concentrations in dry deposition with time at Tula de Allende.

Tula de Allende sampling site constitutes a complex mosaic of rural, agricultural and industrialized areas. At the edges of the City, agriculture activities prevail, making evident the influence of ammonium on neutralization process. Furthermore, it can be observed that SO_4^{2-} and NO_3^- showed the same behavior, suggesting that both ions possibly had a common source.

4.3 Annual deposition fluxes

4.3.1 Los Petenes Biosphere Reserve

The average dry deposition fluxes (in Kg ha-1 yr-1) ranged between 0.39 and 7.19 for Los Petenes.

Figure 13 illustrates that dry deposition fluxes for ammonium and nitrate were the highest. During dry periods, the reduced mixing layer concentrates only ionic species of local origin; it supports the fact of high levels of nitrate and ammonium had a local origin, in this case, forest fires occurring during the mid-summer drought and emissions from "Lerma" Power Plant.

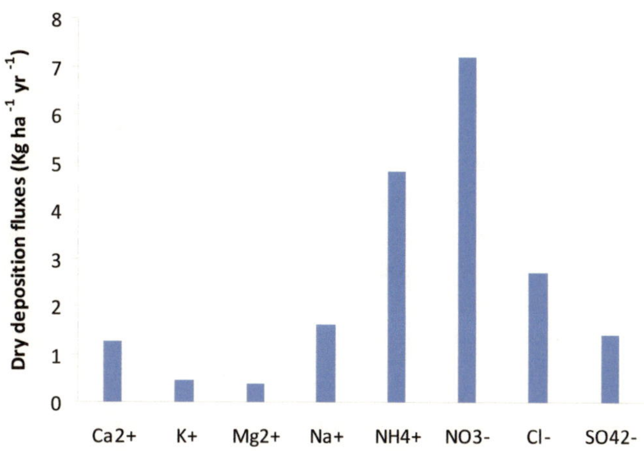

Fig. 13. Dry deposition fluxes for Los Petenes Biosphere Reserve.

4.3.2 Tula de Allende

The average dry deposition fluxes (in Kg ha-1 yr-1) ranged between 0.34 and 8.04 for Tula de Allende. It can be observed, from Figure 14, that dry deposition fluxes for calcium, sulfate and nitrate were the highest. In this case, anthropogenic emissions from Tula-Vito-Apaxco complex contributed to high levels of calcium, sulfate and nitrate, having in mind that, during dry periods, the reduced mixing layer only concentrates ionic species of local origin.

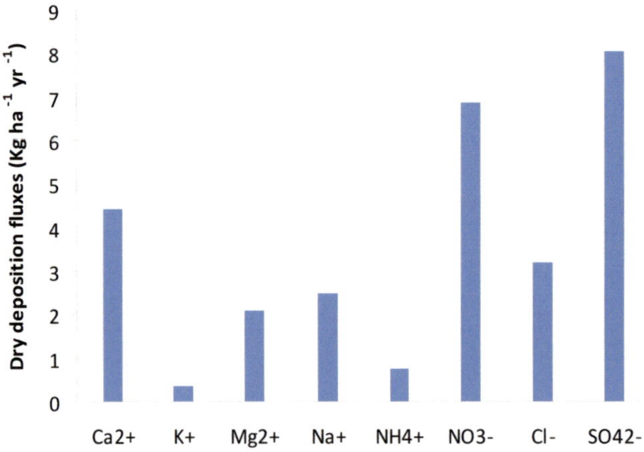

Fig. 14. Dry deposition fluxes for Tula de Allende.

4.4 Deposition fluxes and critical loads

A critical load of 10 Kg N ha^{-1} yr^{-1} has been proposed in the United Kingdom (Jones, 2005), however, given that the sensitivity varies from one specie to other, Power et al (2006) proposed a value of 8 Kg N ha^{-1} yr^{-1}. In addition, it has been proposed a value of 5 Kg N ha^{-1} yr^{-1} for sensitive ecosystems. For the impacted ecosystems in U.S, the critical load reported ranges between 3 and 8 Kg N ha^{-1} yr^{-1}. In the case of S, for natural forests, some authors propose values between 2 and 5 Kg S ha^{-1} yr^{-1} (Nilsson & Grennfelt, 1988).

Data on dry deposition inputs of N and S in Mexican sites are scarce, taking in account the critical values reported for Europe and United States of America, we can observe that in Los Petenes Biosphere Reserve, N input (11.99 Kg ha^{-1} yr^{-1}) is slightly above the critical loads, whereas S input (1.4 Kg ha^{-1} yr^{-1}) did not exceeded the critical values, this supports the proposal that this site is only subject to local emissions. These results indicate that critical loads of nitrogen can increase in a short time period if strategic measures are not taken. Our results are sub estimated considering that we are not reporting data for wet deposition, since this site is a natural reserve, it can constitute a serious threat to flora and fauna species that live in this ecosystem.

On the other hand, N input (7.6 Kg ha^{-1} yr^{-1}) for Tula de Allende, exceeds proposed value for sensitive ecosystems, but it is below of critical load reported for American and European sites. However, in the case of S, we can observe that S input (8.04 Kg ha^{-1} yr^{-1}) exceeded the critical value for various eco-regions in the world. It is clear that emissions released by sources in Tula-Vita-Apaxco industrial complex contributed in a significant way to this input. Alkaline particles emitted from cement plants contributed to neutralize the acidity, therefore, pH values resulting from this neutralization process were not in the acid range. However, pH itself it is not a conservative tracer of ecological effects of acid deposition; for that reason, we propose to improve the current environmental policies and the promulgation of new regulations based on a conservative tracer and focused to decrease sulfate levels in this region.

5. Conclusion

It can be concluded that in Los Petenes site, alkaline particles from the calcareous soil played an important role in the acidity neutralization process derived from high emissions of NO_3^- from the power plant and due to biomass burning in the vicinity of this sampling site. N input in Los Petenes exceeded slightly the critical values reported for other sites of the world. Considering the ecological importance of this site, local strategic policies must be applied focused on protecting the wildfire life in this Biosphere Reserve.

On the other hand, in spite of high sulfate and nitrate levels found in Tula de Allende (attributed to a power plant), acidity neutralization process was improved by the presence of cement plants near to this sampling site. However, S input exceeded critical values reported for several sites around the world. In spite of Tula de Allende is not an eco-region, since is already an industrialized site, S deposition could be a threat to historic wealth located in the vicinity of this place.

6. Acknowledgement

This work was financially co-supported by Sectorial Fund CONACYT-SEMARNAT (project 107948) and by National Institute of Ecology (projects INE/A1-047/2008; INE/A1-007/2009; INE/A1-031/2010). Authors gratefully acknowledge their help to students that supported this work.

7. References

Alonso, R., Bytnerowicz, A., & Boarman, W.I. (2005). Atmospheric dry deposition in the vicinity of the Salton Sea, California - I: Air pollution and deposition in a desert environment. *Atmospheric Environment,*Vol 39, No. 26, August 2005, pp. 4671-4679, ISSN 1352-2310.

Balestrini, R., Galli, L., & Tartari, G. (2000). Wet and dry atmospheric deposition at prealpine and alpine sites in northern Italy. *Atmospheric Environment*, Vol. 34, No. 9, pp. 1455-1470, ISSN 1352-2310.

Dasch, M.J. (1985). Direct measurement of dry deposition to a polyethylene bucket and various surrogate surfaces. *Environmental, Science and Technology,* Vol. 19, No. 8, August 1985, pp. 721-725, ISSN 0013-936X.

Davies Inc (n.d.). Davies Instruments. Vantage Pro2. Weather Link. Data Logger & Software.

Flues,M., Hama, P., Lemes, M.J.L., Dantas, E.S.K., & Fornaro, A. (2002). Evaluation of the rainwater acidity of a rural region due to a coal-fired power plant in Brazil. *Atmospheric Environment*, Vol. 36, No. 14, May 2002, pp. 2397-2404, ISSN 1352-2310.

Galbally, I.E. & Gillel, R.W. (1988). Acidification in Tropical Countries. Chapter 3. In: *Processes regulating nitrogen compounds in the tropical atmosphere.* H. Rodhe & R. Herrera. Pp. 1-405, John Wiley & Sons, ISBN 047 91870 9, Great Britain.

Galloway, J.N. & Likens, G.E. (1976). Calibration of collection procedures for the determination of precipitation chemistry. *Water, Air & Soil Pollution*, Vol. 6, No. 2-4, pp. 241-258, ISSN 0049-6979.

Galloway, J.N., Thornton, J.D., Norton, S.A., Volchok, H.L., & McLean, R.A.N. (1982)Trace metals in atmospheric deposition: a review and assessment. *Atmospheric Environment*, Vol. 16, No. 7, pp. 1677-1700, ISSN 1352-2310.

Hegg, D.A., Radke, L.F., Hobbs, P.V., & Riggon, P.J. (1988). Ammonia emissions from biomass burning. *Geophys. Res. Letter*, Vol. 15, No. 4, pp. 335-337, ISSN 0094-8276.

Jones, M.L.M. (2005). Nitrogen deposition in upland grasslands: Critical loads, management and recovery. Thesis. University of Sheffield. U.K.

Lakes Environmental Inc. (2011). Lakes Environmental Software. WRPLOT View v.6.5.2. Freeware Wind Rose Plots for Meteorological Data.

Nilsson, J. & Grennfelt, P. (1988). Critical loads for sulphur and nitrogen. Report from a workshop held at Skokloster, Sweden. March 19-24. The Nordic Council of Ministers. Report 1988: 15, Copenhagen, Denmark. ISBN 87-7303-248-4.

NOAA (n.d.). National Oceanic and Atmospheric Administration. U.S. Department of Commerce. HYSPLIT -Hybrid Single Particle Lagrangian Integrated Trajectory Model, In: *Air Resources Laboratory http://ready.arl.noaa.gov/HYSPLIT.php.*

Power, S.A., Green, E.R., Barker, C.G., Bell, J.N.B., & Ashmore, M.R. (2006). Ecosystem recovery: heathland response to a reduction in nitrogen deposition. *Global Change Biology*, Vol. 12, No. 7, July 2006, pp. 1241-1252, ISSN 1354-1013.

SAS Inc. (1998). SAS/STAT, Guide for personal computers. Version 8. United States of America.

Tsitouridou, R. & Anatolaki, Ch. (2007). On the wet and dry deposition of ionic species in the vicinity of coal-fired power plants, northwestern Greece. *Atmospheric Research*, Vol 83, pp. 93-105, ISSN 0169-8095.

Zuk, M., Garibay, V., Iniestra, R., López, M.T., Rojas, L., & Laguna, I. (2006). *Introducción a la evaluación de los impactos de las termoeléctricas de México (Primera Edición Octubre 2006)*. Secretaría del Medio Ambiente y Recursos Naturales, Instituto Nacional de Ecología, ISBN 968-817-804-7, México D.F, México.

9

Some Aspects of the Sentinel Method for Pollution Problems

A. Omrane*

Laboratoire CEREGMIA, EA 2440, I.E.S Guyane, Campus Universitaire Trou-Biran, Route de Baduel BP 792, 97337 Cayenne Cedex French Guiana

1. Introduction

Modelling environmental problems leads to mathematical systems with missing data: Weather problems have generally missing initial conditions. This chapter is concerned with identifying pollution terms arising in the state equation of some dissipative system with incomplete initial condition.

To this aim the so-called sentinel method is used. We explain how the problemn of determining a sentinel is equivalent to a null-controllability problem for which Carleman inequalities are revisited.

In a second part of the chapter, we use the same techniques to discuss of how to get instantaneous information (at fixed $t = T \in [0, +\infty[$) on pollution terms in distributed systems of incomplete data in some ecology and/or meteorology problems.

We consiter a fixed final time $T > 0$, and Ω an open subset of $\mathbb{R}^d$ of smooth boundary $\partial\Omega$, and we denote by $Q = \Omega \times]0, T[$ the space-time cylinder. We are intersted with systems partially known; we consider here the state equation :

$$\begin{cases} y' - \Delta y + f(y) = \xi + \lambda\hat{\xi} & \text{in } \; Q, \\ \quad y(0) \quad = y^0 + \tau\hat{y}^0 & \text{in } \; \Omega, \\ \quad y \quad = 0 & \text{on } \; \Sigma_1, \\ \quad \dfrac{\partial y}{\partial \nu} \quad = 0 & \text{on } \; \Sigma_2, \end{cases} \tag{1.1}$$

where $y = y(x, t; \lambda, \tau)$, and where Σ_1 is a piece of the boundary $\Sigma = \partial\Omega \times]0, T[$ and $\Sigma_2 = \Sigma \backslash \Sigma_1$. We assume here that $f : \mathbb{R} \to \mathbb{R}$ is of class C^1, the functions ξ and y^0 are known with $\xi \in L^2(Q)$ and $y^0 \in L^2(\Omega)$. But, the terms : $\lambda\hat{\xi}$ (so-called pollution term) and $\tau\hat{y}^0$ (so-called perturbation term) are unknown, $\hat{\xi}$ and $\hat{y}^0$ are renormalized and represent the size of pollution and perturbation

$$\|\hat{\xi}\|_{L^2(Q)} \leq 1, \quad \|\hat{y}^0\|_{L^2(\Omega)} \leq 1, \qquad \text{so that the reals} \quad \lambda, \tau \quad \text{are small enough.}$$

*This projet is supported by the CPER AGROECOTROP 2007-2013, Région Guadeloupe-État-Europe

Some non empty open subset $\mathcal{O} \subset \Omega$, is called observatory set. The observation is y on $\mathcal{O}$, for the time T. We denote by y_{obs} this observation

$$y_{obs} = m_o \in L^2(\mathcal{O} \times (0,T)). \tag{1.2}$$

We suppose that (1.1) has a unique solution denoted by $y(\lambda,\tau) := y(x,t;\lambda,\tau)$ in some relevant space. The question is

$$\left| \begin{array}{l} \text{how to calculate the pollution term } \lambda\hat{\xi} \text{ in} \\ \text{the state equation, independently from the} \\ \text{variation } \tau\hat{y}^0 \text{ around the initial data ?} \end{array} \right. \tag{1.3}$$

Least squares. Question (1.3) is natural and leds to some developments; some answer is given by the least squares method. The method consists in considering the unknowns $\{\lambda\hat{\xi}, \tau\hat{y}^0\} = \{v, w\}$ as control variables, then the state $y(x,t;v,w)$ has to be driven as close as possible to m_o.

This comes to some optimal control problem. By this way we look for the pair (v,w), there is then no real possibility to find v or w independently.

Sentinels. The sentinel method of Lions Lions J.-L. (1992) is a particular least squares method which is adapted to the identification of parameters in ecosystems with incomplete data; many models can be found in litterature. The sentinel concept relies on the following three objects : some state equation (for instance (1.1)), some observation function (1.2), and some control function w to be determined.

Many papers use the definition of Lions in the theoretical aspect (see for example Bodart Bodart O. (1994), Bodart-Fabre Bodart O. & Fabre C. (1993)Bodart O. & Fabre C. (1995)), as well as in the numerical one (see Bodart-Demeestere Bodart O. & Demeestere P. (1997), Demeestere Demeestere P. (1997) and Kernevez Kernevez J.-P. (1997)).

We use the techniques in Miloudi Y. et al. (2007) to give an answer to the question (1.3). Let h_0 be some function in $L^2(\mathcal{O} \times (0,T))$. Let on the other hand ω be some open and non empty subset of Ω. For a control function $\mathrm{w} \in L^2(\omega \times (0,T))$, we define the functional

$$S(\lambda,\tau) = \int_0^T \int_{\mathcal{O}} h_0 \, y(\lambda,\tau) \, dxdt + \int_0^T \int_{\omega} \mathrm{w} \, y(\lambda,\tau) \, dxdt. \tag{1.4}$$

We say that S defines a sentinel for the problem (1.1) if there exists w such that S is insensitive (at first order) with respect the to missing terms $\tau\hat{y}^0$, which means

$$\frac{\partial S}{\partial \tau}(0,0) = 0 \tag{1.5}$$

for any $\hat{y}^0$ where here $(0,0)$ corresponds to $\lambda = \tau = 0$, and if w minimizes the norm $\|\mathrm{v}\|_{L^2(\omega\times(0,T))}$.

Important Remark. The Lions sentinels S assume $\omega = \mathcal{O}$. In this case, the observation and the control share the same support, and the solution $\mathrm{w} = -h_0$ is trivial.

The definition (1.4) extends the one by Lions to the case where the observation and the control have different supports. This point of view (where $\omega \neq \mathcal{O}$) has been considered for the first time in Nakoulima O. (2004) and Miloudi Y. et al. (2007).

1.1 The informations given by the sentinel

Because of (1.5) we can write

$$S(\lambda,\tau) \simeq S(0,0) + \lambda \frac{\partial S}{\partial \lambda}(0,0), \qquad \text{for} \quad \lambda,\tau \quad \text{small}.$$

In (1.4), $S(\lambda,\tau)$ is observed and using (1.2),

$$S(\lambda,\tau) = \int_Q (h_0\chi_O + \mathrm{w}\chi_\omega) m_o \, dxdt$$

so that (1.5) becomes

$$\lambda \frac{\partial S}{\partial \lambda}(0,0) \simeq \int_Q (h_0\chi_O + \mathrm{w}\chi_\omega)(m_o - y_0)\, dxdt, \tag{1.6}$$

with

$$\frac{\partial S}{\partial \lambda}(0,0) = \int_Q (h_0\chi_O + \mathrm{w}\chi_\omega) y_\lambda \, dxdt,$$

where here χ_O and χ_ω denote the characteristic functions of O and ω respectively.

The derivative $y_\lambda = (\partial y/\partial\lambda)\,(0,0)$ only depends on $\hat{\xi}$ and other known data. Consequently, the estimates (1.6) contains the informations on $\lambda\hat{\xi}$ (see for details remark 1 below).

2. Null-controllability problem

The existence of a sentinel is equivalent to a null-controllability property. Indeed, we begin by transforming the insensibility condition (1.5).

Denote by

$$y_\tau = \frac{d}{d\tau} y(\lambda,\tau)\Big|_{\lambda=\tau=0}.$$

Then the function y_τ is solution of

$$\begin{cases} y'_\tau - \Delta y_\tau + f'(y_0) y_\tau = 0 & \text{in } Q, \\ y_\tau(0) = \hat{y}^0 & \text{in } \Omega, \\ y_\tau = 0 & \text{on } \Sigma_1, \\ \dfrac{\partial y_\tau}{\partial \nu} = 0 & \text{on } \Sigma_2, \end{cases} \tag{2.1}$$

where $y_0 = y(0,0)$. Problem (2.1) is linear and has a unique solution y_τ under mild assumptions on f.

The insensibility condition (1.5) holds if and only if

$$\int_Q \left(h_0\chi_O + \mathrm{w}\chi_\omega\right) y_\tau \, dxdt = 0. \tag{2.2}$$

We can transform (2.2) by introducing the classical adjoint state. More precisely, we define the function $q = q(x,t)$ as the solution of the backward problem :

$$\begin{cases} -q' - \Delta q + f'(y_0)q = h_0\chi_O + \mathrm{w}\chi_\omega & \text{in } Q, \\ q(T) = 0 & \text{in } \Omega, \\ q = 0 & \text{on } \Sigma_1, \\ \dfrac{\partial q}{\partial \nu} = 0 & \text{on } \Sigma_2. \end{cases} \tag{2.3}$$

As for the problem (2.1), the problem (2.3) has a unique solution q (under mild assumptions on $f'(y_0)$). The function q depends on the control w that we shall determine :
Indeed, if we multiply the first equation in (2.3) by y_τ, and we integrate by parts over Q, we obtain

$$\int_Q \Big(h_0\chi_O + \mathrm{w}\chi_\omega\Big)\, y_\tau \, dxdt \;=\; \int_\Omega q(0)\hat{y}^0 dx \quad \forall \hat{y}^0, \quad \|\hat{y}^0\|_{L^2(\Omega)} \leq 1.$$

So, the condition (1.5) (or (2.2)) is equivalent to

$$q(0) \;=\; 0. \tag{2.4}$$

This is a null-controllability problem.

Remark 1. *The knowledge of the optimal control* w *provides informations about the pollution term* $\lambda\hat{\xi}$*. Indeed, denote by*

$$L = \frac{\partial}{\partial t} - \Delta + f'(y_0)\, I_d$$

and let $y_\lambda = \dfrac{\partial y}{\partial \lambda}(0,0)$ *be the solution of*

$$\begin{cases} L y_\lambda = \hat{\xi} \ in \ Q, \\ y_\lambda(0) = 0 \ in \ \Omega, \\ y_\lambda = 0 \ on \ \Sigma_1, \\ \dfrac{\partial y_\lambda}{\partial \nu} = 0 \ on \ \Sigma_2. \end{cases}$$

Integrating by parts, we then obtain

$$\int_Q y_\lambda\, L^* q \; dxdt = \int_Q q\hat{\xi}\, dxdt$$

with

$$L^* = -\frac{\partial}{\partial t} - \Delta + f'(y_0)\, I_d.$$

So that from (2.3) and (1.6) we deduce

$$\int_Q \lambda\hat{\xi}\, dxdt = \int_Q (h_0\chi_O + \mathrm{w}\chi_\omega)(m_o - y_0)\, dxdt.$$

3. Existence of a sentinel

We begin with some observability inequality, which will be proved in detail in the last section. Denote by

$$\mathcal{V} = \Big\{ v \in \mathcal{C}^\infty(\overline{Q}) \text{ such that : } v_{|\Sigma_1} = \frac{\partial v}{\partial t}\Big|_{\Sigma_1} = 0 \text{ and } \frac{\partial v}{\partial \nu}\Big|_{\Sigma_2} = 0 \Big\}. \tag{3.1}$$

Then we have :

Theorem 1. *Let be $u \in \mathcal{V}$, then there exists a positive constant $C = C(\Omega, \omega, O, T, f'(y_o))$ such that*

$$\int_Q \frac{1}{\theta^2} |u|^2 \, dxdt \leq C \Big[\int_Q |L u|^2 \, dxdt + \int_0^T \int_\omega |u|^2 \, dxdt \Big], \tag{3.2}$$

where $\theta \in \mathcal{C}^2(Q)$ positive with $\frac{1}{\theta}$ bounded.

According to the RHS of (3.2), we consider the space $\mathcal{V}$ endowed with the bilinear form $a(.,.)$ defined by :

$$a(u,v) \quad = \quad \int_Q Lu \, Lv \, dxdt + \int_0^T \int_\omega u \, v \, dxdt.$$

Let V be the completion of $\mathcal{V}$ with respect to the norm

$$v \mapsto \|v\|_V = \sqrt{a(v,v)},$$

then, V is a Hilbert space for the scalar product $a(v, \hat{v})$ and the associated norm.

Remark 2. *We can precise the structure of the elements of V. Indeed, let $H_\theta(Q)$ be the weigthed Hilbert space defined by*

$$H_\theta(Q) = \Big\{ v \in L^2(Q) \quad \text{such that :} \quad \int_Q \frac{1}{\theta^2} |v|^2 \, dxdt < \infty \Big\},$$

endowed with the natural norm $\|v\|_\theta = (\int_Q \frac{1}{\theta^2} |v|^2 \, dxdt)^{\frac{1}{2}}$. This shows that V is imbedded continuously $\|v\|_\theta \leq C \|v\|_V$.

Now if $h_0 \in L^2(Q)$ and $\theta h_0 \in L^2(Q)$ (i.e. : $h_0 \in L^2_\theta(Q)$), then from (3.2) and the Cauchy-Schwartz inequality, we deduce that the linear form defined on V by

$$v \mapsto \int_Q h_0 \chi_O \, v \, dxdt$$

is continuous. Therefore, from the Lax-Milgram theorem there exits a unique u in V solution of the variational equation :

$$a(u,v) = \int_Q h_0 \chi_O \, v \, dxdt \qquad \forall v \in V.$$

Theorem 2. *Assume that $h_0 \in L^2_\theta(Q)$, and let u be the unique solution of (3.3). We set*

$$\mathrm{w} = -u \chi_\omega$$

and

$$q = Lu.$$

Then, the pair (w, q) is such that (2.3)-(2.4) hold (i.e there is some insensitive sentinel defined by (1.4)-(1.5)).

4. Proof of theorem 1

The proof for the observability inequality in theorem 1 will hold from Carleman estimates that we carefully show in the following results.

Lemma 1. *Let be ω_0 an open set such that $\overline{\omega_0} \subset \omega$. Then there is $\psi \in C^2(\overline{\Omega})$ such that*

$$\begin{cases} \psi(x) & > 0 \ \forall \ x \in \Omega, \\ \psi(x) & = 0 \ \forall \ x \in \Gamma, \\ |\nabla\psi(x)| & \neq 0 \ \forall \ x \in \overline{\Omega - \omega_0}. \end{cases}$$

See Imanuvilov O. Yu. Imanuvilov (1995)

We now use a function ψ as given by the previous lemma, to define convenient *weight* functions. For $\lambda > 0$, we set

$$\varphi(x,t) = \frac{e^{\lambda\psi(x)}}{t(T-t)},$$

and

$$\eta(x,t) = \frac{e^{2\lambda|\psi|_\infty} - e^{\lambda\psi(x)}}{t(T-t)}.$$

Then

$$\nabla\varphi = \lambda\varphi\nabla\psi, \ \nabla\eta = -\lambda\varphi\nabla\psi.$$

We also notice the following properties :

$$\left|\frac{\partial\varphi}{\partial t}\right| \leq T\varphi^2, \ \left|\frac{\partial^2\varphi}{\partial t^2}\right| \leq T^2\varphi^3, \tag{4.1}$$

$$\left|\frac{\partial\eta}{\partial t}\right| \leq T\varphi^2, \ \left|\frac{\partial^2\eta}{\partial t^2}\right| \leq T^2\varphi^3. \tag{4.2}$$

Remark 3. *Note that η increases to $+\infty$ when $t \to T$ or $t \to 0$, but η is uniformly bounded on $\Omega \times [\delta, T-\delta]$ for any $\delta > 0$.*
On the other hand, for fixed $s > 0$ the function $e^{-s\eta(x,t)}$ goes to 0 when $t \to T$ or $t \to 0$.

The following theorem states the Carleman inequalities concerning (3.1) :

Proposition 1. *There exist constants $s_0 > 0$, $\lambda_0 > 0$ and $C > 0$ depending on Ω, ω, ψ, and T, such that for all $s \geq s_0$, $\lambda \geq \lambda_0$, and for any function $u \in \mathcal{V}$ given by (3.1), we have*

$$\begin{aligned} & 2s^3\lambda^4 \int_Q \varphi^3 e^{-2s\eta} |u|^2 \, dxdt + 4s^2\lambda \int_{\Sigma_2} \varphi \frac{\partial\eta}{\partial t}\frac{\partial\psi}{\partial\nu} e^{-2s\eta} |u|^2 \, d\gamma dt \\ & -4s^3\lambda^3 \int_{\Sigma_2} \varphi^3 |\nabla\psi|^2 \frac{\partial\psi}{\partial\nu} e^{-2s\eta} |u|^2 \, d\gamma dt - 4s^2\lambda^3 \int_{\Sigma_2} \varphi^2 |\nabla\psi|^2 \frac{\partial\psi}{\partial\nu} e^{-2s\eta} |u|^2 \, d\gamma dt \\ & -2s\lambda \int_{\Sigma_2} \varphi \frac{\partial\psi}{\partial\nu} e^{-2s\eta} \frac{\partial u}{\partial t} u \, d\gamma dt - 4s\lambda \int_{\Sigma_1} \varphi \nabla\psi e^{-2s\eta} \nabla u \frac{\partial u}{\partial\nu} \, d\gamma dt \\ & +2s\lambda \int_{\Sigma} \varphi \frac{\partial\psi}{\partial\nu} e^{-2s\eta} |\nabla u|^2 \, d\gamma dt \\ \leq \ & C\left(\int_Q e^{-2s\eta} \left| \frac{\partial u}{\partial t} - \Delta u \right|^2 dxdt + s^3\lambda^4 \int_0^T \int_\omega \varphi^3 e^{-2s\eta} |u|^2 \, dxdt \right). \end{aligned} \tag{4.3}$$

We use the method by Fursikov and Imanuvilov A. Fursikov & O. Yu. Imanuvilov (1996), Imanuvilov O. Yu. Imanuvilov (1995), Lebeau and Robbiano Lebeau G. & Robbiano L. (1995), and Puel Puel J.-P. (2001) (case $\Sigma_2 = \emptyset$ and $\Sigma_1 = \Sigma$).
For $s \geq s_0$ and $\lambda \geq \lambda_0$, we define

$$w(x,t) = e^{-s\eta(x,t)} u(x,t).$$

We easily notice that

$$w(x,0) = w(x,T) = 0.$$

Calculating $g = (\partial_t - \Delta)(e^{s\eta} w)$, with notation (4), we get

$$P_1 w + P_2 w = g_s,$$

where

$$\begin{aligned} P_1 w &= \frac{\partial w}{\partial t} + 2s\lambda\varphi\, \nabla\psi\, \nabla w + 2s\lambda^2 \varphi |\nabla\psi|^2 w, \\ P_2 w &= -\Delta w - s^2\lambda^2\varphi^2 |\nabla\psi|^2 w + s\frac{\partial \eta}{\partial t} w, \\ g_s &= e^{-s\eta} g + s\lambda^2 \varphi |\nabla\psi|^2 w - s\lambda\varphi\, \Delta\psi\, w. \end{aligned}$$

Taking the L^2 norm we get :

$$\int_Q |P_1 w|^2\, dxdt + \int_Q |P_2 w|^2\, dxdt + 2\int_Q P_1 w P_2 w\, dxdt = \int_Q |g_s|^2\, dxdt.$$

We shall now calculate $\int_Q P_1 w P_2 w\, dxdt$. This will give 9 terms $I_{k,l}$.

In order to organize our calculus, we denote by A and B the quantities such that A contains all the terms which can be upper bounded by

$$c\left(s\lambda + \lambda^2\right) \int_Q \varphi\, |\nabla w|^2\, dxdt,$$

and by B all those which can be bounded by

$$c\left(s^2\lambda^4 + s^3\lambda^3\right) \int_Q \varphi^3\, |w|^2\, dxdt.$$

We denote by ν the outer normal on Γ. Note down that ψ cancels on Γ. We then have the following results :

$$\begin{aligned} I_{1,1} &= -\int_Q \frac{\partial w}{\partial t} \Delta w\, dxdt \\ &= -\int_\Sigma \frac{\partial w}{\partial t}\frac{\partial w}{\partial \nu}\, d\gamma dt + 0, \\ I_{1,2} &= -s^2\lambda^2 \int_Q \frac{\partial w}{\partial t} \varphi^2 |\nabla\psi|^2 w\, dxdt = B, \end{aligned}$$

$$I_{1,3} = s\int_Q \frac{\partial w}{\partial t}\frac{\partial \eta}{\partial t} w\, dxdt = \frac{s}{2}\int_Q \frac{\partial \eta}{\partial t}\frac{\partial}{\partial t}\left(|w|^2\right) dxdt$$
$$= -\frac{s}{2}\int_Q \frac{\partial^2 \eta}{\partial t^2}|w|^2\, dxdt = B.$$

And,

$$\begin{aligned} I_{2,1} &= -2s\lambda \int_Q \varphi \nabla\psi \nabla w \Delta w\, dxdt \\ &= -2s\lambda \int_\Sigma \varphi\nabla\psi.\nabla w \frac{\partial w}{\partial \nu}\, d\gamma dt + 2s\lambda^2 \int_Q \varphi|\nabla\psi.\nabla w|^2\, dxdt \\ &\quad + s\lambda \int_\Sigma \varphi \frac{\partial \psi}{\partial \nu}\,|\nabla w|^2\, d\gamma dt - s\lambda^2 \int_Q \varphi |\nabla\psi|^2|\nabla w|^2\, dxdt + A, \end{aligned}$$

$$\begin{aligned} I_{2,2} &= -2s^3\lambda^3 \int_Q \varphi^3|\nabla\psi|^2\nabla\psi\nabla w.w = -s^3\lambda^3\int_Q \varphi^3|\nabla\psi|^2\nabla\psi\nabla|w|^2 \\ &= -s^3\lambda^3 \int_\Sigma \varphi^3\,|\nabla\psi|^2\,\frac{\partial\psi}{\partial \nu}|w|^2 + 3s^3\lambda^4\int_Q \varphi^3\,|\nabla\psi|^4\,|w|^2 + B, \end{aligned}$$

$$\begin{aligned} I_{2,3} &= 2s^2\lambda \int_Q \varphi\nabla\psi\,\nabla w\,\frac{\partial\eta}{\partial t} w\, dxdt \\ &= s^2\lambda \int_\Sigma \varphi \frac{\partial \eta}{\partial t}\frac{\partial\psi}{\partial\nu}|w|^2\, dxdt + B. \end{aligned}$$

Finally :

$$\begin{aligned} I_{3,1} &= -2s\lambda^2 \int_Q \varphi|\nabla\psi|^2\Delta w.w\, dxdt \\ &= -2s\lambda^2 \int_\Sigma \varphi\,|\nabla\psi|^2\,\frac{\partial w}{\partial\nu}.w + 2s\lambda^2\int_Q \varphi|\nabla\psi|^2|\nabla w|^2 + A + B, \end{aligned}$$

$$I_{3,2} = -2s^3\lambda^4 \int_Q \varphi^3|\nabla\psi|^4|w|^2\, dxdt,$$

$$I_{3,3} = 2s^2\lambda^2 \int_Q \varphi\frac{\partial\eta}{\partial t}|\nabla\psi|^2|w|^2\, dxdt = B.$$

Summing all the terms, it follows :

$$\begin{aligned} 2\int_Q P_1 w P_2 w\, dxdt = A + B &+ 2s\lambda^2\int_Q \varphi|\nabla\psi|^2|\nabla w|^2\, dxdt \\ &+ 2s^3\lambda^4\int_Q \varphi^3|\nabla\psi|^4\,|w|^2\, dxdt + 4s\lambda^2\int_Q \varphi|\nabla\psi.\nabla w|^2\, dxdt \\ &- 2\int_\Sigma \frac{\partial w}{\partial t}\frac{\partial w}{\partial\nu}\, d\gamma dt - 4s\lambda\int_\Sigma \varphi\nabla\psi.\nabla w\frac{\partial w}{\partial\nu}\, d\gamma dt \\ &+ 2s\lambda\int_\Sigma \varphi\frac{\partial\psi}{\partial\nu}|\nabla w|^2\, d\gamma dt - 2s^3\lambda^3\int_\Sigma \varphi^3|\nabla\psi|^2\frac{\partial\psi}{\partial\nu}|w|^2\, d\gamma dt \\ &+ 2s^2\lambda\int_\Sigma \varphi\frac{\partial\eta}{\partial t}\frac{\partial\psi}{\partial\nu}|w|^2\, d\gamma dt - 4s\lambda^2\int_\Sigma \varphi\,|\nabla\psi|^2\,\frac{\partial w}{\partial\nu}.w\, d\gamma dt. \end{aligned}$$

But $|\nabla\psi| \neq 0$ on $\overline{\Omega - \omega_0}$, hence there is $\delta > 0$ such that

$$|\nabla\psi| \geq \delta \quad \text{on} \quad \overline{\Omega - \omega_0}.$$

On the other hand

$$\int_Q |g_s|^2 \, dxdt \leq \int_Q e^{-2s\eta} \, |g|^2 \, dxdt + B,$$

so that

$$\int_Q |P_1 w|^2 \, dxdt + \int_Q |P_2 w|^2 \, dxdt + 2 \int_Q P_1 w P_2 w \, dxdt$$
$$\leq \int_Q e^{-2s\eta} \, |g|^2 \, dxdt + B.$$

Consequently :

$$\int_Q |P_1 w|^2 \, dxdt + \int_Q |P_2 w|^2 \, dxdt + 2s\lambda^2\delta^2 \int_Q \varphi \, |\nabla w|^2 \, dxdt$$
$$+2s^3\lambda^4\delta^4 \int_Q \varphi^3 \, |w|^2 \, dxdt - 2 \int_\Sigma \frac{\partial w}{\partial t} \frac{\partial w}{\partial \nu} \, d\gamma dt - 4s\lambda \int_\Sigma \varphi \nabla\psi . \nabla w \frac{\partial w}{\partial \nu} \, d\gamma dt$$
$$+2s\lambda \int_\Sigma \varphi \frac{\partial \psi}{\partial \nu} |\nabla w|^2 \, d\gamma dt - 2s^3\lambda^3 \int_\Sigma \varphi^3 |\nabla\psi|^2 \frac{\partial \psi}{\partial \nu} |w|^2 \, d\gamma dt$$
$$+2s^2\lambda \int_\Sigma \varphi \frac{\partial \eta}{\partial t} \frac{\partial \psi}{\partial \nu} |w|^2 \, d\gamma dt - 4s\lambda^2 \int_\Sigma \varphi |\nabla\psi|^2 \frac{\partial w}{\partial \nu} . w \, d\gamma dt + A + B$$
$$\leq \int_Q e^{-2s\eta} \, |g|^2 \, dxdt + B + 2s\lambda^2\delta^2 \int_0^T \int_{\omega_0} \varphi \, |\nabla w|^2 \, dxdt + 2s^3\lambda^4\delta^4 \int_0^T \int_{\omega_0} \varphi^3 \, |w|^2 \, dxdt.$$

We can eliminate A and B by choosing s and λ large enough. And we observe that :

$$\int_0^T \int_\omega \varphi\theta^2 \, |\nabla w|^2 \, dxdt$$
$$\leq C \left(\int_Q \varphi P_2 w \theta^2 w dxdt + \int_0^T \int_\omega \varphi^{\frac{1}{2}} \theta^2 \, |\nabla w| \, \varphi^{\frac{1}{2}} w \, dxdt + s^2\lambda^2 \int_0^T \int_\omega \varphi^3 w^2 dxdt \right),$$

for $\theta \in \mathcal{D}(\omega)$ such that $0 \leq \theta \leq 1$ and $\theta(x) = 1$ on ω_0 gives

$$2s\lambda^2\delta^2 \int_0^T \int_{\omega_0} \varphi \, |\nabla w|^2 \, dxdt \leq \frac{1}{2} \int_0^T \int_\Omega |P_2 w|^2 \, dxdt + cs^3\lambda^4 \int_0^T \int_\omega \varphi^3 \, |w|^2 \, dxdt,$$

Now, we should write the inequality below in terms of the solution u, since

$$|w|^2 = e^{-2s\eta} |u|^2.$$

So,

$$
\begin{aligned}
&\frac{1}{2}\int_Q |P_1 w|^2\,dxdt + \frac{1}{2}\int_Q |P_2 w|^2\,dxdt \\
&+2s\lambda^2\int_Q \varphi\,|\nabla w|^2\,dxdt + 2s^3\lambda^4\int_Q \varphi^3 e^{-2s\eta}\,|u|^2\,dxdt \\
&+2s^2\lambda\int_\Sigma \varphi\frac{\partial\eta}{\partial t}\frac{\partial\psi}{\partial \nu}e^{-2s\eta}|u|^2\,d\gamma dt - 2s^3\lambda^3\int_\Sigma \varphi^3|\nabla\psi|^2\frac{\partial\psi}{\partial\nu}e^{-2s\eta}|u|^2\,d\gamma dt \\
&-2\int_\Sigma \frac{\partial w}{\partial t}\frac{\partial w}{\partial \nu}\,d\gamma dt - 4s\lambda\int_\Sigma \varphi\nabla\psi.\nabla w\frac{\partial w}{\partial\nu}\,d\gamma dt \\
&+2s\lambda\int_\Sigma \varphi\frac{\partial\psi}{\partial\nu}|\nabla w|^2\,d\gamma dt - 4s\lambda^2\int_\Sigma \varphi|\nabla\psi|^2\frac{\partial w}{\partial\nu}.w\,d\gamma dt \\
&\leq C\left(\int_Q e^{-2s\eta}\,|g|^2\,dxdt + s^3\lambda^4\int_0^T\int_\omega \varphi^3 e^{-2s\eta}\,|u|^2\,dxdt\right).
\end{aligned}
$$

Now from

$$\nabla u = e^{s\eta}\left(\nabla w - s\lambda\varphi\nabla\psi w\right),$$

we deduce

$$\int_Q \varphi e^{-2s\eta}\,|\nabla u|^2\,dxdt \leq C\left(\int_Q \varphi\,|\nabla w|^2\,dxdt + s^2\lambda^2\int_Q \varphi^3\,|w|^2\,dxdt\right).$$

We then use the explicit form of $P_1 w$ and $P_2 w$, and get

$$\frac{1}{s}\int_Q \frac{1}{\varphi}\left|\frac{\partial w}{\partial t}\right|^2 dxdt \leq C\left(\int_Q e^{-2s\eta}|g|^2 dxdt + s^3\lambda^4\int_0^T\int_\omega \varphi^3|w|^2 dxdt\right),$$

and

$$\frac{1}{s}\int_Q \frac{1}{\varphi}|\Delta w|^2\,dxdt \leq C\left(\int_Q e^{-2s\eta}|g|^2 dxdt + s^3\lambda^4\int_0^T\int_\omega \varphi^3|w|^2 dxdt\right).$$

We sum up to finally have

$$
\begin{aligned}
&2s^3\lambda^4\int_Q \varphi^3 e^{-2s\eta}\,|u|^2\,dxdt + 4s^2\lambda\int_\Sigma \varphi\frac{\partial\eta}{\partial t}\frac{\partial\psi}{\partial\nu}e^{-2s\eta}|u|^2\,d\gamma dt \\
&-4s^3\lambda^3\int_\Sigma \varphi^3|\nabla\psi|^2\frac{\partial\psi}{\partial\nu}e^{-2s\eta}|u|^2\,d\gamma dt - 4s^2\lambda^3\int_\Sigma \varphi^2|\nabla\psi|^2\frac{\partial\psi}{\partial\nu}e^{-2s\eta}|u|^2\,d\gamma dt \\
&+\int_\Sigma\left(2s\frac{\partial\eta}{\partial t} - 4s\lambda^2\varphi|\nabla\psi|^2 - 4s^2\lambda^2\varphi^2|\nabla\psi|^2\right)e^{-2s\eta}\frac{\partial u}{\partial\nu}u \\
&-2s\lambda\int_\Sigma \varphi\frac{\partial\psi}{\partial\nu}e^{-2s\eta}\frac{\partial u}{\partial t}u\,d\gamma dt - 4s\lambda\int_\Sigma \varphi\nabla\psi e^{-2s\eta}\nabla u\frac{\partial u}{\partial\nu}\,d\gamma dt \\
&+2s\lambda\int_\Sigma \varphi\frac{\partial\psi}{\partial\nu}e^{-2s\eta}|\nabla u|^2\,d\gamma dt - 2\int_\Sigma e^{-2s\eta}\frac{\partial u}{\partial t}\frac{\partial u}{\partial\nu}\,d\gamma dt \\
&\leq C\left(\int_Q e^{-2s\eta}\,|g|^2\,dxdt + s^3\lambda^4\int_0^T\int_\omega \varphi^3 e^{-2s\eta}\,|u|^2\,dxdt\right).
\end{aligned}
$$

Using the fact that $u = \left.\frac{\partial u}{\partial t}\right|_{\Sigma_1} = 0$, and $\left.\frac{\partial u}{\partial\nu}\right|_{\Sigma_2} = 0$, we obtain (4.3).

Now we proceed as the following. We define

$$\widetilde{\varphi}(x,t) = \frac{e^{-\lambda\psi(x)}}{t(T-t)},$$

and

$$\widetilde{\eta}(x,t) = \frac{e^{2\lambda|\psi|_\infty} - e^{-\lambda\psi(x)}}{t(T-t)}.$$

Then

$$\nabla\widetilde{\varphi} = -\lambda\widetilde{\varphi}\nabla\psi,\ \nabla\widetilde{\eta} = \lambda\widetilde{\varphi}\nabla\psi;$$

we still have the properties (4.1) and (4.2).

For $s \geq s_0$ and $\lambda \geq \lambda_0$, we define

$$\widetilde{w}\,(x,t) = e^{-s\widetilde{\eta}(x,t)}u\,(x,t)\,. \tag{4.4}$$

We easily notice that

$$\widetilde{w}\,(x,0) = \widetilde{w}\,(x,T) = 0. \tag{4.5}$$

Calculating $\widetilde{P}\widetilde{w} = e^{-s\widetilde{\eta}}g = e^{-s\widetilde{\eta}}\left[(\partial_t - \Delta)(e^{s\widetilde{\eta}}\widetilde{w})\right]$, we set

$$\widetilde{P}_1\widetilde{w} + \widetilde{P}_2\widetilde{w} = \widetilde{g}_s, \tag{4.6}$$

where

$$\begin{aligned}
\widetilde{P}_1\widetilde{w} &= \frac{\partial\widetilde{w}}{\partial t} - 2s\lambda\widetilde{\varphi}\,\nabla\psi\,\nabla\widetilde{w} + 2s\lambda^2\widetilde{\varphi}|\nabla\psi|^2\widetilde{w},\\
\widetilde{P}_2\widetilde{w} &= -\Delta\widetilde{w} - s^2\lambda^2\widetilde{\varphi}^2|\nabla\psi|^2\widetilde{w} + s\frac{\partial\widetilde{\eta}}{\partial t}\widetilde{w},\\
\widetilde{g}_s &= e^{-s\widetilde{\eta}}g + s\lambda^2\widetilde{\varphi}|\nabla\psi|^2\widetilde{w} + s\lambda\widetilde{\varphi}\,\Delta\psi\,\widetilde{w}.
\end{aligned}$$

We easily deduce the following result :

Proposition 2. *There exists $s_0 > 0$, $\lambda_0 > 0$ and C a positive constant depending on Ω, ω, ψ, and T, such that for all $s \geq s_0$, $\lambda \geq \lambda_0$, and for any fuction u of (3.1), we have*

$$\begin{aligned}
&2s^3\lambda^4\int_Q \widetilde{\varphi}^3 e^{-2s\widetilde{\eta}}\,|u|^2\,dxdt - 4s^2\lambda\int_{\Sigma_2}\widetilde{\varphi}\frac{\partial\widetilde{\eta}}{\partial t}\frac{\partial\psi}{\partial\nu}e^{-2s\widetilde{\eta}}|u|^2\,d\gamma dt\\
&+4s^3\lambda^3\int_{\Sigma_2}\widetilde{\varphi}^3|\nabla\psi|^2\frac{\partial\psi}{\partial\nu}e^{-2s\widetilde{\eta}}|u|^2\,d\gamma dt + 4s^2\lambda^3\int_{\Sigma_2}\widetilde{\varphi}^2|\nabla\psi|^2\frac{\partial\psi}{\partial\nu}e^{-2s\widetilde{\eta}}|u|^2\,d\gamma dt\\
&+2s\lambda\int_{\Sigma_2}\widetilde{\varphi}\frac{\partial\psi}{\partial\nu}e^{-2s\widetilde{\eta}}\frac{\partial u}{\partial t}\,u\,d\gamma dt + 4s\lambda\int_{\Sigma_1}\widetilde{\varphi}\nabla\psi e^{-2s\widetilde{\eta}}\nabla u\,\frac{\partial u}{\partial\nu}\,d\gamma dt\\
&-2s\lambda\int_{\Sigma}\widetilde{\varphi}\frac{\partial\psi}{\partial\nu}e^{-2s\widetilde{\eta}}|\nabla u|^2\,d\gamma dt\\
&\leq C\left(\int_Q e^{-2s\widetilde{\eta}}\left|\frac{\partial u}{\partial t} - \Delta u\right|^2 dxdt + s^3\lambda^4\int_0^T\!\!\int_\omega \widetilde{\varphi}^3 e^{-2s\eta}\,|u|^2\,dxdt\right).
\end{aligned} \tag{4.7}$$

The proof is similar to the one of Proposition 1, we let it to the reader. We obtain from the above Propositions 1 and 2 the following observability inequality :

Corollary 1. *There is a positive constant $C = C(\Omega,\omega,\psi,T)$ such that we have*

$$\int_Q \frac{1}{\theta^2}\,|u|^2\,dxdt \leq C\Big[\int_Q\left|\frac{\partial u}{\partial t} - \Delta u\right|^2 dxdt + \int_0^T\!\!\int_\omega |u|^2\,dxdt\Big], \tag{4.8}$$

where $\dfrac{1}{\theta^2} = \varphi^3 e^{-2s\eta} + \widetilde{\varphi}^3 e^{-2s\widetilde{\eta}}$ is a bounded weight function.

Summing the terms in (4.3) and (4.7) we get the following :

$$
\begin{aligned}
&2s^3\lambda^4 \int_Q \left(\varphi^3 e^{-2s\eta} + \widetilde{\varphi}^3 e^{-2s\widetilde{\eta}}\right) |u|^2 \, dxdt \\
&-4s^2\lambda \int_{\Sigma_2} \left(\varphi \frac{\partial \eta}{\partial t} e^{-2s\eta} - \widetilde{\varphi} \frac{\partial \widetilde{\eta}}{\partial t} e^{-2s\widetilde{\eta}}\right) \frac{\partial \psi}{\partial \nu} |u|^2 \, d\gamma dt \\
&-4s^3\lambda^3 \int_{\Sigma_2} \left(\varphi^3 e^{-2s\eta} - \widetilde{\varphi}^3 e^{-2s\widetilde{\eta}}\right) |\nabla \psi|^2 \frac{\partial \psi}{\partial \nu} |u|^2 \, d\gamma dt \\
&-4s^2\lambda^3 \int_{\Sigma_2} \left(\varphi^2 e^{-2s\eta} - \widetilde{\varphi}^2 e^{-2s\widetilde{\eta}}\right) |\nabla \psi|^2 \frac{\partial \psi}{\partial \nu} |u|^2 \, d\gamma dt \\
&+2s\lambda \int_{\Sigma_2} \left(\varphi e^{-2s\eta} - \widetilde{\varphi} e^{-2s\widetilde{\eta}}\right) \frac{\partial \psi}{\partial \nu} \frac{\partial u}{\partial t} u \, d\gamma dt \\
&-4s\lambda \int_{\Sigma_1} \left(\varphi e^{-2s\eta} - \widetilde{\varphi} e^{-2s\widetilde{\eta}}\right) \nabla \psi \nabla u \frac{\partial u}{\partial \nu} \, d\gamma dt \\
&+2s\lambda \int_{\Sigma} \left(\varphi e^{-2s\eta} - \widetilde{\varphi} e^{-2s\widetilde{\eta}}\right) \frac{\partial \psi}{\partial \nu} |\nabla u|^2 \, d\gamma dt \\
&\leq C \Big[\int_Q \left(e^{-2s\eta} + e^{-2s\widetilde{\eta}}\right) \left| \frac{\partial u}{\partial t} - \Delta u \right|^2 dxdt \\
&+s^3\lambda^4 \int_0^T \int_\omega \left(\varphi^3 e^{-2s\eta} + \widetilde{\varphi}^3 e^{-2s\widetilde{\eta}}\right) |u|^2 \, dxdt \Big].
\end{aligned}
$$

Now, it suffices to notice that $\varphi = \widetilde{\varphi}$ and $\eta = \widetilde{\eta}$ on Σ.

5. Instantaneous sentinels

In this part, we discuss of how to get instantaneous information (at fixed $t = T \in [0, +\infty[$) on pollution terms in systems of incomplete data in ecology and/or meteorology problems.

Here, the ecological system is affected by pollution in the boundary of the domain (a border of an air pollution cloud for example). We verify that if the initial data is completely unknown, the sentinel is nul. If there is some information on the initial data, the *instantaneous* sentinel naturally exists if the control set is bigger than the one where the observation may be defined.

Finally, we give the characterization of the instantaneous sentinel with some remarks.

Using the techniques of the sections below and those in Miloudi Y. et al. (2009). We prove the existence and characterization of a sentinel, which permits to identify the pollution parameters at fixed time T. The problem we consider now is the following :

$$
\frac{\partial y}{\partial t} + Ay + f(y) = 0 \qquad \text{in } Q :=]0, T[\times \Omega, \tag{5.1}
$$

where $\Omega \subset \mathbb{R}^d$ is an open domain of regular boundary $\Gamma = \partial\Omega$ for instance, A represents a second order elliptic operator such that $a_{kl} = a_{lk}$, $a_{kl} \in C^2(\bar{\Omega})$, and $\sum_{i,j=1}^d a_{ij}\eta_i\eta_j \geq c^2|\eta|^2$, and $f : \mathbb{R} \to \mathbb{R}$ is a (nonlinear) C^1 function.

To (5.1) we should add initial and boundary conditions. It is in these conditions that we can meet incomplete data. We assume that we know that $y_{|t=0}$ is in a ball of $L^2(\Omega)$ of center y^o, say $\|y(0,.) - y^o(.)\|_{L^2(\Omega)} \leq \tau$, that we should write :

$$
y(0) = y^o + \tau\hat{y}^o, \tag{5.2}
$$

where $\|\hat{y}^o\| \leq 1$, and where $\tau\hat{y}^o$ is the missing data. To the boundary condition ξ is added a pollution term $\lambda\hat{\xi}$ which is unknown, and which *we want* to identify. It appears on a part $\Sigma_0 = \Gamma_0 \subset \Sigma$ of the total boundary $\Sigma =]0,T[\times\Gamma$, as we have :

$$y = \begin{cases} \xi + \lambda\,\hat{\xi} & \text{on } \Sigma_0, \\ 0 & \text{on } \Sigma\backslash\Sigma_0. \end{cases} \tag{5.3}$$

The question is to obtain information on the pollution $\lambda\hat{\xi}$, not affected by the missing term $\tau\hat{y}^o$ of the initial data.

Still, one can use the least square method : In the context of the above problem, it consists in considering the unknowns $\{\lambda\hat{\xi}, \tau\hat{y}^o\} = \{v, w\}$ as two control variables. At time T, the state is then $y(T,x;v,w)$, and we want this solution to be *as close as possible* to some measurement m_o. We consider then the distance $J(v,w) = \|y(v,w) - m_o\|$ in an appropriate norm and we search for

$$\inf_{\{v,w\}} J(v,w)$$

where v and w are arbitrary. As we know, with this method, we can not seperate v and w. The more suitable is then the Sentinel method (see Lions J.-L. (1992), Miloudi Y. et al. (2009), Kernevez J.-P. (1997) and the references therein).

The state solution $y(\lambda,\tau) := y(t,x;\,\lambda\hat{\xi},\tau\hat{y}^o)$. The observation is given in an observatory $\mathcal{O} \subset \Omega$, but here we observe y at fixed *instant* T :

$$y_{\text{obs}} = m_o. \tag{5.4}$$

Now, we are given

$$h_o \in L^2(\mathcal{O}) \tag{5.5}$$

and

$$\mathrm{w} \in L^2(\omega) \tag{5.6}$$

where $\omega \subset \Omega$ the open set of controls. We search for w, such that the mean function

$$\begin{aligned} \mathcal{S}(\lambda,\tau) &= \int_{\mathcal{O}} h_o\, y(T,x;\lambda,\tau)\,dx + \int_{\omega} \mathrm{w}\, y(T,x;\lambda,\tau)\,dx \\ &= \int_{\Omega} \Big(h_o\chi_{\mathcal{O}} + \mathrm{w}\chi_{\omega}\Big) y(T,x;\lambda,\tau)\,dx \end{aligned} \tag{5.7}$$

(here $\chi_{\mathcal{O}}$ and χ_ω denote the characteristic functions of $\mathcal{O}$ and ω respectively), is insensitive to the missing data at the first order; i.e.

$$\frac{\partial \mathcal{S}}{\partial \tau}(0,0) = 0,$$

for $\lambda = \tau = 0$, and such that

$$\|\mathrm{w}\|_{L^2(\omega)} \quad \text{is of minimal norm.} \tag{5.8}$$

The integral (5.7) is called *sentinel*, following the definition of Lions Lions J.-L. (1992).
The pollution sources in (5.3) can be considered as functionals whose position and nature are known, but their amplitudes are however unknown. The goal is to estimate these amplitudes; the other missing terms (in the initial data (5.2)) do not interest us.

Remark 4. *In Lions Lions J.-L. (1992), $\omega = \mathcal{O}$, so there exists always a sentinel (at least the one where $\mathsf{w} = -h_o$). Then the problem is only to find w solution to (5.8). At last, we have to be sure that $\mathsf{w} \neq -h_o$.*

Below, we use the sentinel method as formulated in Miloudi Y. et al. (2007) and Miloudi Y. et al. (2009) and Nakoulima O. (2004), where the observation and the control have their support in two different open sets. Indeed, one can observe somewhere and control elsewhere !

6. Instantaneous sentinel. Case of no information on the missing data

In this section, we extend the method of sentinels to the case of observation and control having their supports in two different open sets. Moreover, we want information at precise time T which is a difficult problem.

Denote by $y(t,x;\lambda,\tau) := y(\lambda,\tau)$, the state solution of (5.1)-(5.3). We begin by noticing that the solution $y = y(\lambda,\tau)$ satisfies the system :

$$\begin{cases} \mathcal{L}y_\tau &= 0 \ \text{ in } Q, \\ y_\tau(0) &= \hat{y}^o \ \text{ in } \Omega, \\ y_\tau &= 0 \ \text{ on } \Sigma, \end{cases}$$

where y_τ denotes the derivative

$$y_\tau = \frac{\partial y}{\partial \tau}(0,0)$$

and where $\mathcal{L}$ is the linear differential operator defined by

$$\mathcal{L} = \frac{\partial}{\partial t} + A + f'(y_o)\, I_d$$

with $y_o = y(t,x;0,0)$ for $\lambda = \tau = 0$. Then

$$\frac{\partial S}{\partial \tau}(0,0) = \int_\Omega \left(h_o\chi_{\mathcal{O}} + \mathsf{w}\chi_\omega \right) y_\tau(T)\, dx,$$

Remark 5. *From above, the insensitive criterion (5) to the missing term $\tau\hat{y}^o$ is given by :*

$$\int_\Omega \left(h_o\chi_{\mathcal{O}} + \mathsf{w}\chi_\omega \right) y_\tau(T)\, dx = 0.$$

Lemma 2. *Let q be the solution to the following well-posed backward problem :*

$$\mathcal{L}^* q = 0 \qquad in \quad Q, \tag{6.1}$$

$$q(T) = h_o\chi_{\mathcal{O}} + \mathsf{w}\chi_\omega \quad in \quad \Omega, \tag{6.2}$$

$$q = 0 \qquad on \quad \Sigma, \tag{6.3}$$

where $\mathcal{L}^* = -\frac{\partial}{\partial t} + A^* + f'(y_o)\, I_d$ is the adjoint operator. Then, the existence of an instantaneous sentinel insensitive to the missing data (i.e. such that (5) holds), is equivalent to the null-controllability problem (6.1)-(6.3) with

$$q(0) = 0 \qquad in \quad \Omega. \tag{6.4}$$

Proof - Multiplying (6.1) by $y_\mathcal{T}$ and integrating by parts we obtain :

$$\int_Q (\mathcal{L}^* q)\, y_\mathcal{T}\, dxdt = \int_Q q\, \mathcal{L} y_\mathcal{T}\, dxdt - \int_\Omega q(T) y_\mathcal{T}(T)\, dx$$
$$+ \int_\Omega q(0) y_\mathcal{T}(0)\, dx - \int_\Sigma \frac{\partial q}{\partial \nu} y_\mathcal{T}\, d\sigma + \int_\Sigma \frac{\partial y_\mathcal{T}}{\partial \nu} q\, d\sigma = 0. \quad (6.5)$$

But, $y_\mathcal{T}$ is solution to (6), and q verifies (6.2) and (6.3). Hence

$$-\int_\Omega \left(h_o \chi_\mathcal{O} + \mathrm{w}\chi_\omega\right) y_\tau(T)\, dx + \int_\Omega q(0)\, \hat{y}^o\, dx = 0.$$

If a sentinel exists, then we have (5). So it remains :

$$\int_\Omega q(0)\, \hat{y}^o\, dx = 0 \qquad \text{for every} \quad \hat{y}^o \in L^2(\Omega),$$

and consequently $q(0) = 0$ in Ω. The converse is obvious. ∎

Corollary 2. *If there is no information on the missing data, then the instantaneous sentinel is nul.*

Proof - The proof is easy and lies on the backward uniqueness property (see Lions-Malgrange Lions J.-L. & Malgrange B. (1960)) : If an instantaneous sentinel exists, then from the above Lemma we have (6.1), (6.3) and (6.4). We deduce that

$$q \equiv 0 \quad \text{ in } \; Q.$$

Thus, in particular for $t = T$, we have $q(T) = 0$. So that $h_o \chi_\mathcal{O} + \mathrm{w}\chi_\omega = 0$ in Ω, and hence -as in Lions Lions J.-L. (1992)- the sentinel is nul. ∎

6.0.0.1 Information given by the sentinel

We briefly show how (5) and (5.8) are sufficient to get information on the pollution term $\lambda\hat{\xi}$. We write

$$S(\lambda, \tau) \simeq S(0,0) + \lambda \frac{\partial S}{\partial \lambda}(0,0), \qquad \text{for} \quad \lambda, \tau \quad \text{small}.$$

Using (5.4), we have

$$\lambda \frac{\partial S}{\partial \lambda}(0,0) = \int_Q (h_0 \chi_\mathcal{O} + \mathrm{w}\chi_\omega) y_\lambda\, dxdt \simeq \int_Q (h_0 \chi_\mathcal{O} + \mathrm{w}\chi_\omega)(m_o - y_0)\, dxdt, \quad (6.6)$$

where the derivative

$$y_\lambda = (\partial y / \partial \lambda)\, (0,0)$$

only depends on $\hat{\xi}$ and other known data. Consequently, the estimates (6.6) contain the information on $\lambda\hat{\xi}$. Indeed, y_λ is solution to the well-posed problem

$$\mathcal{L} y_\lambda = 0 \text{ in } Q, \quad y_\lambda(0) = 0 \text{ in } \Omega, \quad \text{and } y_\lambda = \hat{\xi} \text{ on } \Sigma. \quad (6.7)$$

Multiplying by q and integrating by parts, we obtain

$$\int_Q \lambda\hat{\xi}\, dxdt = \int_Q (h_0 \chi_\mathcal{O} + \mathrm{w}\chi_\omega)(m_o - y_0)\, dxdt.$$

Remark 6. *From the previous corollary, we see that in order to get a non nul sentinel, we need more information (on the structure) of the missing initial data. We will need here the assumption (7) below.*

7. Instantaneous sentinel. Case of partial information on the missing data

In this section, we consider the case of the *incomplete* initial data :

$$y(0) = y^o + \sum_{i=1}^{N} \tau_i \hat{y}_i^o \tag{7.1}$$

where $\hat{y}_i^o, 1 \leq i \leq N$ are linearly independent in $L^2(\Omega)$, and belong to a vector subspace of N dimension, which we denote by $Y = \langle \hat{y}_1^o, \hat{y}_2^o, \cdots, \hat{y}_N^o \rangle$. The parameters τ_i are *unknown* and are supposed small.

If we denote by $y_{\tau_i} = \dfrac{\partial y}{\partial \tau_i}(0,0)$ for $\lambda = \tau_i = 0, 1 \leq i \leq N$, then y_{τ_i} is solution with $y_{\tau_i}(0) = \hat{y}_i^o$. Now, we assume the following :

$$\omega \cap \mathcal{O} \neq \emptyset$$

and we define the instantaneous sentinel by :

$$\mathcal{S}(\lambda, \tau) = \int_\Omega \Big(h_o \chi_{\mathcal{O}\cap\omega} + \mathrm{w}\chi_\omega \Big) y(T; \lambda, \tau)\, dx = 0$$

with $\tau = (\tau_1, \cdots, \tau_N)$. Hence, looking for w such that the sentinel $\mathcal{S}$ is insensitive to the missing terms $\tau_i \hat{y}_i^o$, is finding w such that :

$$\frac{\partial \mathcal{S}}{\partial \tau_i}(0, \tau_i = 0) = \int_\Omega \Big(h_o \chi_{\mathcal{O}\cap\omega} + \mathrm{w}\chi_\omega \Big)\, y_{\tau_i}(T)\, dx = 0, \quad 1 \leq i \leq N. \tag{7.2}$$

Lemma 3. *The existence of an instantaneous sentinel insensitive to the missing terms is equivalent to the existence of a unique pair* (w, q) *such that we have :*

$$\begin{cases} \mathcal{L}^* q = 0 & \text{in } Q, \\ q(T) = h_o \chi_{\mathcal{O}\cap\omega} + \mathrm{w}\chi_\omega & \text{in } \Omega, \\ q = 0 & \text{on } \Sigma. \end{cases} \tag{7.3}$$

and such that

$$q(0) \in Y^\perp. \tag{7.4}$$

Proof - Multiplying the first equation in (7.3) by y_{τ_i} and integrating by parts, we find :

$$-\int_\Omega q(T)\, y_{\tau_i}(T)\, dx + \int_\Omega q(0) \hat{y}_i^o\, dx = 0,$$

since y_{τ_i} is solution of (6). And so,

$$-\int_\Omega \Big(h_o \chi_{\mathcal{O}\cap\omega} + \mathrm{w}\chi_\omega \Big)\, y_{\tau_i}(T)\, dx + \int_\Omega q(0) \hat{y}_i^o\, dx = 0, \qquad 1 \leq i \leq N.$$

But the sentinel should satisfy (7.2). Hence,

$$\int_\Omega q(0) \hat{y}_i^o\, dx = 0, \quad \text{for all} \quad 1 \leq i \leq N.$$

So that

$$q(0) \perp \hat{y}_i^o, \qquad 1 \leq i \leq N \qquad \text{if and only if} \qquad q(0) \in Y^\perp,$$

where $Y^\perp$ is the orthogonal of Y in $L^2(\Omega)$. ■

Remark 7. *The controllability problem (7.3)-(7.4) has at least a solution: Given $h_o \in L^2(\mathcal{O})$, the control 'solution'* w *is given by*

$$\mathrm{w} = \begin{cases} -h_o & \text{in } \mathcal{O} \cap \omega, \\ 0 & \text{in } \omega \setminus (\mathcal{O} \cap \omega), \end{cases} \tag{7.5}$$

and so, the set of solutions to (7.3)-(7.4) is not empty.

8. Penalization

Here we are interested in the problem (5.8). We consider the optimization problem :

$$(\mathcal{P}) \qquad \min_{(\mathrm{w},z)\in\mathcal{A}} \|\mathrm{w}\|_{L^2(\omega)}$$

with

$$\mathcal{A} = \left\{ (\mathrm{w},q) \text{ such that} \left| \begin{array}{rl} \mathcal{L}^* q = 0 & \text{in } Q,\ q(0) \perp \widehat{y}_i^o\ 1 \le i \le N, \\ q(T) = h_o\chi_{\mathcal{O}\cap\omega} + \mathrm{w}\chi_\omega & \text{in } \Omega, \quad q = 0 \quad \text{on } \Sigma \end{array} \right. \right\}.$$

Theorem 3. *There is a unique pair $(\widehat{\mathrm{w}}, \widehat{q})$ solution to the problem $(\mathcal{P})$.*

Proof - The domain $\mathcal{A}$ is no empty and is closed. The mapping : $\mathrm{w} \longrightarrow \|\mathrm{w}\|_{L^2(\omega)}$ is continuous, coercitive and strictly convex. We deduce that there exists a unique solution to $(\mathcal{P})$ denoted by $(\widehat{\mathrm{w}}, \widehat{q}) \in \mathcal{A}$, which satisfies

$$\|\widehat{\mathrm{w}}\|_{L^2(\omega)} \le \|\mathrm{w}\|_{L^2(\omega)} \qquad \forall\ (\mathrm{w}, q) \in \mathcal{A}.$$

■

We now use the penalization method in order to characterize the optimal control $(\widehat{\mathrm{w}}, \widehat{q})$. Let be $\varepsilon > 0$, we introduce the function

$$J_\varepsilon(\mathrm{w}, q) = \frac{1}{2}\|\mathrm{w}\|^2_{L^2(\omega)} + \frac{1}{2\varepsilon}\|\mathcal{L}^* q\|^2_{L^2(Q)},$$

and we consider the problem $(\mathcal{P}_\varepsilon)$ given by

$$(\mathcal{P}_\varepsilon) \qquad \min_{(\mathrm{w},q)\in\mathcal{U}} J_\varepsilon(\mathrm{w}, q)$$

with

$$\mathcal{U} = \left\{ (\mathrm{w},q) \text{ such that} \left| \begin{array}{rl} \mathcal{L}^* q \in L^2(Q) & \text{in } Q,\ q(0) \perp \widehat{y}_i^o\ 1 \le i \le N, \\ q(T) = h_o\chi_{\mathcal{O}\cap\omega} + \mathrm{w}\chi_\omega & \text{in } \Omega, \quad q = 0 \quad \text{on } \Sigma \end{array} \right. \right\}.$$

The following proposition gives the existence of a solution to the penalized problem $(\mathcal{P}_\varepsilon)$.

Proposition 3. *The problem $(\mathcal{P}_\varepsilon)$ has a unique solution denoted by $(\mathrm{w}_\varepsilon, q_\varepsilon)$.*

Proof - We have $\mathcal{A} \subset \mathcal{U}$. Moreover, $\mathcal{A}$ is nonempty by the previous theorem. Consequently, $\mathcal{U}$ is nonempty and closed. The cost function J_ε is continuous, coercitive and strictly convex. Hence, problem $(\mathcal{P}_\varepsilon)$ has a unique solution. ■

We now prove the following proposition.

Proposition 4. *Let be $(w_\varepsilon, q_\varepsilon)$ the unique solution of $(\mathcal{P}_\varepsilon)$, then for $\varepsilon \longrightarrow 0$ we obtain*

$$\begin{cases} w_\varepsilon \rightharpoonup \widehat{w} & \text{weakly in } \; L^2(\omega), \\ q_\varepsilon \rightharpoonup \widehat{q} & \text{weakly in } \; \mathcal{W}(0,T), \end{cases}$$

where $\mathcal{W}(0,T) = \left\{ q \in C(0,T); \quad q(t) \in L^2(\Omega) \right\}$.

Proof - Since $(w_\varepsilon, q_\varepsilon)$ is solution to $(\mathcal{P}_\varepsilon)$ then

$$J_\varepsilon(w_\varepsilon, q_\varepsilon) \le J_\varepsilon(w, q) \;\; \forall (w, q) \in \mathcal{A}.$$

In particular $(\widehat{w}, \widehat{q}) \in \mathcal{A} \subset \mathcal{U}$ is solution to $(\mathcal{P}_\varepsilon)$. Then :

$$\|w_\varepsilon\|_{L^2(\omega)} \le \|\widehat{w}\|_{L^2(\omega)} \le C \qquad \text{and so} \qquad \|\mathcal{L}^* q_\varepsilon\|_{L^2(Q)} \le C\sqrt{\varepsilon},$$

where C is a positive constant which is not the same at each time.

Knowing that $(w_\varepsilon, q_\varepsilon) \in \mathcal{U}$, we deduce

$$\|q_\varepsilon\|_{H^{2,1}(Q)} \le C.$$

Then, there is a subsequent $(w_\varepsilon, q_\varepsilon)$, and two functions $w_0 \in L^2(\omega)$ and $q_0 \in H^{2,1}(Q)$ such that

$$\begin{cases} w_\varepsilon \rightharpoonup w_0 & \text{in } L^2(\omega), \\ q_\varepsilon \rightharpoonup q_0 & \text{in } H^{2,1}(Q). \end{cases}$$

Now as $H^{2,1}(Q) \hookrightarrow L^2(Q)$ with compact injection, the pair (w_0, q_0) satisfies the following :

$$\mathcal{L}^* q_0 = h + k_0 \chi_\omega \text{ in } Q, \quad q_0(T) = q_0 = 0 \text{ in } \Omega, \quad q_0 = 0 \text{ in } \Sigma. \tag{8.1}$$

But,

$$\frac{1}{2}\|w_0\|^2_{L^2(\omega)} \le \liminf_{\varepsilon \to 0} J_\varepsilon(w_\varepsilon, q_\varepsilon) \le J_\varepsilon(\widehat{w}, \widehat{q}) \le \frac{1}{2}\|\widehat{w}\|_{L^2(\omega)}.$$

Since $(\widehat{w}, \widehat{q})$ is the unique solution of $(\mathcal{P})$, then $\widehat{w} = w_0$. Finally, as q_0 satisfies (8.1), we obtain $\widehat{q} = q_0$. ■

9. Concluding remarks

From the numerical aspect, there where some results in the litterature. In B.-E. Ainseba et al. Ainseba B.E. et al. (1994a)Ainseba B.-E. et al. (1994b), the authors compare numerically the least square method and the sentinel one during time $[0, T[$ and find that they are equivalent in the linear case $f(y) = Cy$. However, the sentinel approach has an advantage in case of several measures -as it is really the case in general-; Indeed, it suffices to calculate a simple integral to identify the parameter each time, since one has to minimize a quadratic cost functional for the least square approach in order to determine both the parameter and the missing terms.

In the nonlinear case and when the observation is noisy, the least square method costs a lot numerically and can fail after a large number of iterations, while the sentinel method is relatively robust face to the perturbations of the observation. Moreover, the calculus of w does not depend on the observation and then makes the method efficient in case of several measures. We refer to Ainseba B.E. et al. (1994a)Ainseba B.-E. et al. (1994b) and to the book of J.-P. Kernevez Kernevez J.-P. (1997) for further information and numerical details.

As we have seen, the existence of an instantaneous sentinel is equivalent to controllability problems. In the case of no supplementary information on the missing data (that we do not want to identify), the sentinel is nul as in Lions Lions J.-L. (1992).

In the case where we have some more information on the structure of the missing data (which becomes incomplete data), we proved that the existence of the instantaneous sentinel is equivalent to a nontrivial controllability problem, which has a solution under the assumption $\mathcal{O} \cap \omega \neq \emptyset$.

10. References

Ainseba B.-E., Kernevez J.-P., Luce R. *Identification de pollution dans une rivière,* ESAIM:Mathematical Modeling and Numerical Analysis M2AN, 1994, pp 819 – 823.

Ainseba B.-E., Kernevez J.-P., Luce R. *Identification de paramètres dans les problèmes non linéaires à données manquantes,* ESAIM:Mathematical Modeling and Numerical Analysis M2AN, 1994, pp 313 – 328.

Bodart O., Application de la méthode des sentinelles à l'identification de sources de pollution dans un système distribué. Thèse de Doctorat, Université de Technologie de Compiègne, 1994.

Bodart O., Demeestere P., Contrôle frontière de l'explosion en temps fini de la solution d'une équation de combustion. C.R. Acad. Sci., Série I, 325(8), 1997, pp 841 – 845.

Bodart O., Fabre C., Contrôle insensibilisant la norme de la solution d'une équation de la chaleur semi-linéaire. C.R. Acad. Sci., Série I, 316(8), 1993, pp 789 – 794.

Bodart O., Fabre C., Controls insensitizing the norm of a semilinear heat equation. Journ. of Math. Analysis and Applications, 195, 1995, pp 658 – 683.

Brezis H., Analyse fonctionnelle, Collection mathématiques appliquées pour la maîtrise, Masson, 1984.

Demeestere P., Méthode des sentinelles : étude comparative et application à l'identification d'une frontière Contrôle du temps d'explosion d'une équation de combustion. Thèse de Doctorat, Université de Technologie de Compiègne, 1997.

A. Fursikov, O. Yu. Imanuvilov, Controllability of evolution equations. Lecture Notes, Research Institute of Mathematics, Seoul National University, Korea (1996).

O. Yu. Imanuvilov Controllability of parabolic equations. Sbornik Mathematics 186 : 6, 1995, pp 879 – 900.

Kernevez J.-P., The Sentinel Method and its Application to Environmental Pollution Problems, CRC. Press, Boca Raton, FL, 1997.

Lebeau G., Robbiano L., Contrôle exacte de l'équation de la chaleur, Comm. P.D.E., 20, 1995, pp 335 – 356.

Lions J.-L., Contrôle optimal de systèmes gouvernés par des équations aux dérivées partielles. Dunod, gauthier-Villars, Paris 1968.

Lions J.-L. Sentinelles pour les systèmes distribués à données incomplètes. Masson, Paris, 1992.

Lions J.-L., Malgrange B. *Sur l'unicité rétrograde dans les problèmes mixtes paraboliques,* Math. Scand., 8, pp 277-286 (1960).

Miloudi Y., Nakoulima O., Omrane A., A method of detection of pollution in dissipative systems of incomplete data, Proc. of the European Society of Appl. and Indust. Math., ESAIM:Proc Vol. 17, 2007, pp 67 – 79.

Miloudi Y., Nakoulima O., Omrane A., Instantaneous sentinels for distributed systems of incomplete data, Inverse Problems in Science and Engineering, Vol. 17, 4, 2009, pp 451 – 459.

Nakoulima O., Contrôlabilité à zéro avec contraintes sur le contrôle. C. R. Acad. Sci. Paris, Ser. I 339, 2004, pp 405 – 410.

Puel J.-P., Contrôlablité approchée et contrôlabilité exacte. Notes de cours de D.E.A. Université Pierre et Marie Curie, Paris, 2001.

10

PM10 and Its Chemical Composition: A Case Study in Chiang Mai, Thailand

Somporn Chantara
Chiang Mai University
Thailand

1. Introduction

This chapter describes the chemical composition and spatial and temporal variations of airborne particulate matter that have a diameter of less than 10 μm (PM10), as well as its possible health effects. A case study of air pollution in the city of Chiang Mai, Thailand was conducted due to its geographical features and meteorological conditions; i.e. temperature inversion, wind velocity and precipitation levels. Chiang Mai, with an altitude of approximately 310 m above sea level, is situated approximately 700 km north of Bangkok. The city covers an area of approximately 20,107 km^2 and is the country's second largest province. The city is situated in the Chiang Mai-Lamphun Basin and is surrounded by mountains. In the dry season, there is a low level of precipitation, and there are calm winds and a vertical temperature inversion, while air pollutants are generated from various sources of mostly anthropogenic activities and accumulate in the lower atmosphere and definitely have an effect on human health and the environment. PM10 is considered the most significant air pollutant that contributes to serious air pollution in the dry season of Northern Thailand. Its major sources are open burning and internal combustion exhaust from traffic. However, traffic density seems to be constant for the whole year, while open burning is mostly performed in the dry season, which coincides with the peak of the annual haze episode in the upper region of Northern Thailand. Large scale open burning in this region consists of forest fires and the burning of agricultural waste. These activities definitely emit a variety of air pollutants in the forms of both particulates and gases.

Air pollution is quite common in most countries with big cities. Atmospheric aerosol particles originate from a wide variety of natural and anthropogenic sources. Primary particles are directly emitted from sources, such as biomass burning, incomplete combustion of fossil fuels and traffic-related suspension of road, soil, dust, sea salt and biological materials. Secondary particles are formed by gas-to-particle conversion in the atmosphere (Pöschl, 2005). Concentration, composition and size distribution of atmospheric aerosol particles are temporally and spatially highly variable. The predominant particle components of air particulate matter (PM) are sulfate, nitrate, ammonium, sea salt, mineral dust, organic compounds and black or elemental carbon (Pöschl, 2005). The effects of aerosols to the atmosphere, climate and public health are among the central topics in the current environmental research. An airborne particulate matter consists of several of inorganic and organic species, many of which can adversely affect human health (Pöschl, 2005; Mauderly

& Chow, 2008). Of these constituents, the classes of polycyclic aromatic hydrocarbons (PAHs) are widespread environmental pollutants which are formed during the incomplete combustion process of organic material emitted from a large variety of industrial processes, motor vehicles, domestic waste burning and regenerated burning in agriculture (Lee, et al., 1995). PAHs have been measured in many media (air, water, food, and soil samples) because some of them are known to be carcinogenic. Specifically prepared foods, such as charcoal broiled or smoked meat, contain higher concentrations of PAHs than city air. Nevertheless, the daily air intake of humans (10–25 m^3) is of comparable mass to the daily intake of food and water (2–4 kg) (Furton & Pentzke, 1998). Therefore, the air polluted by airborne PAHs is of significant concern.

Numerous particulates and gaseous compounds (e.g. CO and volatile organic compounds) that come from biomass burning are known to be hazardous to human health (Torigoe et al., 2000). Mass concentrations of CO, NO_2, $PM_{2.5}$, organic carbon (OC) and elemental carbon (EC) in $PM_{2.5}$ were significantly associated with emergency department visits at hospitals due to cardiovascular diseases (Metzger et al., 2004). The most important gasses which have affected the acidic dry deposition are HNO_3, HCl and H_2SO_4. Those gasses can be transformed into aerosols by neutralization reactions (Seinfeld & Pandis, 1998). Semi-volatile NH_4NO_3 and NH_4Cl are formed via reversible phase equilibrium with NH_3, HNO_3 and HCl (Pio & Harrison, 1987). Understanding the formation of secondary particulate species SO_4^{2-}, NO_3^- and NH_4^+ formation, requires investigation of the atmospheric concentration levels for HNO_3, HCl and SO_2 and related aerosol ions. The inorganic component is made up of some insoluble dust and ash material, and soluble salts, in which potassium, ammonium, sulfate and nitrate are the most important species (Fuzzi et al., 2007).

Toxic metals in air are considered a global problem and a growing threat to the environment. Heavy metals are part of a large group of air pollutants called air toxins, which upon inhalation or ingestion can be responsible for a range of health effects such as cancer, neurotoxicity, immunotoxicity, cardiotoxicity, reproductive toxicity, teratogenesis and genotoxicity. For most metals, the bound or complex species are less toxic than the elemental forms, hence there is a loss in specificity when they are broadly classified as metals and their compounds. When inhaled, very small particles containing metals or their compounds deposit beyond the bronchial regions of the lungs into the alveoli region. Epidemiological studies have established relationships between inhaled PM and morbidity or mortality in populations. Studies in occupational or community settings have established the health effects of exposure to heavy metals, such as lead and their compounds. The localised release of some heavy metals from inhaled PM have been hypothesised to be responsible for lung tissue damage (U.S. EPA, 1999c). Metals occur in air in different phases as solids, gases or adsorbed to particles having aerodynamic sizes ranging from below 0.01 to 100 μm and larger. Two major categories of PM are fine particles (PM2.5) and coarse particles (PM10). Heavy metals such as As, Ga, Eu, W, Au, Cd, Ga, Mo, Pb, Sb, W, Zn, Fe, V, Cr, Co, Ni, Mn, Cu and Se exist in both coarse and fine fractions in ambient air. Ca, Al, Ti, Mg, Sc, La, Hf and Th exist predominantly in the coarse fraction. Metals such as Ba, Cs and Se enrich the fine fraction of PM. The degree to which the metals are absorbed depends on the properties of the metals concerned, i.e. valence, radius, degree of hydration and co-ordination with oxygen, physicochemical environment, nature of the adsorbent, other metals present and their concentrations and the presence of soluble ligands in the surrounding fluids (Valerie et al., 2003). The metals that are of high environmental priority,,due to their toxicity from char and ash samples of biomass, could be released to the atmosphere when biomass is burnt. The elemental compositions of biomass are complex. They involve 6 major elements (C, Cl, H, N, O and S) in the organic phase and at least 10 other elements (Al, Ca, Fe, K, Mg, Na, P, Si, Sr and Ti) in the inorganic phase called ash elements, which is important to ash characterization. Heavy metals (As, Cd, Co, Cr, Hg, Ni, Pb, Sb and Se) in both phases of biomass are trace levels (Demirbas et al., 2008).

In attempt to identify the major emission sources responsible for adverse health effects, several comprehensive surveys of atmospheric contaminants from a variety of sources have been performed worldwide (Furton & Pentzke, 1998; Pengchai, et al., 2009). In Thailand, only a few studies have been conducted to determine atmospheric PAHs in big cities, such as Bangkok and Chiang Mai. In particular, the northern part of Thailand has been facing air pollution in the dry season almost every year. It is important to know the levels of air pollutants, including PAHs and their variations, throughout the year. Moreover, sources of air pollutants are also needed to be identified.

2. PM10 situation in the city of Chiang Mai

Ambient PM10 concentrations have been automatically measured at the Air Quality Monitoring (AQM) station set up by the Pollution Control Department (PCD) of Thailand. They were measured with the tapered element oscillating microbalance (TEOM) Series 1400a (Rupprecht & Patashnick, USA). TEOM mass detectors or microbalances utilize an inertial mass weighing principle. A TEOM detector consists of a substrate (usually a filter cartridge) placed on the end of a hollow tapered tube. The tube with the filter on the free end was oscillated in a clamped-free mode at its resonant frequency. This frequency depends on the physical characteristics of the tube and the mass on its free end. A particle laden air steam was drawn through the filter where the particles were deposited and then through the hollow tube. The frequency of oscillation was measured and recorded by the microprocessor; the change in frequency was used to calculate the mass of particulate matter deposited on the filter (Teflon coated with glass fiber filter surface). The air-flow rate at 16.67 L min^{-1} was sampled through the sampling head and divided between the filter flow (3 L min^{-1}) and an auxiliary flow (13.67 L min^{-1}). The filter flow was sent to the instrument's mass transducer, which contained the analyte fraction of the particulate. The inlet was heated to 50°C prior to the particles being deposited onto the filter in order to eliminate the effect of condensation or evaporation of particle water (Washington State Department of Ecology, 2004; Patashnick et al., 2002).

Twenty-four hour PM10 concentrations from 2005-2009 monitored at Chiang Mai AQM station located at the city center were plotted and are illustrated in Fig.1. The trend of PM10 concentrations are almost the same every year. They were high in the dry season especially in February and March. The level decreased in the middle of April due to rain precipitation. During the first 2 weeks of March 2007, air pollution levels in Chiang Mai and the surrounding provinces rose steadily above the safety limit and produced haze that significantly cut visibility down to less than 1 km. PM10 reached their peak concentration on

March 14th 2007at 383 µg/m³ three times higher than 24 hrs PM10 standard of Thailand (120 µg/m³). About 500,000 people were reported by the Public Health Ministry to be affected by the heavy air pollution. Hospitals and clinics across the affected area reported a surge in the number of patients with respiratory problems in March 2007, approximately a 20% increase compared to the same period in 2006 (Pengchai, et al., 2009).

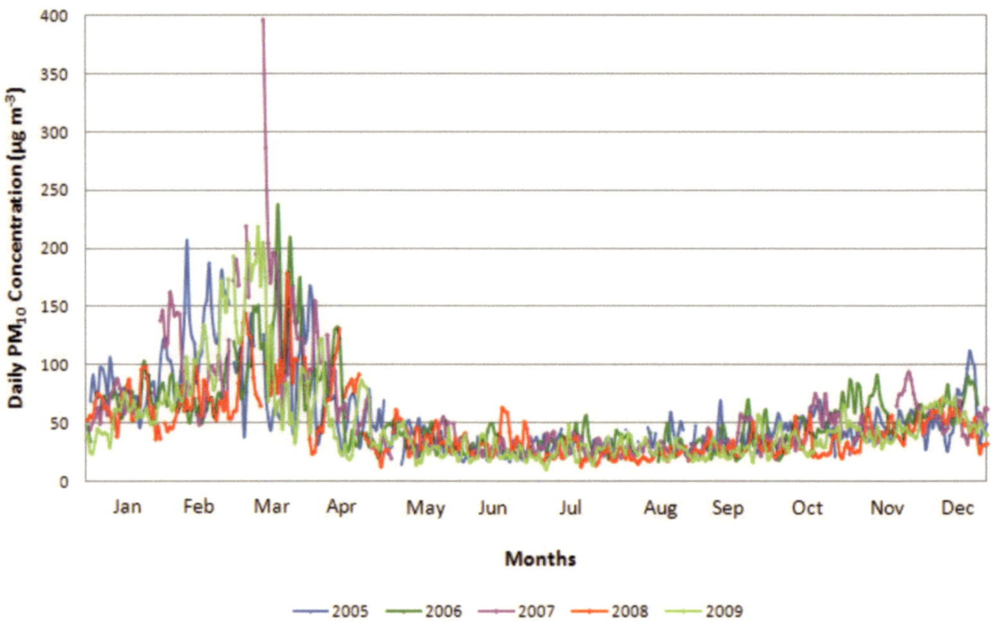

Fig. 1. PM10 concentrations in Chiang Mai's ambient air from 2005-2009.

PM10 concentrations and rain precipitation from late February 2010 to the beginning of March 2011 are illustrated in Fig. 2. PM10 samples were collected by mini volume air samplers (Airmetric, USA) at the rooftop of a 9-storey building (ScB1), Faculty of Science, Chiang Mai University (CMU). This station was set up to monitor ambient PM10 variation by avoiding the direct effects of traffic emissions. Another two PM10 data series were obtained from the AQM Station belonging to PCD located at Provincial Hall (PH) and Yupparaj Wittayalai School (YP). Regional area air quality data was downloaded from the PCD website (Pollution Control Department, 2011). All mentioned sampling sites are shown in Fig. 3. Data of rain precipitation was obtained from website of Chiang Mai meteorological station (http://www.cmmet.tmd.go.th/index1.php).

In the dry season of the year 2010, PM10 concentrations obtained from those 3 stations were much closer to each other. This is probably due to the effect from the large scale of open burning, which is mostly performed in the dry season. In the rainy season of the year 2010 and at the beginning of the dry season in the year 2011, PM10 in the air was generated from other anthropogenic sources, such as traffic, etc. Due to the fact that the station at CMU was on the rooftop of high building, it had a minimized influence from traffic emissions. Therefore, during the non-burning season PM10 concentrations at this sampling site were lower than the other two sites, where increased human activities existed.

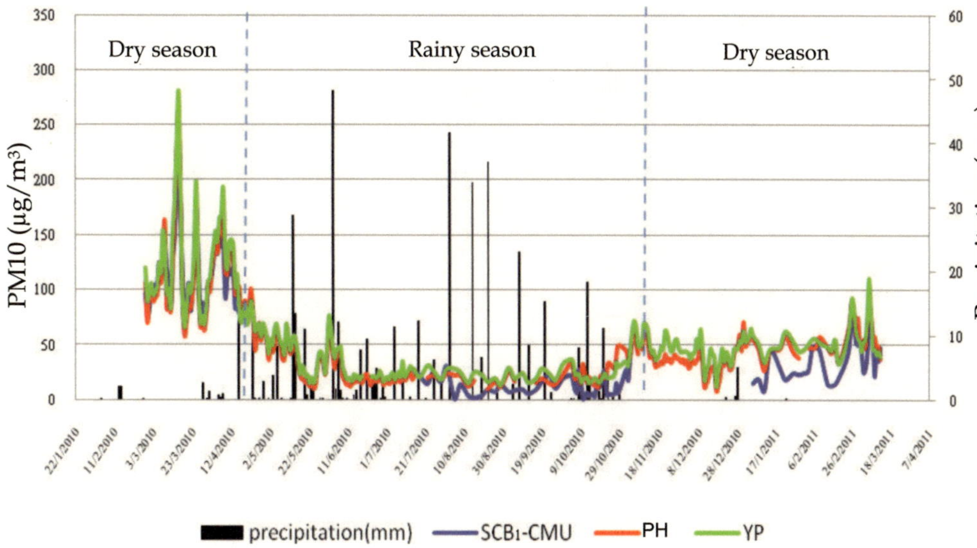

Fig. 2. PM10 concentrations (μg/m³) at Chiang Mai University (CMU), Provincial Hall (PH) and Yupparaj Wittayalai School (YP) and precipitation (mm) from 2010-2011

3. Sampling site selection for the study of PM10 and its chemical composition

This study aims to investigate PM10 concentrations and PM10 chemical composition in the ambient air of the cities located in the Chiang Mai-Lamphun Basin in order to assess the air quality and predict possible health effects. Three study sites were located in Chiang Mai Province: Yuparaj Witayalai School (YP), Municipality Hospital (HP), Sarapee District (SP) and one site was located in Muang District, Lamphun Province (LP). A map of the sampling sites is illustrated in Fig. 3. YP site is located at the centre of Chiang Mai City. It is a residential area surrounded by government buildings, schools and temples with quite high traffic density. HP site is a densely populated commercial area. It has been reported to possess particulate concentrations exceeding the standard throughout the year (Pengchai, et al., 2009 as cited in Vinitketkumnuen et al., 2002). SP site is surrounded by government buildings and is considered an area experiencing dense traffic within the countryside. There have been significant reports of people suffering from respiratory disease (Pengchai, et al., 2009 as cited in Vinitketkumnuen et al., 2002). LP site is located in a residential area at the centre of Lamphun with light traffic density.

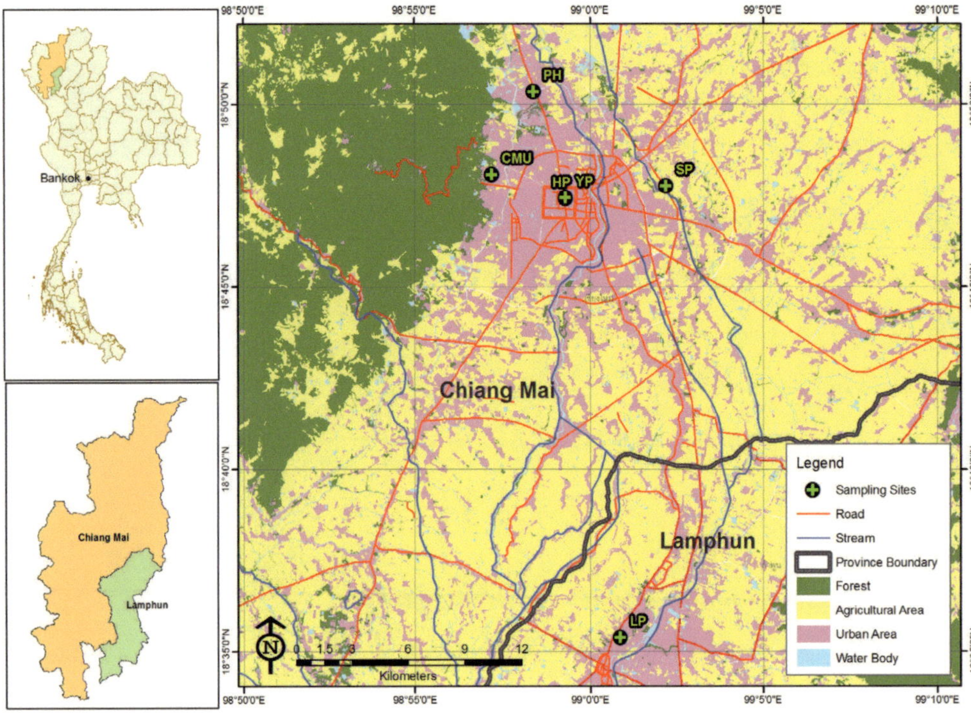

Fig. 3. Map of sampling sites

3.1 Ambient PM10 concentrations

Twenty-four hour PM10 samples were collected on quartz fiber filters every 3 days at each sampling site using a High Volume PM10 Air Sampler (Wedding & Associates Inc., USA) at the flow rate of 1,130 L/min. Sampling had been done for 1 year from June 2005 to June 2006 to cover 3 seasonal periods, which were the wet season (June - September), the dry season (December - March) and the transition periods (October - November and April - May). PM10 samples were weighed using the five digit-scale balance (Mettler Toledo AG285, USA) covered by the temperature and humidity controlling cabinet (DE-300).

PM10 concentrations of all sampling stations from June 2005 - June 2006 are illustrated in Fig. 4. Their concentrations for all seasons were almost the same. They were lower than 80 $\mu g/m^3$ in the rainy season and getting higher in the dry season. The highest concentration was found in March (~150 $\mu g/m^3$) at all three sampling sites except at the LP site, which was uncommonly high in April. This was probabaly due to local activities, such as open burning of community garbage.

Seasonal and spatial variations of mean PM_{10} concentrations and their standard deviation (SD) are illustrated in Table 1. It was found that concentrations of PM10 collected in the dry season (Dec - Mar) were significantly higher than those of the other seasons, while the lowest concentration was found in the rainy season (Jun - Sep). Average PM_{10} concentrations in each sampling site were found to be very much similar. However, the mean concentration of the Sarapee (SP) site was higher than those of the other sites (Chantara et al., 2009).

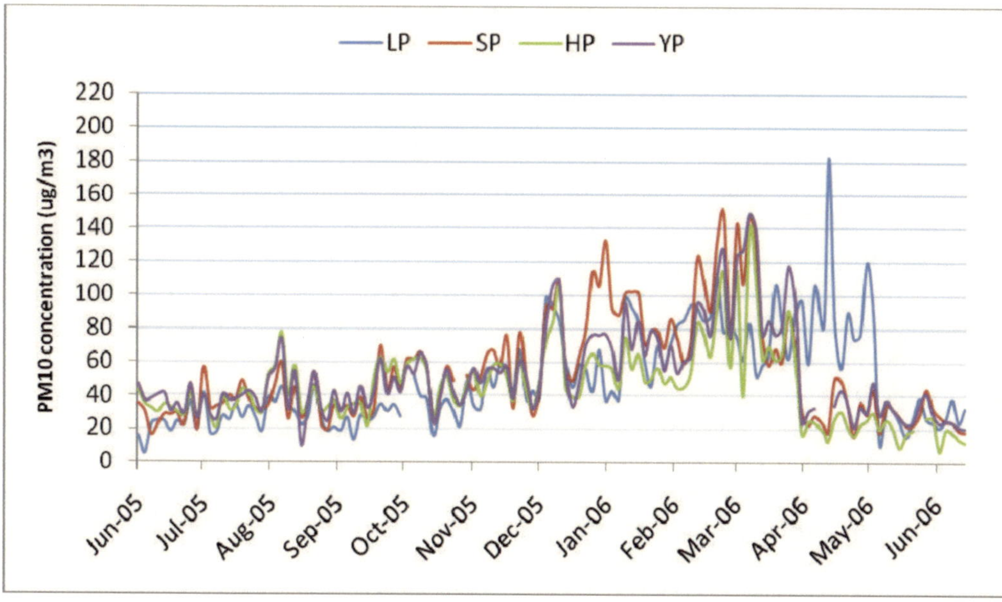

Fig. 4. PM_{10} concentrations of all sampling stations from June 2005 – June 2006

Season[a,b]	PM10 concentration	Sampling site[a,b]	PM10 concentration
Dry: Dec – Mar (n = 160)	73.64 ± 25.34 /7	LP (n=127)	46.53 ± 29.30 /1
Trans I: Apr – May (n = 84)	*40.44 ± 25.33* /2	SP (n=127)	**55.94 ± 36.64** /2
Rainy: Jun – Sep (n = 184)	*33.17 ± 21.39* /1	HP (n=127)	44.08 ± 20.60 /3
Trans II: Oct – Nov (n = 80)	**47.19 ± 12.36** /2	YP (n=127)	**50.30 ± 23.26** /6

[a]/n number of non detected or incomplete data
[b] Groups of number in *italic* and/or **bold** are not significant different ($p > 0.05$)

Table 1. Mean PM10 concentrations (μg/m³) ± SD for 4 periods and 4 sampling sites

4. Analysis of PM10 chemical composition

The post-weighed quartz fiber paper (8 x 10 inches) of the PM10 sample was equally divided into eight pieces (2.5 x 4 inches) by a stainless steel roller blade cutter. Two of the eight pieces were grouped for the analysis of carbon, PAHs, soluble ions and elements. All fiber samples, except the samples for carbon analysis, were cut into small pieces by stainless steel scissors prior to analysis.

4.1 Analysis of total carbon in PM10 samples

4.1.1 Analysis method of PM10 carbon content by CHN-S/O analyzer

A tin capsule (1 x 1 cm) as a sample container was weighed by MX5 Automated-S Microbalance (Mettler Toledo). The quartz filter contains PM10 was punched by a puncher with a diameter of 0.55 cm. The punched filter was put into a prepared capsule, wrapped and weighed. The sample was then analysed for total carbon (TC) by Elemental Analyzer (PE 2400 series II CHN-S/O analyzer, Perkin Elmer Cooperation) (Perkin Elmer, 1991) using L-cystine and BS Slag 2 as standards and an unused quartz filter paper as a blank.

4.1.2 Carbon content of PM10 samples

Spatial and temporal variations of mean carbon concentrations are shown in Fig. 5. It was found that the distribution pattern of carbon contents in all stations were almost the same. However, the PM10 samples collected at the SP station had higher carbon content than the others. A high concentration of carbon was found in the dry season, especially in March, while in the rainy season the carbon content was low. A correlation of carbon content with PM10 weight (Table 2) revealed that they were strongly correlated in the dry season and the transition period (Apr-May) for all sampling sites. It can be concluded that carbon content was positively correlated to PM_{10} weight. However, this consumption relationship can not be applied to the rainy season, in which the correlation was low. It can be explained that low concentrations of PM10 and carbon in the rainy season gave a relatively high variation of values. Moreover, ratios of carbon and PM10 (μg/mg PM10) at all sites were higher in the rainy season than the other seasons. Therefore a correlation was not obviously observed.

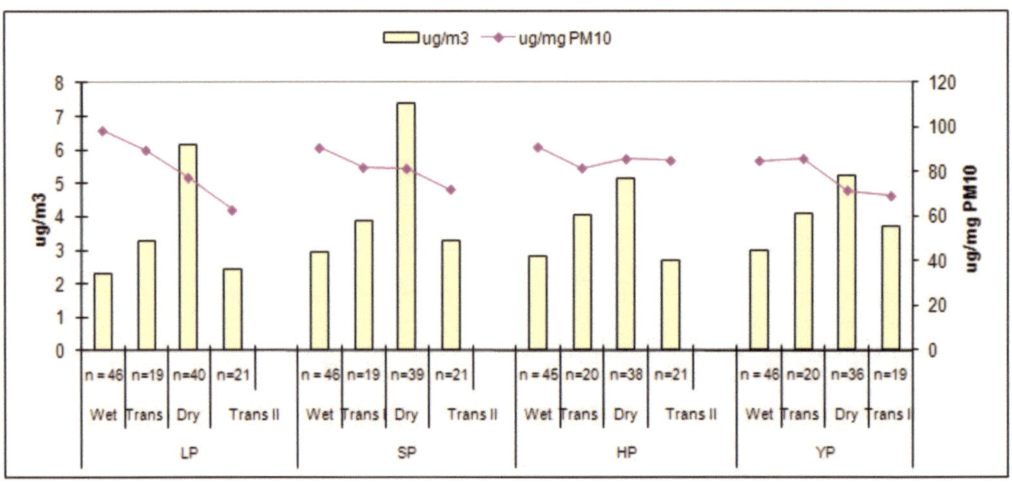

Fig. 5. Spatial and temporal variations of mean carbon concentrations

Carbon concentrations found in this study were compared with the study in Bangkok (Chuersuwan et. al., 2008), in which carbon content in PM2.5 was determined. The carbon concentrations were 38.5±19.3, 21.7±12.7, 21.9±13.3 and 17.6±11.0 μg/m^3 in Din Daeng, Chankasaem, Ban Somdej and Bangna, respectively. The contents were 4-8 times higher than those found in the PM10 samples in this study. This is probably due to a much higher number of carbon sources, especially in terms of the traffic in Bangkok.

Sampling site	r	Season	r
LP	0.861	Dry (Dec-Mar)	0.849
SP	0.682	Transition (Apr-May)	0.836
HP	0.784	Rainy (Jun-Sep)	0.208
YP	0.798	Transition (Oct-Nov)	0.445

Table 2. Correlation of carbon ($\mu g/m^3$) and PM_{10} weight (mg)

4.2 PM10-bound PAHs

4.2.1 Validation of analysis method for PM10 bound PAHs using GC-MS

Gas Chromatography - Mass Spectrometry (GC-MS) (Hewlett Packard, USA) was used for investigation of PAHs in this study. One micro liter of a mixed solution of 16-PAH standards (2,000 μg/ml in methylenechloride, Restek, USA) and internal standards (D10-ACE, D10-PHE and D12-PYR from Supelco, USA) was injected in splitless mode onto an HP-5MS GC capillary column of 30 m length, 0.25 mm diameter and 0.25 μm thickness internally coated with 5% phenyl methyl polysiloxane stationary phase. Standard solutions of the 16 PAHs with concentrations ranging from 0.005 to 0.500 μg/ml were combined with 0.10 μg/ml internal standards and were injected onto the same GC capillary column of the GC-MS employed under an optimum condition in order to obtain a suitable calibration curve in which peak area ratios were plotted against concentrations. The GC condition was as follows. The injector temperature was 275°C. The GC oven was programmed with an initial temperature of 70°C, held for 2 min, then increased to 290°C at 8°C/min. The MS was operated in SIM mode. Characteristic ions of selected PAHs were used for quantification, which was based on the peak area of the PAH standard relative to the deuterated internal standard closest in molecular weight to the analyte. Detection limits of GC-MS for PAHs and recoveries of PAHs from extraction are illustrated in Table 3. Detection limit of GC-MS for an individual PAH was calculated as three times of the standard deviation of blank concentration (Miller & Miller, 1993). Detection limits for 16-PAHs obtained were ranged from 0.003-0.007 ng/m^3. Recoveries of PAHs from the extraction of spiked filters ranged from 58-120%. Of all the 16 PAHs investigated, 12 PAHs were found to yield the recovery higher than 70%.

Optimization for the extraction procedure of PAHs in PM 10 samples was conducted using ultrasonication to obtain an appropriate extraction volume of acetonitrile (HPLC grade, Lab-Scan Analytical Science) and evaporation conditions. The quartz filter was cut into small pieces using a pair of stainless steel scissors and the cut filter was put into a 60-ml amber-bottle wrapped with aluminium foil and mounted by paraffin film. The 0.3 μg/ml of mixed 16-PAH standards was spiked onto the paper, covered and left for 10 min. The spiked samples were extracted in 35 ml acetonitrile by ultrasonicator (T710DH, Elma, Germany) at 100% ultrasound power for 30 min under controlled temperature (approximately 20°C). The solution was filtered through a 0.45 μm nylon filter and evaporated by low-pressure evaporator at 30°C until it became nearly dry. The pellet was re-extracted in 30 ml acetronitrile by repeating the previous step. The solution was transferred into a 1 ml volumetric flask and a mixed solution of internal standards was added, with the final volume adjusted to 1 ml with acetronitrile. The solution was then analyzed for 16 PAHs by GC-MS. Percent recoveries of individual PAHs were obtained.

Compound	Number of ring	DL (ng/m^3)	Recovery±SD (%)
Naphthalene (NAP)	2	0.006	64±5
Acenaphthylene (ACY)	3	0.005	76±6
Acenaphthene (ACE)	3	0.006	66±6
Fluorene (FLU)	3	0.004	80±7
Phenanthrene (PHE)	3	0.006	60±1
Anthracene (ANT)	3	0.004	77±2
Fluoranthene (FLA)	4	0.003	76±2
Pyrene (PYR)	4	0.003	72±1
Benzo(a)anthracene* (BaA)	4	0.007	120±4
Chrysene* (CHR)	4	0.005	78±2
Benzo(b)fluoranthene* (BbF)	5	0.006	112±7
Benzo(k)fluoranthene* (BkF)	5	0.007	84±4
Benzo(a)pyrene* (BaP)	5	0.005	81±12
Indeno(1,2,3-cd)pyrene* (IND)	5	0.004	73±14
Dibenzo(a,h)anthracene* (DBA)	6	0.005	77±16
Benzo(g,h,i)perylene (BPER)	6	0.007	58±15

* carcinogen suspects

Table 3. Detection limits (DL) of GC-MS for PAHs and recoveries of PAHs from extraction

4.2.2 Analysis of 16 PAHs in PM10 samples by GC-MS

After sampling, the weighed quartz filter was cut into small pieces by a pair of stainless steel scissors and put into a 60-ml amber-bottle wrapped with aluminium foil and mounted with paraffin film. The samples were extracted using the same conditions as the standards described above. Each of the sample solutions was then analyzed for 16 PAHs by GC-MS.

Fig. 6 illustrates spatial and temporal variations of PAHs concentrations (Chantara et al., 2010). The highest concentrations of an individual and total PAHs, as well as carcinogenic PAHs, were found in the dry season and were higher than those found in other seasons.

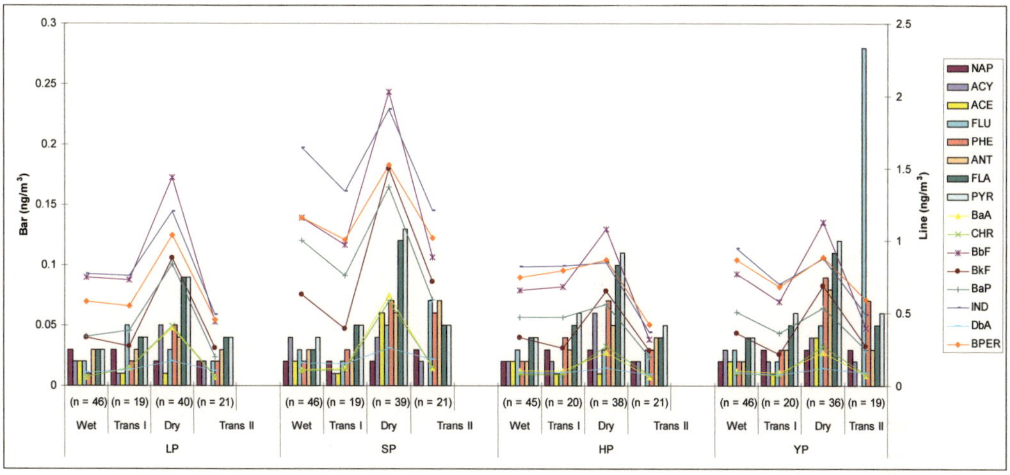

Fig. 6. Spatial and temporal variations of PAHs concentrations

Seasonal variation of PM_{10}-bound PAHs can be observed in Table 4 . Mean concentrations of various groups of PAHs including those with 2-3 rings, 4-6 rings, carcinogenic and non-carcinogenic PAHs and sums of 16 PAHs were statistically tested based on the season basis. The results revealed that all PAH groups presented a similar pattern. The highest concentration was found in the dry season (Dec-Mar) and were significantly higher (p<0.05) than the other seasons, in which PAHs concentrations were not significantly different. In terms of the sampling site (Table 5), high molecular weight (4-6 rings) and carcinogenic PAHs were significantly higher at the SP site than the other sites (Chantara et al., 2010).

PAHs*†	Dec – Mar (n = 160)	Apr – May (n = 84)	Jun – Sep (n = 184)	Oct – Nov (n = 80)
2-3 rings	**0.15 ± 0.13 /1**	0.10 ± 0.16 /2	0.07 ± 0.04 /5	0.08 ± 0.06 /4
4-6 rings	**6.58 ± 3.20 /1**	2.76 ± 2.05 /2	*4.06 ±2.96 /1*	3.59 ±1.69 /2
Carcinogenic‡	**5.29 ± 2.72 /1**	2.05 ± 1.65 /2	*3.15 ±2.42 /1*	2.73 ±1.41 /2
Non carcinogenic	**0.36 ± 0.20 /1**	0.20 ± 0.18 /2	0.13 ± 0.06 /1	0.18 ± 0.09 /2
Total 16 PAHs	**6.72 ± 3.20 /1**	2.87 ± 2.08 /2	*4.13 ±2.97 /1*	3.67 ±1.71 /2

* /n number of non detected or incomplete data

† Groups of number in *italic*, underline and/or **bold** are not significant different (p > 0.05)

‡ Carcinogenic PAHs are BaA, CHR, BbF, BkF, BaP, DBA and IND

Table 4. Mean concentrations (ng/m³) ± SD of PM_{10}-bound PAHs in the four seasons

PAHs*†	LP (n=127)	SP (n=127)	HP (n=127)	YP (n=127)
2-3 rings	0.07 ± 0.06 [/5]	0.10 ± 0.09 [/3]	0.10 ± 0.08 [/2]	**0.14 ± 0.17** [/2]
4-6 rings	3.97 ± 2.53 [/1]	**6.93 ± 4.16** [/2]	3.58 ± 2.03 [/1]	3.82 ± 1.81 [/2]
Carcinogenic‡	3.18 ± 2.12 [/1]	**5.57 ± 3.51** [/2]	2.71 ± 1.61 [/1]	2.88 ± 1.46 [/2]
Non carcinogenic	0.17 ± 0.11 [/1]	*0.23 ± 0.16* [/2]	**0.22 ± 0.15** [/1]	*0.28 ± 0.22* [/2]
Total 16 PAHs	4.04 ± 2.55 [/1]	**7.03 ± 4.18** [/2]	3.68 ± 2.06 [/1]	3.96 ± 1.84 [/2]

* [/n] number of non detected or incomplete data

† Groups of number in *italic* and/or **bold** are not significant different ($p > 0.05$)

‡ Carcinogenic PAHs are BaA, CHR, BbF, BkF, BaP, DBA and IND

Table 5. Mean concentrations (ng/m^3) ± SD of PM_{10}-bound PAHs from four sampling sites

Table 6 illustrates the ratios of non-carcinogenic and carcinogenic PAHs found in PM_{10} samples from each sampling site, which were 1.0:3.7 (LP), 1.0:3.8 (SP), 1.0:2.7 (YP) and 1.0:2.8 (HP). SP and LP ratios revealed that carcinogenic PAHs were obviously higher than non-carcinogenic ones. The highest carcinogenic PAHs content was found at the SP site and was significantly higher than other sites (Table 5). The carcinogenic ratios of all sites found in this study were higher than those found along the roadsides of the inner part of Bangkok (1.0:2.5) from November 02 to April 03 (Norramit et al., 2005). This reveals that our study sites presented higher concentrations of carcinogenic PAHs but lower PM_{10} content in comparison to the Bangkok area. This fact should be noted and brought into the attention of relevant authorities for subsequent air quality management.

Sampling site	% Non-carcinogenic PAHs	% Carcinogenic PAHs	non-carcinogenic : carcinogenic PAHs
LP	21.3	78.7	1.0 : 3.7
SP	20.8	79.2	1.0 : 3.8
HP	26.4	73.6	1.0 : 2.8
YP	27.3	72.7	1.0 : 2.7
Bangkok (Norramit et al., 2005)	28.6	71.4	1.0 : 2.5

Table 6. Ratios of non-carcinogenic to carcinogenic PM_{10}-bound PAHs

Mean total PAHs concentrations detected in this study were almost the same as values found in the community area of the previous study (Chantara & Sangchan, 2009) carried out in 2004 (Table 6). However, the concentrations found in this study were obviously lower than those found in Bangkok and Pathumthanee, even under the same category (roadside area) (Boonyatumanond et al., 2007; Norramit et al., 2005; Chetwittayachan et al., 2002; Kim Oanh et al., 2000), but the values were similar to the rural area (Boonyatumanond et al., 2007) and the general area (Chetwittayachan et al., 2002) as is shown in Table 7. This is attributed to the transport pattern and the traffic volume that exist near the sampling site.

Based on the ratios of non-carcinogenic and carcinogenic PAHs found in PM_{10} samples from each sampling site (Table 6), the highest carcinogenic PAHs content was found at the SP site.

The carcinogenic ratios of all sites found in this study were higher than those found in roadsides of the inner part of Bangkok from November 02 to April 03 (Norramit et al., 2005) (Table 7). This reveals that our study sites presented higher concentrations of carcinogenic PAHs but lower PM_{10} contents in comparison to the Bangkok area. This should be noted and of significant concern for air quality management.

City	Study period	Location	Mean (ng/m^3)	Reference
Chiang Mai	Jun 05 - Jun 06	Roadside area	1.3-12.2	This study
Lamphun	Jun 05 - Jun 06	Roadside area	1.7-7.4	This study
Chiang Mai	Jun - Nov 04	Traffic area Community area Rural area	7.6-16.6 3.9-9.1 2.7-8.4	Chantara & Sangchan, 2009
Bangkok [a]	Nov 03 - Jan 04	Rural area Roadside area	7.2-10.0 10.1-28.1	Boonyatumanond et al., 2007
Bangkok	Nov 02 - Apr 03	Roadsides	12.59±0.94	Norramit et al., 2005
Bangkok [b]	Mar 01	Roadside area General area	50-53 11-12	Chetwittayachan et al., 2002
Pathumthani (40 km north of Bangkok)	Jun 1996 - Apr 1997	Road site	17.4	Kim Oanh et al., 2000

[a] Sum of concentrations of 18 PAHs: PHE, ANT, 1-methyl PHE, 2-methyl PHE, 3-methyl PHE, 9-methyl PHE, FLA, PYR, BaA, CHR, BbF, BjF, BkF, BeP, BaP, IND, BPER and coronene.
[b] Sum of concentrations of 11 PAHs: ANT, PHE, PYR, BaA, CHR, BbF, BkF, BaP, IND, DBA and BPER

Table 7. Comparison of mean PAH concentrations in Chiang Mai, Lamphun and Bangkok, Thailand

4.2.3 Correlation of PM10 and PAHs

Correlations between PM_{10} content and total PAH concentrations for each sampling site in each season are shown in Table 8. Correlations between PM_{10} content and PAH concentrations for all seasons were found to be in the range 0.618 - 0.731. The highest correlation was found in the LP site, while the lowest one was in the YP site. The correlations for all sites were relatively high in the dry season and low in the wet season, respectively. The correlations for the SP and HP sites were high throughout the year; they were also high in the transition periods and were obviously higher than those for the LP and YP sites. This is probably due to local sources, which distributed PM_{10}-bound PAHs into the atmosphere during the sampling time period. In terms of PAH source, both SP and HP sampling sites are surrounded by markets with high traffic volume and human activities. The difference between these two places is that the SP site is a place where small scale factories are situated. The fuels used in such factories could have contributed to the level of pollutants emitted to the air.

Season	LP	SP	HP	YP
Wet (Jun-Sep)	0.530	0.549	0.703	0.595
Dry (Dec- Mar)	0.704	0.572	0.601	0.590
Transition period (Apr-May, Oct-Nov)	0.257	0.650	0.690	0.365
All Season	0.731	0.686	0.659	0.618

Table 8. Pearson correlation of PM_{10} and PAHs

In comparison to previous studies, the correlation found in this study is almost the same as that reported by Chantara & Sangchan (2009) in which the mass of PM_{10} collected by a Minivol air sampler in traffic areas was found to be slightly correlated ($R^2 = 0.45$) with the total PAH concentration of the solution obtained from extraction of the collected particulate. The afore-mentioned correlation was higher than the value in the study by Koyano et al. (1998) in which airborne particulate samples were collected from commercial and residential sites of Chiang Mai from July to December 1989 and their results revealed that there was not a large differnce in PAHs concentrations between different sampling sites. However, the airborne particulates and large amounts of rough particulates encountered in their study prompted them to conclude that there was wide spread air pollution throughout the city.

4.2.4 BaP and BaP-equivalent carcinogenic power (BaPE)

The limit value of BaP recommended by WHO is 1 ng/m^3 due to its high carcinogenic property. Other PAHs such as BaA, BbF, BkF and DBA also have carcinogenic potential. To combine these potential carcinogenic compounds, the benzo(a)pyrene-equivalent carcinogenic power (BaPE) is calculated. The BaPE is an index that has been introduced for better denoting aerosol carcinogenicity related to the whole PAH fraction instead of the BaP (Liu et al., 2007). The BaPE was calculated in this study using the following equation:

$$BaPE = (BaA \times 0.06) + ((BbF + BkF) \times 0.07) + BaP + (DBA \times 0.6) + (IND \times 0.08) \quad (1)$$

Values of monthly mean BaPE and PM_{10} concentrations for each sampling site are illustrated in Fig. 7. It can be obviously seen that the BaPE values of the SP site are the highest, while those of the other sites are approximately two times lower. It can therefore be concluded that the air quality of the SP site was worse than the other sampling sites.

Patterns of the BaPE values from June 05 to June 06 were found to be similar for all sites. Fig. 7 shows two peaks of BaPE values. The first peak was obtained in the rainy season (Aug 05) while the second one was obtained in the dry season (Jan-Mar 06). Based on the fact that traffic is the main PAHs source in the urban area, it presents almost the same level of pollutants to the air throughout the year. Therefore, the BaPE values contributed from the traffic should be a background value for all sites. It is quite clear that in the dry season, open burning, i.e. burning of agricultural waste and community garbage, as well as forest fires, mainly contributed pollutants to the atmosphere. In the rainy season, the high BaPE values might have come from other sources. The air pollutants in this season could be from various local activities apart from open burning. One of the possibilities is from small scale factories in the area, whereby seasonal agricultural products are processed.

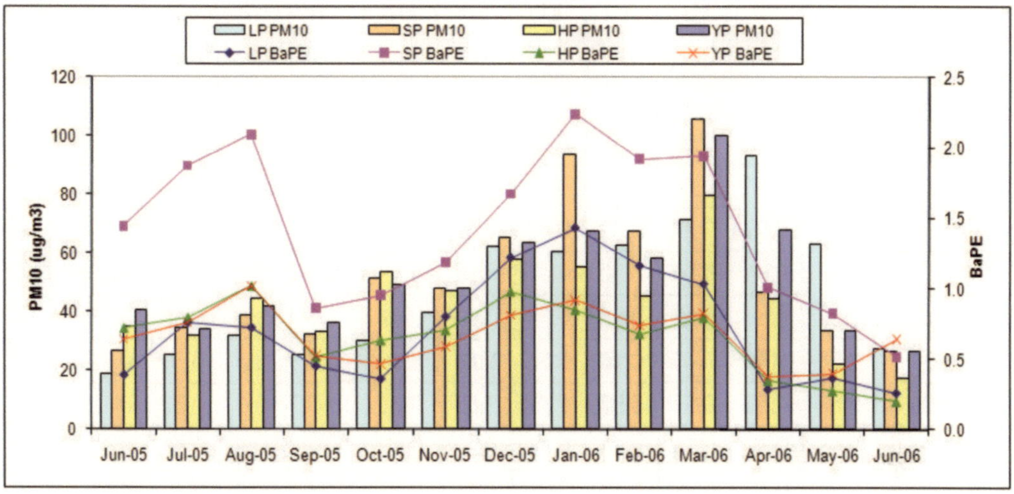

Fig. 7. Monthly mean BaPE and PM10 concentrations from all sampling sites

4.2.5 Diagnostic ratios

Dianostic ratio can be used for the identification of possible emission sources. The diagnostic ratios of this study were calculated by separating sampling sites and months as shown in Table 9 (Chantara et al., 2010). The IND/(IND+BPER) ratio can be used to identify traffic sources. The value in a range from 0.35-0.70 indicates that the PAHs were emitted from diesel engines (Lui et. al, 2007 and Kavouras et. al, 2001). In this study, the average ratios of all sites were 0.50-0.56. Therefore, the PAHs found in this study could have originated from diesel engine emissions. The BaP/BPER ratio was also used for characterization of the PAHs sources. The low BaP/BPER ratio (<0.60) is an evidence of greater emission of BPER from traffic sources (Pandey et. al., 1999). The mean ratios of the 4 sampling sites were 0.56-0.80. The mean ratios of HP and YP sites were fitted with the low ratio value, therefore traffic might be the main source for these two sites. However, the ratio values in some months, especially in the dry season were greater than 0.60, which indicated some other PAH sources. The SP and LP sites had a higher mean ratio value (0.80 and 0.64, respectively) as well as monthly values, depicted as *italic* values in Table 9. This could indicate that the PAHs in these two sites were not only from the traffic but were also from some other sources on a similar basis described for the BaPE above. Apart from these two ratios, the BaP/(BaP + CHR) ratio was also considered. The ratio values of 0.57 and 0.89 could be related to diesel and gasoline exhaust, respectively (Khalili et al., 1995). The BaP/(BaP + CHR) ratio values obtained in this work were 0.68-0.77, which could be reckoned as somewhere between 0.57 and 0.89. Based on this ratio alone, it could not be concluded what the major source of PAHs was in the areas studied in this work. However, the IND/(IND+BPER) ratio of 0.50-0.56 obtained in this study fits within the range of 0.35-0.70 which suggests that the diesel engine source plays a more important role than the gasoline source.

	IND/(IND+BPER)				BaP/BPER				BaP/(BaP+CHR)			
	LP	SP	HP	YP	LP	SP	HP	YP	LP	SP	HP	YP
Jun-05	0.54	0.57	0.50	0.48	*0.61*	*0.87*	*0.63*	0.50	0.66	0.57	0.58	0.60
Jul-05	0.53	0.55	0.49	0.48	*0.70*	*1.06*	*0.66*	*0.61*	0.86	0.79	0.83	0.83
Aug-05	0.61	0.62	0.56	0.55	*0.61*	*0.95*	*0.70*	*0.67*	0.90	0.83	0.86	0.86
Sep-05	0.57	0.60	0.52	0.52	0.46	0.59	0.48	0.46	0.77	0.71	0.71	0.72
Oct-05	0.60	0.58	0.52	0.52	0.48	*0.62*	0.51	0.45	0.83	0.78	0.79	0.80
Nov-05	0.56	0.55	0.50	0.50	*0.81*	*0.94*	*0.69*	*0.61*	0.75	0.77	0.83	0.86
Dec-05	0.52	0.54	0.47	0.48	*0.83*	*1.03*	*0.64*	*0.63*	0.68	0.66	0.73	0.76
Jan-06	0.52	0.56	0.45	0.46	*0.86*	*0.86*	*0.66*	*0.64*	0.67	0.56	0.81	0.78
Feb-06	0.57	0.58	0.53	0.52	*0.89*	*1.01*	*0.71*	*0.63*	0.65	0.62	0.81	0.78
Mar-06	0.55	0.55	0.54	0.54	*0.65*	*0.79*	0.59	0.54	0.68	0.55	0.67	0.67
Apr-06	0.52	0.55	0.48	0.48	0.53	*0.66*	0.50	0.48	0.72	0.49	0.65	0.64
May-06	0.52	0.53	0.45	0.43	0.40	0.50	0.45	0.38	0.79	0.73	0.81	0.77
Jun-06	0.52	0.52	0.47	0.54	0.44	0.56	0.47	*0.61*	0.88	0.80	0.88	0.80
Avg	**0.55**	**0.56**	**0.50**	**0.50**	**0.64**	**0.80**	**0.59**	**0.56**	**0.76**	**0.68**	**0.77**	**0.76**
SD	0.03	0.03	0.03	0.03	0.17	0.20	0.10	0.09	0.09	0.11	0.09	0.08

Table 9. Diagnostic ratios of PAHs for four sampling sites in different months

5. Analysis of dissolved ions in PM10 samples by Ion Chromatography

5.1 Ion analysis method by Ion Chromatography

Dissolved ions including Na^+, NH_4^+ , K^+, Ca^{2+}, Mg^{2+}, Cl^-, NO_3^- and SO_4^{2-} were analyzed following a method developed from the Technical Document for Filter Pack Method (Acid Deposition Monitoring Network in East Asia, 2003) using Ion Chromatograph (IC). Samples were put into 100 ml beakers followed by 50 ml of deionized water. The beakers were then covered by paraffin film and ultrasonicated at 100% ultrasound power (T490DH, Elma, Germany) for 30 min. The solution was filtered through a cellulose acetate filter (pore size 45 µm, diameter 13 mm) into a plastic bottle for further analysis by IC (Metrohm, Switzerland). Analytical columns were Metrosep A Supp4 (4 x 250 mm) and Metrosep C2 (4 x 100 mm) for anions and cations, respectively.

5.2 pH and ion contents of PM10

The extracted samples were divided into 2 parts for pH measurement and ion analysis. The pH distribution for all sampling stations was plotted as shown in Fig. 8. More than 50% of the samples from all sampling sites except the YP station (45%) had pH value less than 5.6. Specifically, half of all the samples from the SP station was in the lower pH range value (<5.0). The seasonal variation of the pH values illustrated that low pH values were inspected in the dry season, particularly in March and April.

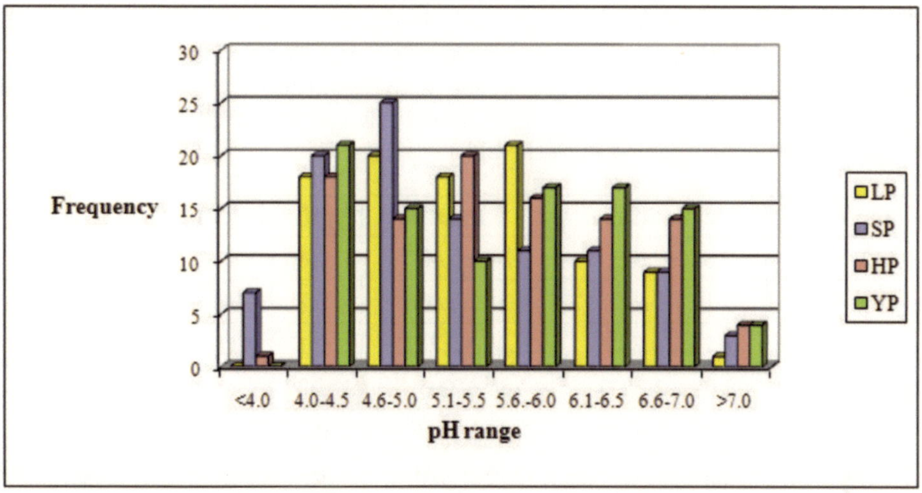

Fig. 8. pH distribution of PM_{10} extractions from four sampling stations

Average (SD) concentrations (ng/m³) of PM_{10}-bound ions in each season and sampling site are shown in Tables 10-11 and Fig. 9. Both anions (SO_4^{2-} , NO_3^-, and Cl^-) and cations (NH_4^+, Na^+, K^+, Mg^{2+} and Ca^{2+}) were significantly higher in the dry period (Dec-Mar) and the transition period I (Oct-Nov) than those found in other seasons (Table 10). The dominant anion and cation were SO_4^{2-} and NH_4^+, respectively. There was a strong correlation between NH_4^+ and SO_4^{2-} (r = 0.953), followed by Na^+ and SO_4^{2-} (r = 0.651) and K^+ and Cl^- (r = 0.606). This suggested that the main acidity of dry deposition in this region was due to H_2SO_4, which was neutralized by NH_4^+ (Chantara & Chunsuk, 2009). Noticeably, K^+ (biomarker of vegetative burning) was significantly 2-3 times higher in the dry season. Therefore, it can be

Ion*†	Dec - Mar (n = 160)	Apr - May (n = 84)	Jun - Sep (n = 184)	Oct - Nov (n = 80)
SO_4^{2-}	**4.02 ± 2.39**	*2.56 ±1.92* $^{/2}$	*1.20 ±0.97* $^{/2}$	**6.04 ± 5.35** $^{/2}$
NO_3^-	**1.84 ± 1.31**	*0.68 ±0.61* $^{/2}$	*0.46 ±0.24* $^{/7}$	*0.71 ±0.34* $^{/2}$
Cl^-	**0.33 ± 0.33**	*0.11 ±0.10* $^{/2}$	*0.15 ±0.11* $^{/1}$	0.20 ± 0.15 $^{/2}$
Total anions	**6.21 ± 3.09**	3.35 ± 2.44 $^{/2}$	*1.79 ±1.10* $^{/1}$	**6.95 ± 5.35** $^{/2}$
NH_4^+	**1.45 ± 1.22**	*0.96 ±0.78* $^{/17}$	0.15 ± 0.16 $^{/7}$	**1.17 ± 1.06** $^{/4}$
Na^+	**0.17 ± 0.09** $^{/3}$	*0.11 ±0.09* $^{/2}$	**0.15 ± 0.11** $^{/4}$	**0.16 ± 0.11** $^{/2}$
K^+	**1.38 ± 0.65**	*0.56 ±0.50* $^{/2}$	0.24 ± 0.13 $^{/1}$	*0.55 ±0.46* $^{/2}$
Mg^{2+}	**0.09 ± 0.17** $^{/3}$	0.03 ± 0.02 $^{/2}$	0.04 ± 0.04 $^{/2}$	**0.09 ± 0.04** $^{/2}$
Ca^{2+}	**1.29 ± 0.52**	0.92 ± 0.39 $^{/2}$	0.79 ± 0.29 $^{/1}$	**1.17 ± 0.43** $^{/2}$
Total cations	**4.38 ± 1.89**	2.41 ± 1.55 $^{/2}$	1.36 ± 0.48 $^{/1}$	**3.11 ± 1.31** $^{/2}$
Total ions	**10.59 ± 4.62**	5.76 ± 3.87 $^{/2}$	3.14 ± 1.49 $^{/1}$	**10.07 ± 6.56** $^{/2}$

* $^{/n}$ number of non detected or incomplete data

† Groups of number in *italic* and/or **bold** are not significant different (p > 0.05)

Table 10. Mean concentrations (ng/m³) ± SD of PM_{10}-bound ions in each season

concluded that biomass burning took place in the area during the dry season. Total ion concentrations including total anions and cations were not significantly different among all sites (Table 11). Standing out was the SP site with significantly higher concentrations of some major ions such as NO_3^- and K^+. This is probably due to a higher frequency or larger scale of open burning happening in the area.

Ion*†	**LP (n=127)**	**SP (n=127)**	**HP (n=127)**	**YP (n=127)**
SO_4^{2-}	2.67 ± 2.62 /2	3.39 ± 3.40 /1	2.86 ± 3.16 /1	3.37 ± 3.45 /2
NO_3^-	*1.07 ±1.21 /4*	**1.16 ± 1.09 /3**	0.78 ± 0.78 /2	*0.91 ±0.87 /2*
Cl^-	**0.27 ± 0.28 /1**	**0.31 ± 0.30 /1**	0.13 ± 0.11 /1	0.13 ± 0.10 /2
Total anions	3.98 ± 3.42 /1	4.84 ± 3.98 /1	3.76 ± 3.44 /1	4.42 ± 3.79 /2
NH_4^+	0.76 ± 0.92 /6	1.00 ± 1.09 /8	0.79 ± 1.00 /7	0.89 ± 1.13 /7
Na^+	**0.17 ± 0.12 /1**	**0.16 ± 0.10 /3**	*0.12 ±0.08 /3*	*0.14 ±0.09 /4*
K^+	*0.69 ±0.62 /1*	**0.84 ± 0.73 /1**	*0.59 ±0.62 /1*	*0.68 ±0.67 /2*
Mg^{2+}	0.07 ± 0.11 /2	0.07 ± 0.08 /2	0.06 ± 0.08 /3	0.07 ± 0.14 /2
Ca^{2+}	*0.91 ±0.48 /1*	*1.00 ±0.43 /1*	*1.02 ±0.39 /1*	*1.19 ±0.52 /2*
Total cations	2.56 ± 1.77 /1	3.02 ± 1.91 /1	2.55 ± 1.69 /1	2.93 ± 2.03 /2
Total ions	6.54 ± 4.99 /1	7.85 ± 5.64 /1	6.31 ± 4.83 /1	7.35 ± 5.51 /2

* /n number of non detected or incomplete data
† Groups of number in *italic* and/or **bold** are not significant different ($p > 0.05$)

Table 11. Mean concentrations (ng/m³) ± SD of PM_{10}-bound ions in each sampling site

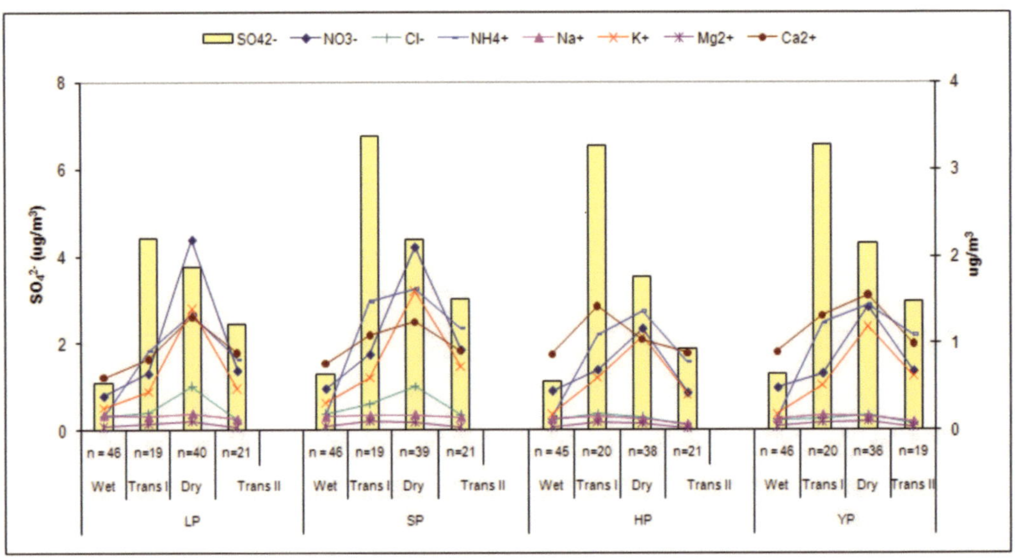

Fig. 9. Spatial and temporal variations of ion concentrations (μg/m³)

Correlations of ion concentrations (µg/m^3) and PM10 weight (mg) were analysed among all sampling stations (Fig. 10a) and among all seasons (Fig. 10b). It was found that almost all ions, except Na^+, Mg^{2+}, were well correlated with PM10 (*p*< *0.001*). K^+ concentrations and had a very strong correlation with PM10 in every sampling site in the dry season (December - May) and the transition period (April - March). As mentioned above, K^+ is often used as a tracer for biomass burning. This figure obviously showed the possibility of a high frequency and huge area of open burning, including forest fires and agricultural burning in Northern Thailand and neighboring countries.

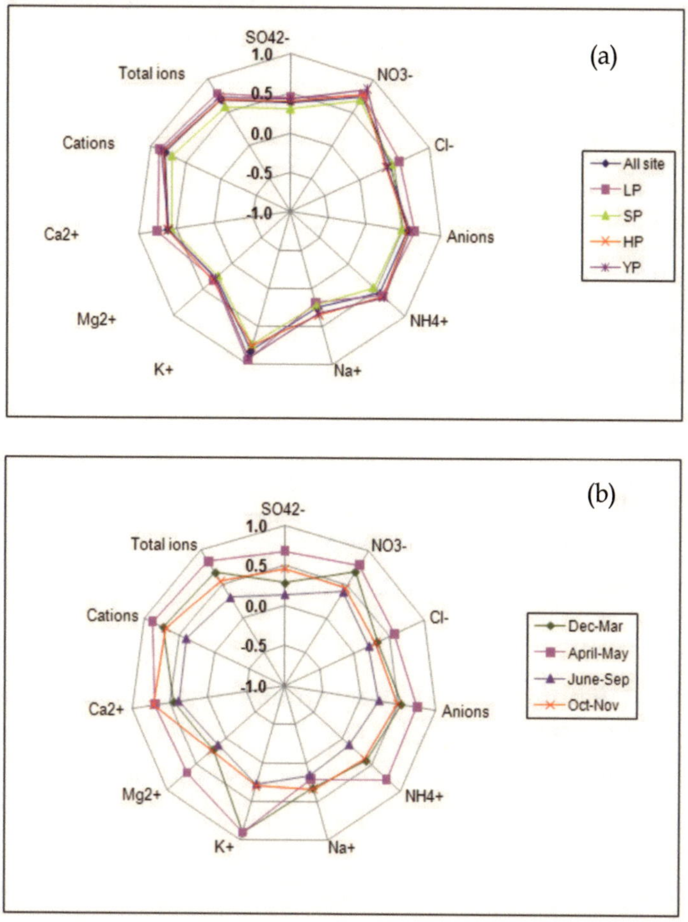

Fig. 10. Correlations between ion and PM10 concentrations; (a) sampling sites and (b) seasons

Table 12 illustrates the major ion composition of PM_{10} samples in this study in comparison with other studies, which were 1) ion composition of dry deposition collected by 4-stages filter pack sampler of the Mae Hia sampling site (approximately 7 km southwest from the city center of Chiang Mai) from Oct 05 - Sep 06 and 2) PM_{10}-bound ions in the Bangkok area from Feb 02 - Jan 03. It was found that concentrations of SO_4^{2-} , NO_3^-, NH_4^+ and Ca^{2+} detected in this study were approximately 2 times higher than those in Mae Hia. This is because Mae Hia is located in a sub-urban area with lower human activities than the stations of this study. When comparing the results of this study with Bangkok, it reveals that NO_3^- and NH_4^+ concentrations were almost the same, while SO_4^{2-} of this study was slightly higher. Unsurprisingly, Cl^- content in the Bangkok area was 7-10 times higher than that which was found in this study. This must be related with the distance from sea.

Ion	This research (Jun 05 – Jun 06)		Mae Hia, Chiang Mai (Oct 05 – Sep 06)*	Bangkok (Feb 02 – Jan 03) (Chuersuwan, et al., 2008)	
	Min±SD	Max±SD		Min±SD	Max±SD
SO_4^{2-}	2.67±2.62	3.37±3.42	1.55±1.52	1.79±0.59	2.37±0.72
NO_3^-	0.73±0.65	1.13±1.04	0.69±1.04	1.16±0.55	1.45±0.66
Cl^-	0.13±0.11	0.31±0.30	0.29±0.76	1.10±0.56	2.00±0.91
NH_4^+	0.76±0.92	1.00±1.09	0.45±0.50	0.62±0.31	0.94±0.62
Na^+	0.12±0.08	0.17±0.12	0.19±0.30	-	-
K^+	0.56±0.57	0.83±0.71	0.86±2.07	-	-
Mg^{2+}	0.06±0.08	0.07±0.14	0.16±0.34	-	-
Ca^{2+}	0.91±0.48	1.18±0.53	0.69±1.06	-	-

* Mean values from filter pack samples; dry deposition

Table 12. Mean concentrations (μg/m³) ± SD of PM_{10}-bound ions in Chiang Mai and Bangkok, Thailand

6. Analysis of PM10 bound elements by Inductively Coupled Plasma – Optical Emission Spectroscopy (ICP-OES)

6.1 Method efficiency for analysis of PM10 bound elements by ICP-OES

Filter samples were extracted based on Compendium Method IO-3.1 (US. EPA, 1999a) for 20 elements including Ca, Al, Si, Fe, Mg, K, Zn, Ti, P, Pb, Ba, Sr, Mn, Ni, Cu, V, Cr, Cd, Hg and As. They were analysed by Inductively Coupled Plasma - Optical Emission Spectroscopy (ICP-OES) (PerkinElmer Optima 3000 operated with WinLab32TM software) based on Compendium Method IO-3.4 (US. EPA, 1999b). Samples were put into Erlenmeyer flasks followed by the addition of 40 ml of mixed acid solution (5.55% HNO_3/16.75% HCl) and refluxed on hot plates for 3 hours. They were left until cool and transferred into 50 ml volumetric flasks. Prior to analysis by ICP, they were filtered through a 0.45 μm syringe nylon filter into plastic bottles.

Percent recoveries of elements were tested by the spiking of 2.50 μg/ml of mixed standards onto the quartz filter paper and followed the analysis method described above. The recoveries ranged between 76.0% (Mg) to 97.2% (Cr), except for Fe which was quite high

(157%). The instrument detection limit (IDL) ranged from between 0.01 ng/m^3 (Sr) to 5.47 ng/m^3 (K).

6.2 PM10 bound elements

Fig. 11 shows the spatial and temporal variation of elements in all sampling sites. Mean concentrations of elements and metals were high in the dry season, except for zinc, for which its concentrations were quite high in the rainy season in every sampling site. Calcium concentration was high all year round in all sampling sites. It was generated from soil re-suspension, which was related to high amounts of particulates being suspended in the air during the dry season. Potassium and silicon concentrations were high in the dry season in every sampling site, especially at the SP site. Both of them were ash elements that were emitted from biomass burning (Demirbas et al., 2008). The result was very well in agreement with the dissolved ion content mentioned earlier.

Toxic metals including lead (Pb), mercury (Hg), cadmium (Cd) and arsenic (As), were detected only in some samples in tracer levels in the dry season as well as in the transition period (Oct-Nov). There was no significant difference of element concentrations in terms of the sampling site.

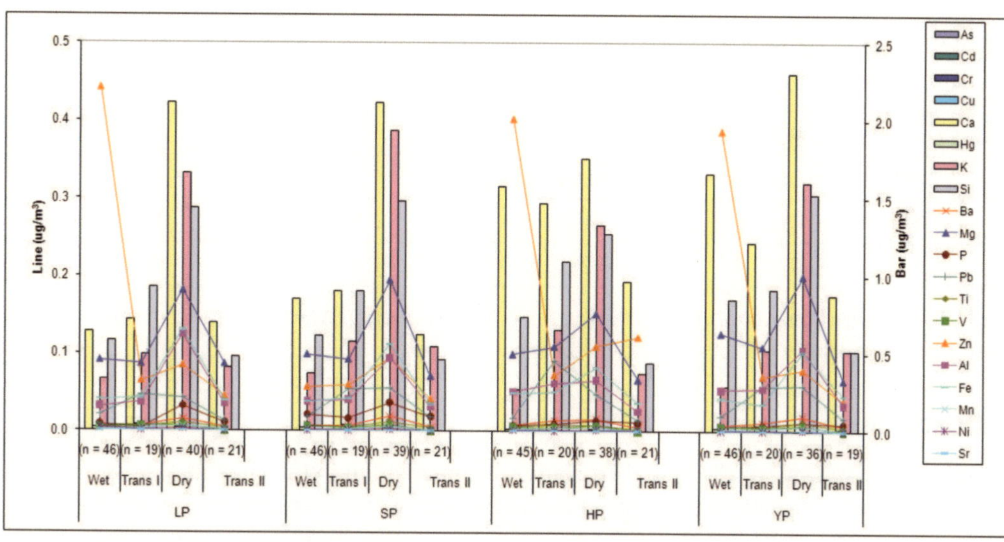

Fig. 11. Spatial and temporal variations of PM10 bound elements

Element concentrations found in this study were compared with the study in the Bangkok area, in which 14 elements were determined from 4 areas including Din Daeng, Chankasaem, Ban Somdej and Bangna, from Feb 02 - Jan 03 (Chuersuwan et al., 2008) (Table 13). It was found that the mean concentration of almost all elements found in the Bangkok area was higher than those in this study. This is probably because sources of elements and heavy metals, especially traffic emissions in Bangkok, were higher both in terms of variations and number.

Element	This research (Jun 05 - Jun 06)		Bangkok (Feb 02 - Jan 03) (Chuersuwan, et al., 2008)	
	Min ± SD	Max ± SD	Min ± SD	Max ± SD
PM10	33.2±21.4	73.6±25.3	57.6±23.9	108.1±35.5
Al	0.27±0.19	0.31±0.43	1.32±0.78	3.02±2.73
As	0.01±0.00	0.02±0.01	0.18±0.13	0.32±0.14
Ca	1.13±1.18	1.67±1.21	1.89±1.50	4.79±2.38
Cr	0.01±0.02	0.02±0.02	0.10±0.05	0.14±0.09
Cu	0.01±0.01	0.02±0.03	0.05±0.02	0.09±0.04
Fe	0.29±0.24	0.37±0.38	1.36±0.69	2.13±0.92
Hg	0.02±0.02	0.03±0.02	-	-
K	0.67±0.75	0.92±0.98	0.55±0.33	1.21±0.72
Mg	0.11±0.09	0.14±0.10	0.49±0.27	0.83±0.32
Mn	0.02±0.01	0.02±0.01	0.07±0.04	0.09±0.09
Ni	0.01±0.01	0.01±0.02	0.03±0.20	0.24±0.31
P	0.01±0.01	0.03±0.03	-	-
Pb	0.04±0.03	0.05±0.08	0.13±0.12	0.22±0.15
Si	0.90±0.68	1.03±0.75	-	-
Sr	0.003±0.003	0.003±0.003	-	-
Ti	0.01±0.01	0.01±0.01	-	-
V	0.01±0.01	0.01±0.01	0.80±0.38	1.13±0.51
Zn	0.07±0.07	0.201±1.023	0.65±0.35	0.84±0.47

Table 13. Min and max concentrations ($\mu g/m^3$) ± SD of mean PM10 and its elemental composition found in this study and in Bangkok, Thailand

7. Conclusion

Sampling and analysis of PM10 and its chemical composition in the Chiang Mai - Lamphun Basin, Thailand revealed interesting patterns of ambient PM10 concentrations in the urban areas. The background level of PM10 was mainly generated from traffic volume, which was quite constant throughout the year. Seasonal variation of PM10 levels was quite distinct between the dry and wet seasons. Rain precipitation was a main factor that reduced concentrations of air pollutants, as both particulate and gas phases. In the wet season, a high amount of precipitation scavenges contaminated atmosphere, and the sky becomes clear. In the dry season, there is a low amount of rain and high air pressure leading to air pollutant accumulation. The geographical characteristics of the basin are that it surrounded by a wall of high mountains. Temperature inversion occurs in the dry season, especially in February and March, acting as a lid covering the basin. Heavy haze occurs almost every year in the basin, especially in the past ten years. It was proof that the main sources are open burning, especially in the way of forest fires and agricultural burning in Northern Thailand and neighbouring countries.

The PM10 chemical composition, including both organic parts (carbon and PAHs) and inorganic parts (dissolved ions and elements & heavy metals) revealed that their concentrations were obviously high in the dry season in all sampling stations. They were well correlated with PM10 concentrations, especially in the dry season. The dominant organic compounds found in the PM10 samples were 4-6 rings PAHs. The ratios of carcinogenic PAHs over non-carcinogenic ones were approximately 3:1-4:1, which was about 1.0-1.5 times higher than those that were reported from the Bangkok area. The BaPE values representing carcinogenic power were highest in the samples taken from the SP site, which is located in the eastern part of the city, and this was due to the higher number of sources, i.e. open burning, street food cooking and small-scale factories. The dominant water-soluble inorganic substances were SO_4^{2-} and NH_4^+, which showed high amounts of fossil fuel combustion. PM10 major elemental composition was Ca, K and Si, which were reported to be emitted from biomass burning. To reduce air pollution in this area, their sources must be reduced. Open burning from both forest fires and agricultural burning need to be managed. Biomass residues have to be removed or reused for other purposes, such as organic fertilizer, green or bio-fuels for cooking in households, etc., to avoid the practice of large-scale burning. Better management of forest areas in the dry season with plenty of leaf litter is also needed. Apart from that, potential sources of air pollution in the urban area also need to be considered. Mobile sources; traffic volume, number and types of vehicles and fuel used, as well as the point source, such as small scale factories, should also be considered. Air pollution management campaigns are only going to be successful and sustained when the issue is brought, as a priority, into the area of public concern. Moreover, increasing the number of air quality monitoring stations is necessary to cover and represent a wider area. This will, in turn, lead to higher precision and accuracy in recording and interpreting the data for solid air quality management.

8. Acknowledgment

Financial support from the Thailand Research Fund (TRF) and research facilities from the Chemistry Department, Faculty of Science, Chiang Mai University are gratefully acknowledged. Special thanks are extended to the Air Project Team, Chemistry Dept., CMU, particularly Assoc. Prof. Dr. Mongkon Rayanakorn, Asst. Prof. Dr. Sunanta Wangkarn and Dr. Urai Tengjaroenkul. Thanks to all member of the Environmental Chemistry Research Laboratory, CMU for their great help and support.

9. References

Acid Deposition Monitoring Network in East Asia (EANET). (2003). *Technical Document for Filter Pack Method in East Asia*, The Third Interim Scientific Advisory Group Meeting of Acid Deposition Monitoring Network in East Asia.

Boonyatumanond, R., Murakami, M., Wattayakorn, G., Togo, A., & Takada, H. (2007). Sources of polycyclic aromatic hydrocarbons (PAHs) in street dust in a tropical Asian mega-city, Bangkok, Thailand. *Science of the Total Environment*, 384, pp. 420-432. ISSN 0048-9697.

Chantara, S., & Chunsuk, N. (2008). Comparison of wet-only and bulk deposition at Chiang Mai (Thailand) based on rainwater chemical composition. *Atmospheric Environment.* 42, pp. 5511-5518. ISSN 1352-2310.

Chantara, S., & Sangchan, W. (2009). Sensitive analytical method for particle-bound polycyclic aromatic hydrocarbons; A case study in Chiang Mai, Thailand. *ScienceAsia.* 35, pp. 42-48. ISSN 0303-8122.

Chantara S., Wangkarn S., Sangchan W. & Rayanakorn M. (2010). Spatial and Temporal Variations of Ambient PM_{10}-bound Polycyclic Aromatic Hydrocarbons in Chiang Mai and Lamphun Provinces, Thailand. *Desalination and Water Treatment.* 19, pp. 17-25. ISSN 1944-3994.

Chantara, S., Wangkarn, S., Tengcharoenkul, U., Sangchan, W. & Rayanakorn, M. (2009). Chemical Analysis of Airborne Particulates for Air Pollutants in Chiang Mai and Lamphun Provinces, Thailand. *Chiang Mai Journal of Science,* 36(2), pp. 001-013. ISSN 0125-2526.

Chetwittayachan, T., Shimazaki, D., & Yamamoto, K. (2002) A comparison of temporal variation of particle-bound polycyclic aromatic hydrocarbons (pPAHs) concentration in different urban environments: Tokyo, Japan, and Bangkok, Thailand. *Atmospheric Environment.* 36, pp. 2027-2037. ISSN 1352-2310.

Chiang Mai Meteorological Station. Statistic of Chiang Mai climate. 2 August 2011. Available from <http://www.cmmet.tmd.go.th/index1.php>

Chuersuwan, N., Nimrat, S., Lekphet, S., & Kerdkumrai, T. (2008). Levels and major sources of PM2.5 and PM10 in Bangkok Metropolitan Region. *Environmental International.* 34, pp. 671-677. ISSN 0160-4120.

Demirbas, A. (2008). Hazardous Emissions from Combustion of Biomass. *Energy Sources, Part A.* 30, pp. 170–178. ISSN 1556-7036.

Furton, K.G., & Pentzke, G. (1998). Polycyclic Aromatic Hydrocarbon, In: *Chromatographic Analysis of Environmental Food Toxicants,* Shibamoto, T. pp. 1-30. Marcel Dekker Inc., ISBN 0824701453, New York.

Fuzzi, S., Decesari, S., Facchini, M.C., Cavalli, F., Emblico, L., Mircea, M., Andreae, M.O., Trebs, I., Hoffer, A., Guyon, P., Artaxo, P., Rizzo, L., Lara, L., Pauliquevis, T., Maenhaut, W., Raes Chi, N.X., Mayol-Bracero, O., Soto-Garcı´a, L.L., Claeys, M., Kourtchev, I., Rissler, J., Swietlicki, E., Tagliavini, E., Schkolnik, G., Falkovich, A., Rudich, Y., Fisch, G., & Gatti L. (2007). Overview of the inorganic and organic composition of size-segregated aerosol in Rondonia, Brazil, from the biomass burning period to the onset of the wet season. *Journal of Geophysical Research,* 112(D1), D01201, doi:10.1029/2005JD006741. ISSN 0148-0227.

Kavouras, I.G., Koutrakis, P., Tsapakis, M., Lagoudaki, E., Stephanou, E., Von Baer, D. & Oyola, P. (2001). Source apportionment of urban particulate aliphatic and polynuclear aromatic hydrocarbons (PAHs) using multivariate methods, *Environmental Science and Technology,* 25, pp. 2288–2294, ISSN 1735 1472.

Khalili, N.R., Scheff, P.A., & Holsen, T.M. (1995). PAH source fingerprints for coke ovens, diesel and gasoline engines, highway tunnels, and wood combustion emissions, *Atmospheric Environment,* 29, pp. 533–542, ISSN 1352-2310.

Kim Oanh, N.T., Bætz Reutergårdh, L., Dung, N.T., Yu, M.H., Yao, W.X. & Co, H.X. (2000). Polycyclic aromatic hydrocarbons in the airborne particulate matter at a location 40 km north of Bangkok, Thailand. *Atmospheric Environment*, 34, pp. 4557-4563, ISSN 1352-2310.

Koyano, M., Endo, O., Goto, S., Tanabe, K., Koottatep, S. & Matsushita, H., (1998). Carcinogenic polynuclear aromatic hydrocarbons in the atmosphere in Chiang Mai, Thailand, *Japanese Journal of Toxicology and Environmental Health*, 44(3), pp. 214–225, ISSN 0013-273X.

Liu, L.B., Hashi, Y., Liu, M., Wei, Y. & Lin, J.M. (2007). Determination of particle-associated polycyclic aromatic hydrocabens in urban air of Beijing by GC/MS. *Analytical Science*, 23, pp. 667-671, ISSN 0910-6340.

Mauderly, J.L. & Chow, J.C. (2008). Health effects of organic aerosols. *Inhalation Toxicology*, 20, pp. 257-288, ISSN 0895-8378.

Metzger, K.B., Tolbert, P.E., Klein, M., Peel, J.L., Flanders, W.D., Todd, K., Mulholland, J.A., Ryan, P.B., & Frumkin, H. (2004). Ambient air pollution and cardiovascular emergency department visits. *Epidemiology* 15, pp. 46-56, ISSN 1044-3983.

Miller J.C. & Miller J.N. (1993). *Statistics for Analytical Chemistry*, 3rd edition, Ellis Horwood, ISBN 0130309907/0-13-030990-7, New York.

Norramit, P., Cheevaporn, V., Itoh, N. & Tanaka, K. (2005). Characterization and carcinogenic risk assessment of polycyclic aromatic hydrocarbons in the respirable fraction of airborne particles in the Bangkok metropolitan area. *Journal of Health Science*, 51, pp. 437-446. ISSN 1347-5207.

Pandey, P.K., Patel, K.S. & Lenicek, J. (1999). Polycyclic aromatic hydrocarbons: Need for assessment of health risks in India? *Environmental Monitoring and Assessment*, 59, pp. 287–319. ISSN 1573-2959.

Patashnick, H., Meyer, M. & Rogers, B. (2002). Tapered element oscillating microbalance technology, *Proceedings of the North American/Ninth U.S. Mine Ventilation Symposium*, Kingston, Ontario, Canada, 8-12 June 2002.

Pengchai, P., Chantara, S., Sopajaree, K., Wangkarn, S., Tengcharoenkul, U. & Rayanakorn, M. (2009). Seasonal variation, risk assessment and source estimation of PM10 and PM10-Bound PAHs in the ambient air of Chiang Mai and Lamphun, Thailand. *Environmental Monitoring and Assessment*, 154, pp. 197-218. ISSN 1573-2959.

Perkin Elmer Cooperation. (1991). *Users Manual PE 2400 Series II CHNS/O Analyser*, Perkin Elmer, Connecticut.

Pio, C.A. & Harrison, R.M. (1987). Vapor-pressure of ammonium-chloride aerosol - effect of temperature and humidity. *Atmospheric Environment* 21, pp. 2711-2715. ISSN 1352-2310.

Pollution Control Department (2011). Regional area air quality data, 2 August 2011, Available from: <http://www.pcd.go.th/AirQuality/Regional/QueryAirThai.cfm?task=default>

Pöschl, U. (2005). Atmospheric aerosols: composition, transformation, climate and health effects. *Angewandte Chemie International Edition*, 44, pp. 7520-7540, ISSN 1521-3773.

Seinfeld, J.H. & Pandis, S.N. (1998). *Atmospheric Chemistry and Physics: From Air Pollution to Climate Change*, John Wiley and Sons Inc., ISBN 0-471-17815-2, New York.

Torigoe, K., Hasegawa, S., Numata, O., Yazaki, S., Matsunaga, M., Boku, N., Hiura, M. & Ino, H. (2000). Influence of emission from rice straw burning on bronchial asthma in children. *Pediatrics International*, 42, pp. 143-150, ISSN 1328-8067.

U.S. EPA (Environmental Protection Agency): Center for Environmental Research Information, Office of Research and Development. (1999a). *Compendium of methods for the determination of inorganic compounds in ambient air.* Compendium Method IO-3.1., Selection, preparation, and extraction of filter material, Retrieved from <www.epa.gov/ttnamti1/files/ambient/inorganic/mthd-3-1.pdf>

U.S. EPA (Environmental Protection Agency):Center for Environmental Research Information, Office of Research and Development. (1999b). *Compendium of methods for the Determination of inorganic compounds in ambient air.* Compendium Method IO-3.4., Determination of metals in ambient particulate matter using inductively coupled plasma (ICP) spectrometry, Retrieved from <www.epa.gov/nrmrl/pubs/625r96010/iocompen.pdf>

U.S. EPA (Environmental Protection Agency). (1999c). *National Air Toxics Program: The Integrated Urban Strategy,* United States Environmental Protection Agency's Federal Register Vol. 64, No. 137,1999, Retrieved from <www.epa.gov/ttnatw01/urban/fr19jy99.pdf>

Valerie, S., Clare, S. & Aoife, D. (2003). Microwave digestion of sediment, soils and urban particulate matter for trace metal analysis. *Talanta.* 60, pp. 715-723, ISSN 0039-9140.

Washington State Department of Ecology, (2004). PM_{10} Tapered Element Oscillating Microbalance Operating Procedure, In: *Air Quality Program,* 1 September 2011, Available from: <www.ecy.wa.gov › Ecology home › Publications & Forms>